AF358697

Thermal Performance of the Building Envelope—Original Methods and Advanced Solutions

Thermal Performance of the Building Envelope—Original Methods and Advanced Solutions

Editors

Gianpiero Evola
Elena Lucchi

Basel • Beijing • Wuhan • Barcelona • Belgrade • Novi Sad • Cluj • Manchester

Editors
Gianpiero Evola
Università degli Studi
di Catania
Catania
Italy

Elena Lucchi
Università degli Studi
di Pavia
Pavia
Italy

Editorial Office
MDPI AG
Grosspeteranlage 5
4052 Basel, Switzerland

This is a reprint of articles from the Special Issue published online in the open access journal *Buildings* (ISSN 2075-5309) (available at: https://www.mdpi.com/journal/buildings/special_issues/thermal_building).

For citation purposes, cite each article independently as indicated on the article page online and as indicated below:

Lastname, A.A.; Lastname, B.B. Article Title. *Journal Name* **Year**, *Volume Number*, Page Range.

ISBN 978-3-7258-2027-6 (Hbk)
ISBN 978-3-7258-2028-3 (PDF)
doi.org/10.3390/books978-3-7258-2028-3

Cover image courtesy of Elena Lucchi

Contents

About the Editors

Gianpiero Evola

Gianpiero Evola is an Associate Professor in Building Physics at the Department of Electric, Electronic and Computer Engineering (DIEEI), University of Catania, Italy. He holds a Master of Science in Mechanical Engineering and a PhD in Building Physics from the University of Palermo, followed by post-doc work at the National Institute of Solar Energy (INES), France. His research and teaching activities address the thermal and acoustic performance of buildings, both at the component scale and at the building scale, including thermal bridges, hygrothermal simulations, energy efficiency assessment, thermal comfort, and renewable energy integration.

Elena Lucchi

Elena Lucchi is a Researcher in Building Physics at the Department of Civil Engineering and Architecture (DICAr), University of Pavia, Italy. She holds a Master of Science in Architecture and a PhD in Building Physics from the Politecnico di Milano, complemented by a post-master's degree in Environmental Education from the Università Cattolica. Her research addresses the connections between energy, environment, and architecture, with a focus on energy efficiency, climate mitigation strategies, and the integration of renewable energies in historic buildings and protected landscapes.

Preface

The following reprint, entitled *"Thermal Performance of the Building Envelope — Original Methods and Advanced Solutions"*, presents advanced research about experimental and numerical techniques to design, assess, and optimize the thermal performance of buildings, both at the component scale and at the building scale. The reprint includes one Editorial by the Guest Editors and ten articles, with nine original research papers and one perspective paper. The articles were authored by 37 different scholars hailing from Italy, Serbia, Portugal, Turkey, Romania, Hungary, Canada, Egypt, India, Vietnam, and Japan.

The papers cover three main topics, i.e., the thermal performance of the building envelope, energy optimization of buildings, and the integration of renewable energies. They were selected and organized to show the importance of a multi-scale approach, including the thermal performance of individual envelope components and the optimization of their overall energy efficiency, implemented through simulations possibly complemented by in situ or laboratory measurements. This reprint is useful to building designers who wish to improve their knowledge and skills regarding the correct thermal design of a building.

Gianpiero Evola and Elena Lucchi
Editors

Editorial

Thermal Performance of the Building Envelope: Original Methods and Advanced Solutions

Gianpiero Evola [1] and Elena Lucchi [2,*]

[1] Department of Electrical, Electronic and Computer Engineering, University of Catania, Via S. Sofia 64, 95123 Catania, Italy; gevola@unict.it

[2] Dipartimento di Ingegneria Civile e Architettura (DICAr), University of Pavia, Via Ferrata 1, 27100 Pavia, Italy

* Correspondence: elena.lucchi@unipv.it

Citation: Evola, G.; Lucchi, E. Thermal Performance of the Building Envelope: Original Methods and Advanced Solutions. *Buildings* **2024**, *14*, 2507. https://doi.org/10.3390/buildings14082507

Received: 11 July 2024
Accepted: 13 August 2024
Published: 14 August 2024

1. Introduction

In the European Union, buildings are responsible for 40% of the final energy demand and approximately 36% of greenhouse gas emissions [1]. The existing building stock's substantial energy demand is mostly due to the poor thermal performance of their envelope components (e.g., roofs, walls, and windows), which is especially relevant in buildings that were built when the current EU regulations addressing energy efficiency in buildings were not in force; at present, it is possible to state that roughly 75% of the EU stock is energy inefficient [2].

In this framework, a correct design of the building envelope, including an informed choice of the materials and their thermal properties, is a key requisite to conjugate energy efficiency, the durability of the envelope, and indoor microclimate. This involves a strategic selection of building materials based on their thermal properties and performance. Designers must adopt a multi-scale approach, evaluating the thermal performance of individual envelope components and optimizing the overall energy efficiency of the building. In situ and laboratory measurements can improve the accuracy of the analysis. Indeed, in situ measurements allow for the assessment of actual building performance under real-life conditions, identifying areas of thermal weakness and potential improvements [3]. Laboratory measurements enable precise control over the testing conditions and the evaluation of material properties and their thermal performance in a controlled environment. These measurements are also useful for dynamic simulation techniques, which account for transient thermal behavior and occupant interaction with the building's performance [4]. This activity should be carried out using reliable tools able to simulate advanced envelope solutions and must be complemented by effective and original evaluation methods to identify the best performing solution.

2. Contributions

This Special Issue titled "The Thermal Performance of the Building Envelope—Original Methods and Advanced Solutions" presents advanced research about the above-mentioned topics. It includes 10 articles, namely 9 original research articles and 1 perspective paper. The 10 articles were authored by 37 scholars from Italy, Serbia, Portugal, Turkey, Romania, Hungary, Canada, Egypt, India, Vietnam, and Japan. The studies concern the following main topics:

- The thermal performance of the building envelope.
- The energy optimization of buildings.
- The integration of renewable energies.

Several different approaches to these topics were followed, as detailed below.

2.1. The Thermal Performance of the Building Envelope

The thermal performance of the building envelope is assessed based on the following:

- The use of different nondestructive techniques.
- The integration of experimental measurements and numerical analyses.
- The development of advanced simulation models.

The integration of different nondestructive techniques allows for a comprehensive analysis of the thermal performance of the building envelope without damaging the structure. Among these techniques, infrared thermography (IRT) detects infrared radiation (IR) emitted by objects to visualize temperature variations, highlighting areas of heat loss, thermal bridges, and insulation deficiencies. IRT provides real-time and accurate data that can be compared with temperature measurements from thermal sensors to understand the thermal properties of materials. Živanović et al. [5] investigated the relationship between measurements performed using IRT and direct temperature measurements using embedded sensors through the testing of an early-age hydration process of a cement-based paste. They prepared two types of cubic samples with different heights using a cement-based paste, with 20% of the cement (by mass) being replaced with fly ash. Infrared (IR) images were obtained using an FLIR E6 IR camera positioned at 0.60 m from the sample, through a tube, to stabilize the environment between the camera and the surface of the samples. The IR camera captures images every half hour during stable temperature periods and every 10 min during periods of intensive temperature fluctuation. The temperature fluctuations were modeled using the asymmetric Gaussian function. The measurements from both the IR camera and surface sensors show excellent agreement with the modeling coefficients, as the temperature differences between the thermo-sensors on the surface and the thermal camera are less than 2 °C. The values obtained from the sensor placed in the central part of the samples are higher than the temperatures measured either through sensors placed on the surface or with the thermal camera. The temperatures measured on the surface with the two different devices show very good accordance.

Integrating experimental measurements and numerical analyses allows for a detailed evaluation of the thermal performance of building envelopes, supporting the validation of theoretical models and the development of a more effective building design. Experimental measurements provide real-world data, ensuring accuracy and reliability. Numerical analyses, on the other hand, allow for the simulation of various conditions and scenarios that may not be feasible to test experimentally. In this direction, Evola and Gagliano [6] studied, both experimentally and numerically, the thermal distribution in a thermal bridge corresponding to a reinforced concrete corner pillar in a building dating back to the 1980s and located in Southern Italy. The authors measured the inner surface temperatures in several points near the corner pillar by using both Pt 1000 temperature probes and thermal imaging techniques. Moreover, 2D finite element simulations were performed based on the same indoor and outdoor conditions as those measured on site during the experimental campaign. The results show that the thermal imaging technique is very useful for obtaining visual information about the presence and extension of the thermal bridging effect, but it is less accurate when quantifying the information.

Similarly, Santos et al. [7] conducted an evaluation of bio-based (pine wood) and recycled (rubber–cork composite) materials to mitigate the thermal bridge effect caused by steel profiles in Lightweight Steel-Framed (LSF) walls. In fact, the high thermal conductivity of steel and the associated thermal bridges can significantly compromise the thermal performance of LSF walls. Their study involved both controlled laboratory measurements and numerical simulation models. The thermal resistance (R-value) of the LSF wall was measured using the Heat Flow Meter (HFM) method in a climatic chamber. These measurements were then compared with calculations from 2D (THERM models) and 3D numerical simulations (ANSYS models), showing a difference of less than ±2%. Both materials performed similarly, with pine wood having a slight advantage due to its greater thickness. Overall, bio-based and recycled materials present an environmentally friendly solution as thermal break strips.

Recent advancements in simulation models for building envelopes could support innovative approaches in environmental design to optimize energy efficiency and thermal performance. These include solar greenhouse technologies, greenery systems, solar chimneys, and ventilated façade systems. Indeed, Kaliakatsos et al. [8] simulated the thermal performance of a bioclimatic greenhouse to assess the influence of features like the type of glazing, thermal mass, size, shading, and ventilation systems, thus providing a general overview of the features that a south-facing solar greenhouse should have when attached to a building in the Mediterranean area. According to their results, the solar greenhouse contributes to greater energy savings in the winter if it has a reduced depth and low-emissivity double glazing. In the summer, the greatest energy savings are achieved by favoring the ventilation of the greenhouse and activating solar shadings to reduce the counterproductive effects of overheating.

Among the possible strategies, the implementation of resilient technologies for the building envelope, such as vertical greenery systems (VGSs), is gaining ground. However, existing models in the literature are not sufficiently detailed in describing all phenomena occurring in a VGS. Nesci et al. [9] tried to overcome this research gap by identifying and improving two mathematical models for green façades and living walls. To this end, a dedicated calculation code to estimate the effect of VGSs on a building's energy performance and indoor thermal comfort was developed and implemented within the EnergyPlus calculation software. Through a BESTest case study selected from ASHRAE 140, it was shown that the shading effect of the vegetation layer and the evapotranspiration process significantly lower the exterior wall surface temperatures during the summer, thus improving the building's energy performance and occupant comfort. Indeed, the shielding effect reduces the incident solar radiation on the back wall, and the evapotranspiration effect of both the vegetation and substrate, based on the technology applied, involves cooling in terms of the external surface temperature and immediate surroundings. Instead, the thermal energy needs for cooling in the summer decrease for green façades and for living walls by 15.2% and 8.5%, respectively. The proposed model can easily be used within a dynamic energy simulation tool.

Similarly, using passive strategies, Nguyen et al. [10] studied solar chimneys to reduce the solar heat gain on a building envelope and enhance natural ventilation. In this work, three configurations of two solar chimneys combined with a heated wall to naturally ventilate a room are proposed: (I) the chimneys are connected serially, (II) the chimneys are parallel and exhaust air at two separate outlets, and (III) the chimneys are parallel but the outlets are combined. The airflow rate achieved with each configuration was predicted with a Computational Fluid Dynamics (CFD) model. The results show the effects of the heat flux in each channel and the geometries of the channels. Configuration (II) shows the highest flow rate. Particularly, the proposed configurations significantly enhance the flow rate by up to 40% compared to the typical setup with a single-channel solar chimney. The findings offer a novel design option for building façades to reduce their solar heat gain and enhance natural ventilation. Lastly, Petresevics and Nagy [11] evaluated the point thermal transmittances created by brackets and anchors in ventilated façade claddings using the 3D Finite Element Method (FEM) to carry out thermal modeling. Ventilated façade systems are popular not only for their esthetic properties, but also because they provide mechanical and acoustic protection for the façade and reduce the building's energy demand. However, point thermal bridges of the fastening system with brackets and anchors are often neglected during simplified energy performance calculations and practical design tasks. To address this, a comprehensive point thermal bridge catalog is created, considering multiple factors of ventilated façades. This research investigates the effects of these parameters on a broader scale than previous studies. Numerical simulations predict the effects of metal fasteners on the thermal performance of the building envelope. The FEM-based results indicate that thermal breaks/isolators can only reduce point thermal transmittances by 2% to 28%

depending on the materials of the brackets and isolators. The material and geometric properties of the brackets can result in up to a 70% difference between the corrected and uncorrected thermal transmittance values. The results of the numerical simulations clearly show that considering only the anchors and doweling for mechanical fixings is insufficient. The effect of the brackets on the point thermal transmittance is also significant. Additionally, significant differences are observed when brackets are applied to different types of masonry or reinforced concrete walls.

2.2. Energy Optimization in Buildings

Several papers discuss models to optimize the energy of entire buildings. Particularly noteworthy is the contribution of Chidiac and Marjaba [12], who introduced a new metric known as the building envelope coefficient of performance (BECOP). This comprehensive metric evaluates the thermal performance of building envelopes by comparing them to an ideal system, ensuring applicability across building types and climate zones. The BECOP captures the combined influence of the thermal resistance, climate zone, and internal heat gains. As the heating degree days (HDDs) increase, the BECOP highlights the enhanced impact of an efficient building envelope. A noted weakness is the range of the BECOP given the low efficiency of the building envelope compared to the ideal system. Nonetheless, this weakness can become a catalyst for designing a more efficient building envelope. Furthermore, the BECOP values underscore the energy saving potential achievable with innovative building envelope systems.

Additionally, Kallioğlu et al. [13] illustrated the effects of different insulation materials and fuel types on the cooling and heating performances of buildings situated in hot and dry, warm and humid, composite, and cold climatic conditions in India. Ten different locations were chosen from diverse climatic regions, and various potential parameters for expanded polystyrene and extruded polystyrene insulation materials were evaluated. Their study demonstrates that applying insulation to buildings' exterior walls results in significant annual savings with a payback period of less than five years, indicating economic feasibility. Additionally, increasing the insulation thickness reduces greenhouse gas (GHG) emissions from fuels. Future research should aim to determine optimal insulation thicknesses for various climate zones and materials.

2.3. The Integration of Renewable Energies

Finally, the perspective paper introduces a new theme: the integration of renewable energy sources (RESs) into the building envelope as a novel research outlook that combines energy efficiency with energy transition. Specifically, the study focuses on the built environment, concentrating on the greatest barriers to the application of these technologies. Particularly, Lucchi [14] explored RES integration within architectural heritage settings, addressing the need to reduce the energy demand and environmental impact without compromising heritage values and esthetic, historical, and material integrity. The perspective study reviews recent studies in the literature to identify key topics, challenges, and advanced solutions for applying solar, wind, geothermal, and bioenergy in heritage contexts. It also highlights acceptability, design criteria, and state-of-the-art technologies through illustrative case studies, offering an understanding of practical implementation strategies. The RES integration criteria in architectural heritage include ensuring conservation compatibility and minimal visual impact to preserve historical and esthetic integrity. Additionally, installations should be reversible, environmentally considerate, and compliant with regulatory standards to balance sustainability with heritage preservation.

3. Conclusions

Adopting a comprehensive design strategy that includes the informed selection of materials with optimal thermal properties and the detailed design of building envelopes is a key action in optimizing the thermal performance of the building envelope. Future research may refer to the following:

- The development of advanced materials, especially those based on bio-based solutions, nanotechnologies, Super Insulating Materials (SIMs), and smart materials.
- The interaction between real-time measurements and advanced simulation tools, thanks to the application of Digital Twins into the building sector.
- The study of hybrid simulation models that combine different tools and techniques (e.g., CFD, FEM, and building energy modeling) to achieve more accurate and comprehensive predictions of a building's performance.

Author Contributions: All authors equally contributed to this editorial. Conceptualization, G.E. and E.L.; formal analysis, G.E. and E.L.; writing—original draft preparation, G.E. and E.L.; writing—review and editing, G.E. and E.L. All authors have read and agreed to the published version of the manuscript.

Conflicts of Interest: The authors declare no conflicts of interest.

References

1. Sajn, N. *Energy Efficiency of Buildings: A Nearly Zero-Energy Future?*; European Parliamentary Research Service; European Union: Brussels, Belgium, 2016. Available online: https://www.europarl.europa.eu/thinktank/en/document/EPRS_BRI(2016)582022 (accessed on 10 July 2024).
2. European Commission, Department: Energy. In Focus: Energy Efficiency in Buildings, Brussels, 17 February 2020. Available online: https://ec.europa.eu/info/news/focus-energy-efficiency-buildings-2020-lut-17_en (accessed on 10 July 2024).
3. Nardi, I.; Lucchi, E. In situ thermal transmittance assessment of the building envelope: Practical advice and outlooks for standard and innovative procedures. *Energies* **2023**, *16*, 3319. [CrossRef]
4. Schito, E.; Lucchi, E. Advances in the Optimization of Energy Use in Buildings. *Sustainability* **2023**, *15*, 13541. [CrossRef]
5. Živanović, N.; Aškrabić, M.; Savić, A.; Stević, M.; Stević, Z. Early-Age Cement Paste Temperature Development Monitoring Using Infrared Thermography and Thermo-Sensors. *Buildings* **2023**, *13*, 1323. [CrossRef]
6. Evola, G.; Gagliano, A. Experimental and Numerical Assessment of the Thermal Bridging Effect in a Reinforced Concrete Corner Pillar. *Buildings* **2024**, *14*, 378. [CrossRef]
7. Santos, P.; Abrantes, D.; Lopes, P.; Mateus, D. Experimental and Numerical Performance Evaluation of Bio-Based and Recycled Thermal Break Strips in LSF Partition Walls. *Buildings* **2022**, *12*, 1237. [CrossRef]
8. Kaliakatsos, D.; Nicoletti, F.; Paradisi, F.; Bevilacqua, P.; Arcuri, N. Evaluation of Building Energy Savings Achievable with an Attached Bioclimatic Greenhouse: Parametric Analysis and Solar Gain Control Techniques. *Buildings* **2022**, *12*, 2186. [CrossRef]
9. Nesci, V.; Ballarini, I.; Rando Mazzarino, P.; Corrado, V. Living Walls and Green Façades: An Implementation Code for Energy Simulation. *Buildings* **2024**, *14*, 2040. [CrossRef]
10. Nguyen, Y.; Nguyen, V.; Tran, L.; Wells, J. CFD Analysis of Different Ventilation Strategies for a Room with a Heated Wall. *Buildings* **2022**, *12*, 1300. [CrossRef]
11. Petresevics, F.; Nagy, B. FEM-Based Evaluation of the Point Thermal Transmittance of Various Types of Ventilated Façade Cladding Fastening Systems. *Buildings* **2022**, *12*, 1153. [CrossRef]
12. Chidiac, S.; Marjaba, G. BECOP: A New Metric for Measuring the Energy Performance of the Building Envelope. *Buildings* **2022**, *12*, 553. [CrossRef]
13. Kallioğlu, M.; Yılmaz, A.; Sharma, A.; Mohamed, A.; Dobrotă, D.; Alam, T.; Khargotra, R.; Singh, T. Optimal Insulation Assessment, Emission Analysis, and Correlation Formulation for Indian Region. *Buildings* **2023**, *13*, 569. [CrossRef]
14. Lucchi, E. Renewable Energies and Architectural Heritage: Advanced Solutions and Future Perspectives. *Buildings* **2023**, *13*, 631. [CrossRef]

 *buildings*

Article

Early-Age Cement Paste Temperature Development Monitoring Using Infrared Thermography and Thermo-Sensors

Nevena Živanović [1], Marina Aškrabić [1,*], Aleksandar Savić [1], Miša Stević [2] and Zoran Stević [3]

1   Faculty of Civil Engineering, University of Belgrade, 11000 Belgrade, Serbia; sasha@grf.bg.ac.rs (A.S.)
2   Elsis, 11070 Belgrade, Serbia
3   Technical Faculty Bor, School of Electrical Engineering Belgrade, University of Belgrade, 11000 Belgrade, Serbia; zstevic@live.com
*   Correspondence: amarina@imk.grf.bg.ac.rs

Abstract: Infrared thermography is an advanced technique usually applied for the assessment of thermal losses through different elements of the building envelope, or as a method for detection of damage (cracks) in reinforced concrete elements, such as bridges. Use of this method for the investigation of temperature development during early cement hydration is still an evolving area of research. For the purpose of verifying the reliability of the method, two types of cubic samples of different heights were prepared using a cement-based paste, with 20% of cement (by mass) replaced with fly ash. Temperature development was measured in two ways: using infrared thermography and thermo-sensors embedded in the samples. Additionally, the obtained results were modeled using the asymmetric Gaussian function. Peak temperatures in the middle of each sample were higher than the peak temperatures measured on the sample surface, with differences ranging between 2 °C and 4 °C. Differences between the temperature measurements of the thermo-sensors placed on the surface of the sample and thermal camera were lower than 2 °C. Very good compliance of the results was obtained for both the camera and the surface sensors measurements, as well as for the modeling coefficients.

Keywords: cement hydration temperature; infrared thermography; asymmetric Gaussian; supplementary cementitious materials

Citation: Živanović, N.; Aškrabić, M.; Savić, A.; Stević, M.; Stević, Z. Early-Age Cement Paste Temperature Development Monitoring Using Infrared Thermography and Thermo-Sensors. *Buildings* **2023**, *13*, 1323. https://doi.org/10.3390/buildings13051323

Academic Editors: Gianpiero Evola and Elena Lucchi

Received: 21 April 2023
Revised: 12 May 2023
Accepted: 14 May 2023
Published: 19 May 2023

1. Introduction

Infrared thermography is an advanced technique of obtaining the intensity of heat radiation invisible to the human eye. The measurements are performed with no contact between the object that is emitting heat and the measuring device (thermal camera). The result of the measurements is an image of temperature distribution on the tested surface, commonly referred to as a thermogram. Surfaces with the same temperature are usually presented using the same color.

As far as the application of this method in the building sector is concerned, its greatest impact lies in the assessment of thermal losses through different elements of the building envelope. It is usually applied in the context of improving the energy efficiency of buildings (for example: detection and identification of thermal bridges, detection of thermal irregularities, determination of building envelope thermal properties) [1,2]. Another application of this method is in the detection of damage (cracks) in reinforced concrete elements, such as bridges [3]. It is also used as a measure of inhomogeneity in certain building materials, as well as for inspection of heating and electrical installations [4].

This paper is focused on a less well known application of infrared thermography, temperature measurement during an early age hydration of cement paste, a known exothermic process. Due to this reaction, the maximum temperature of concrete can reach up to 60 °C to 70 °C in the core, while the high temperature differential between the concrete's surface and core can result in tensile stresses followed by cracking [5,6].

One of the methods used for reducing the temperature development in the early phases of the hydration process is the addition of supplementary cementitious materials (SCMs) as a partial replacement of pure Portland cement (PC). The most common mineral admixtures used for this purpose are fly ash (FA), steel slag (SS), ground granulated blast-furnace slag (GGBS), and natural pozzolanic materials [7,8]. Zhang et al. [9] investigated the effects of different SCMs on hydration heat and temperature of cement pastes. Four types of pastes were tested, one ordinary Portland cement paste and three pastes where 40% of Portland cement was replaced with FA, SS, and GGBS. As expected, the highest temperature was reached for the sample containing only pure cement paste (54 °C), while the temperatures for three other pastes were lower, reaching 39.8 °C for the paste containing GGBS, 38.4 °C for the paste containing SS, and 37 °C for the paste containing FA. The authors also concluded that addition of these SCMs did not change the type of hydration products of the paste system at the early ages.

Liu et al. [10] developed a hydration temperature test device to evaluate the effects of various steel slags with potential hydraulic activity on different properties of cement pastes. The hydration temperature results showed that when the slag content was higher than 30%, the maximum hydration temperature of the cement paste decreased linearly with increasing slag content.

Almattarneh et al. [11] followed the development of the hydration process of two cement pastes—one with pure Portland cement and the other where 20% of the cement was replaced with FA—by measuring dielectric properties during the first 24 h. Through this method, four different stages of the hydration process were detected in both pastes: stage I lasted for approximately 3 h, stage II for 7 h, stage III for 8.3 h, and finally stage IV until the end of measurements. Bentz et al. [12] showed that the heat release curves for the first 24 h of hydration, measured using isothermal calorimetry, are very similar for cement pastes with water-to-cement ratios ranging from 0.325 to 0.425. The maximum heat flow was reached approximately 7 h after the production of samples. However, a change in the water/cement (w/c) ratio had a more distinctive influence on the semi-adiabatic temperature curves of the pastes. The peak temperature was increased when the w/c ratio was lowered from 0.425 to 0.400 to 0.350, but it did not increase when it was further lowered to 0.325. Greater differences between the heat releases of different pastes were noticed during the first 7 days of the hydration process.

Xu and Chung [13] measured the changes in specific heat of several cement pastes using admixture surface treatments. The w/c ratio of the tested pastes was 0.35, and Portland cement type I was used in the mixtures. The authors tested 12 different pastes with the addition of as-received silica fume and silane-treated silica fume, together with carbon fibers. The reference paste containing only cement and water had a w/c ratio of 0.45 and a measured specific heat of 0.736 J/gK. The addition of silica fume influenced the increase in the cement paste specific heat by 12% and the reduction of thermal conductivity of the paste by 40% when compared to addition of silane-treated silica fume.

Azenha et al. [14] used infrared thermography and embedded sensors to monitor the surface and internal temperatures, respectively, of a $0.4 \times 0.4 \times 0.4 \text{ m}^3$ concrete cube since casting. Emissivity was assumed constant, with an adopted value of 0.88. The cement hydration heat release was characterized with the isothermal conduction calorimetry method. The obtained results of the temperature measurements were compared with numerically calculated values using a 3D finite element model. When the measurements using a sensor placed on the surface of the tested cube and thermography in the same point were compared, they showed a very similar trend, with the sensor measuring slightly higher temperatures during the whole experiment. The highest temperature was reached at about 8.3 h after the casting of the concrete. The authors concluded that the coherence between the data collected using two different measuring techniques allowed for strong confidence in the results obtained through thermography for hardening concrete. The authors also emphasized that infrared thermography offers greater information richness regarding the temperature distribution in concrete surfaces, which cannot be obtained using other conventional temperature sensors. The drawback of the method is that it needs "visual contact" with the object in question.

Apart from the high temperature development in massive concrete structures, temperature measurements in the early age concrete are of great interest. They are applied in maturity methods in estimating the concrete strength. These methods are based on the calibration performed in laboratory conditions, as well as in situ measurements of concrete temperature development in time [15]. Measurements of concrete and ambient temperatures are for this purpose usually conducted through embedded thermo-sensors. The thermo-sensors used for the concrete temperature measurements often cannot be reused on another element of structure, which makes these kinds of measurements very costly. Another issue connected to the use of thermo-sensors is that they measure the temperature in one point of the sample (element) volume. Development of a method that would enable measurements of temperature in a higher number of points is an object of interest both for the scientific and engineering communities. Infrared thermography is one of the proposed solutions, due to the possibility of performing measurements in many points and in this way obtaining additional information on the homogeneity of the mass.

This paper aimed to investigate the relationship between measurements performed using infrared thermography and direct temperature measurements using embedded sensors through the testing of an early age hydration process of one cement-based paste, where fly ash was used as an SCM, replacing 20% of the cement (by mass).

2. Materials and Methods

Testing the applications of a thermal camera for the measurement of temperature development in the early stages of cement paste hydration was performed on a paste consisting of pure Portland cement CEM I 42.5 N and fly ash, using water from the city pipeline.

Standard testing of the cement paste (PCP) and the cement–fly ash paste (PCFP) with 20% mass replacement of the cement was performed first. The measurement of standard consistency and setting time, according to SRPS EN 196-3:2017 [16], was included. These results were used as initial data for the planning of the temperature measurements, since standard consistency paste was used for the preparation of samples. Moreover, in order to define the binder properties, two types of standard cement mortar (designated as PCM and PCFM) were prepared for testing of the flexural and compressive strength, according to SRPS EN 196-1:2017 [17], at the age of 7 and 28 days. Measurements were performed on three prismatic specimens, 4 × 4 × 16 cm in size, for each type and age of testing. The breaking force for flexural strength was measured using a range between 0 and 6 kN, with the precision of 0.1 kN, and for compressive strength using range between 0 and 200 kN, with the precision of 0.25 kN. The first mortar mix consisted of pure cement, standard sand, and water (PCM), while in the second mix, 20% of cement was replaced with fly ash (PCFM).

For measurements of cement hydration temperature at the early ages, cubic concrete molds with a 10 cm edge were used. For two samples, the mold was completely filled in, while for two samples, the mold was filled up to the height of 5 cm. Cement–fly ash paste (PCFP) of a water-to-binder ratio 0.25 was used for all of the sample preparations. Two sensors were placed in each of the four samples, one in the middle (at the cross section of the cube diagonals) and another in the center of the only free side. In order not to disturb the placing of the sensors, the samples were placed and compacted by hand. The room temperature and humidity were held constant (22 °C, 50%).

2.1. Testing of Cement

The chemical composition of the cement and fly ash used in the test is presented in Table 1, while the specific and bulk densities of the cement, fly-ash, and sand used are presented in Table 2. Composition of the pastes and mortars tested is presented in Table 3.

Table 1. Chemical composition of cement and fly ash used (%).

Oxide	Cement (PC)	Fly Ash (F)
CaO	63.12	6.12
SiO_2	20.32	58.60
Al_2O_3	4.60	21.92
Fe_2O_3	3.30	5.97
MgO	1.92	1.77
K_2O	0.61	1.50
Na_2O	0.26	0.37
TiO_2	0.44	0.49
MnO	0.07	-
P_2O_5	0.35	-
SO_3	3.20	-
Loss on ignition	1.69	3.26

Table 2. Basic properties of the component materials used.

	Cement (PC)	Fly Ash (F)	Sand
Specific density (g/cm^3)	3.040	2.190	2.740
Bulk density (g/cm^3)	0.890	0.690	1.610

Table 3. Composition of the pastes and mortars.

	PCP	PCFP	PCM	PCFM
Mass of cement (g)	500	400	450	360
Mass of fly ash (g)	-	100	-	90
Mass of water (g)	125	125	225	225
Mass of sand (g)	-	-	1350	1350

The results of the basic properties of the cement and cement–fly ash pastes used in this experiment are presented in Table 4. The mechanical properties of the tested mortars are presented in Table 5. The addition of fly ash led to an increase in both flexural and compressive strength at the age of 28 days. Setting time was prolonged with the addition of fly ash, which was an important information for the temperature measurements.

Table 4. Properties of the tested pastes.

Property	Property/Time	PCP	PCFP
	Water/cement factor	0.25	0.28
Setting time of the paste	Beginning, h:min	02:45	03:48
	Ending, h:min	04:00	06:08
Standard consistency of the paste	Water/cement factor	0.25	0.28
	Vicat's apparatus (mm)	6.5	6.0

Table 5. Properties of the tested mortars.

Property	Age/Time	PCM	PCFM
Flexural strength of mortar (MPa)	7 days	6.88	6.44
	28 days	9.38	10.13
Compressive strength of mortar (MPa)	7 days	34.69	32.42
	28 days	47.66	53.28

2.2. Temperature Measurement Setup

The effects of the heat of hydration were recorded simultaneously on all four samples using a thermal imaging camera (FLIR E6) to monitor the temperature on the surface of the samples, as well as two thermo-sensors (IC LM35) installed in the middle and on the surface of each sample. To measure, display, and record the temperature in the center and on the surface of the four samples, we developed a custom LabVIEW application connected to an eight-channel system of thermo-sensors (see details below). The ambient temperature (thermostatic laboratory) was 22 °C. Temperature monitoring of the fresh concrete lasted for 30 h, and the sampling period was 10 min.

Figure 1 shows the experimental setup used, and Figure 2 presents the positions of the sensors in the samples. A thermal imaging camera (with an accuracy of ±2 °C) recorded a thermogram of the upper surface of the sample. To prevent the reflection of IR radiation from the environment, PVC pipes with diameters of 200 mm were used.

Figure 1. Experimental setup.

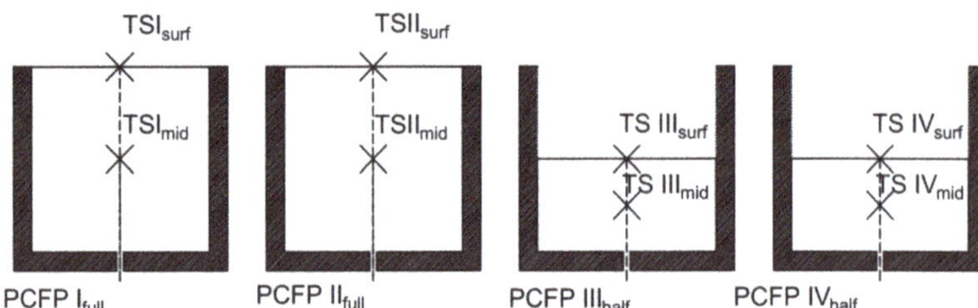

Figure 2. Disposition of the sensors in the samples.

On the basis of the expected temperature range and the required accuracy, an integrated LM35 circuit [18] was used as the thermo-transducer. An additional resistor (100 k) to the negative source (-5 V) enabled the measurement of negative temperatures (Figure 3). After filtering (capacitors 10 nF) and adjustment (resistors 10 kΩ), the signals from the thermo-sensor were fed to the input channels (AI CH0 to AI CH7) of the analog-to-digital converter (NI USB 6009) [19]. The obtained digital signals were fed into a computer (PC) via the USB port and further processed by software. The temperature transmitters themselves were factory-calibrated in the range of $-40\,°C$ to $110\,°C$, so no additional adjustments were required. The accuracy of the measurement was ±0.2 °C. Due to the physical dispersion of the thermal sensors, shielded conductors were used for the transmission of analog signals over a relatively large distance (several meters) for each channel separately.

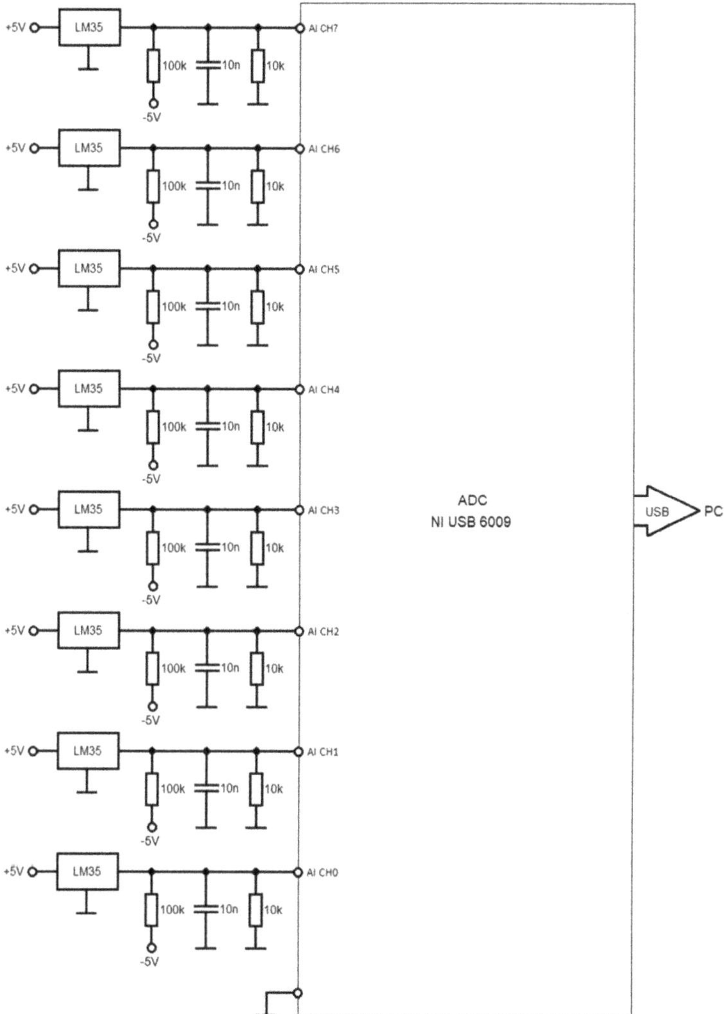

Figure 3. Scheme of the hardware.

An application for measuring, displaying, and storing temperature in 8 points was developed in the software package LabVIEW 2016 [20]. The graphic code (block diagram) of the application is shown in Figure 4. DAQ Assistant, a standard module of the LabVIEW package for data acquisition, was used to measure analog signals from thermo-transducers. The data were then separated into eight channels, averaged (Statistics), and multiplied by 100 to transform the temperature into °C (output from LM35 is 10 mV/°C). The obtained temperatures are shown on the displays Ti (°C) and diagrams (XY Ti) and finally saved in the form of a text file whose name is specified before starting the measurement. The program was able to be stopped at any time by clicking the STOP button [21].

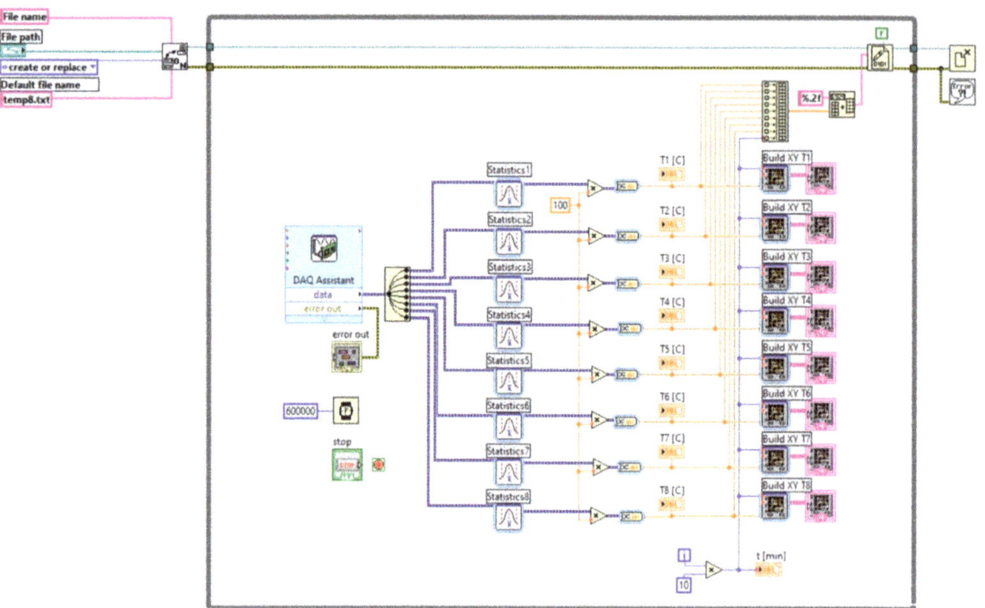

Figure 4. The LabVIEW application.

For clarity, separate diagrams for each of the 8 channels were included on the front panel of the virtual instrument. Numerical displays for temperature (Ti (°C)) and time (t (min)), a file name input field, the STOP button, and an error code display field were also included.

2.3. Thermal Camera Measurement Setup

Images were obtained using a FLIR E6 thermal camera positioned at a distance of 60 cm from the sample, through a tube (as presented in Figure 1). This was done to stabilize the environment between the camera and the surface of the samples. Images were always taken in the same order, after the stabilization of the temperature.

After the finalization of the measurements (30 h), images were processed using the FLIR Tools Plus V12 software, and the reports for each sample were extracted. The processing included defining the surface from which the temperatures should be calculated and inputting the values for emissivity, reflected temperature, and atmospheric temperature, as presented in Figure 5.

The emissivity was measured using two focus points for each sample. The first point was prepared by placing black tape on the surface of the sample. The measurement of temperature on the middle point of this black surface with known emissivity 0.95 [14] is shown in Figure 6a. The second point of measurement was placed on the similar position, but on the sample surface. During the image processing of the second image, it was brought to the same level of the temperature by changing emissivity (as shown in Figure 6b). A value of 0.85 was adopted for the tested paste.

The pictures using the FLIR camera were taken every half an hour in periods of stable temperatures and every 10 min in periods of intensive temperature fluctuation.

Results from the embedded thermo-sensors were plotted in LabVIEW using temperature–time diagrams. Similar diagrams were generated from the test results of the thermal camera images, after the appropriate image analysis.

These diagrams were then analyzed and compared with the setting time results measured on the cement–fly ash paste samples.

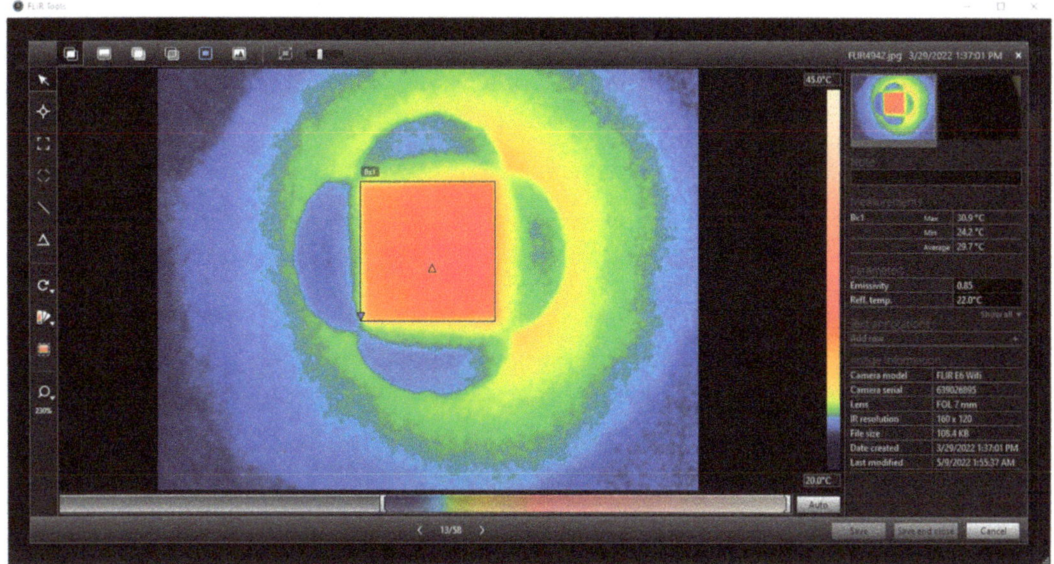

Figure 5. Example of image processing using the FLIR Tools Plus software (the square in the middle represents the surface of interest).

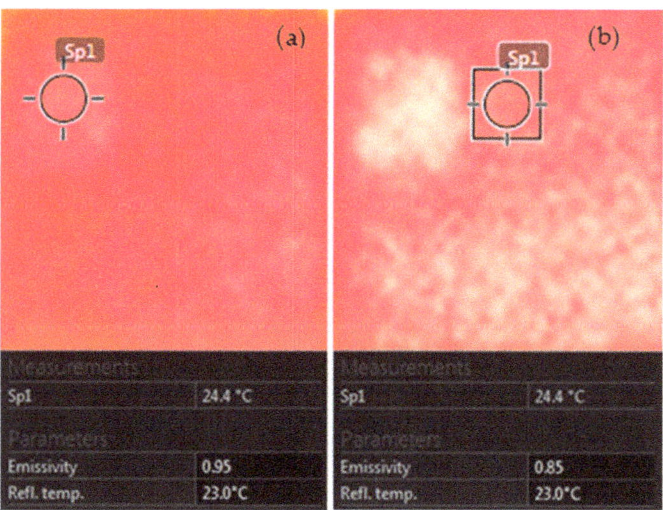

Figure 6. Testing of paste emissivity (**a**) measurement on the surface of known emissivity; (**b**) processed image of the second measurement point.

2.4. Modelling

Modelling of the diagrams was performed for the periods where intensive temperature change was noted. Since the heating of the samples was faster than the cooling, the asymmetric Gaussian was used to describe the process, defined by the following parameters:

$$A\left(z_i, \mu_i^z, \sigma_i^2, r_i\right) \equiv \frac{2}{\sqrt{2\pi}} \frac{1}{\sqrt{\sigma_i^2(r_i+1)}} \begin{cases} exp\left(-\frac{(z_i-\mu_i^z)^2}{2\sigma_i^2}\right) if\ z_i > \mu_i^z. \\ exp\left(-\frac{(z_i-\mu_i^z)^2}{2r_i^2\sigma_i^2}\right) otherwise. \end{cases} \tag{1}$$

This model was referred to as univariate asymmetric Gaussian by the authors Kato et al. [22]. This function returns normalized results ranging from 0 to 1. In order to fit the results of the temperature measurements, the parameters μ_i^z, σ_i^2, r_i were adopted for every diagram. This necessitated the addition of the additional parameters A_1, A_2, N_1, and N_2 in order to correlate the obtained results with the actual temperature measurements, which is explained in more detail in Section 3. The adopted model was then

$$A\left(z_i, \mu_i^z, \sigma_i^2, r_i, A_1, A_2, N_1, N_2\right) \equiv \frac{2}{\sqrt{2\pi}} \frac{1}{\sqrt{\sigma_i^2}(r_i+1)} \begin{cases} A_1 \cdot exp\left(-\frac{(z_i - \mu_i^z)^2}{2\sigma_i^2}\right) + N_1 \ if \ z_i > \mu_i^z. \\ A_2 \cdot exp\left(-\frac{(z_i - \mu_i^z)^2}{2r_i^2\sigma_i^2}\right) + N_2 \ otherwise. \end{cases} \tag{2}$$

The moment when the temperature began to rise was adopted as t = 0, while the moment when the temperature began to stabilize was taken as final.

3. Results

3.1. Temperature Measurements

The results of the temperature measurements from the embedded sensors are presented in Figures 7 and 8, while the results of the thermal camera measurements are presented in Figures 9 and 10. On each of the figures, an additional vertical line was added to mark the beginning of the setting time, as presented in Table 4.

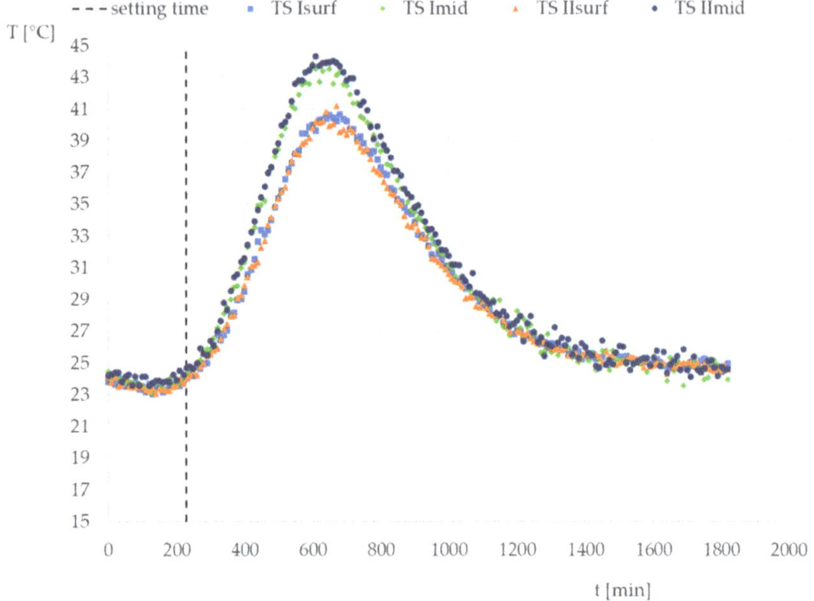

Figure 7. Temperature measurements of the samples PCFP I$_{full}$ and PCFP II$_{full}$ (using sensors).

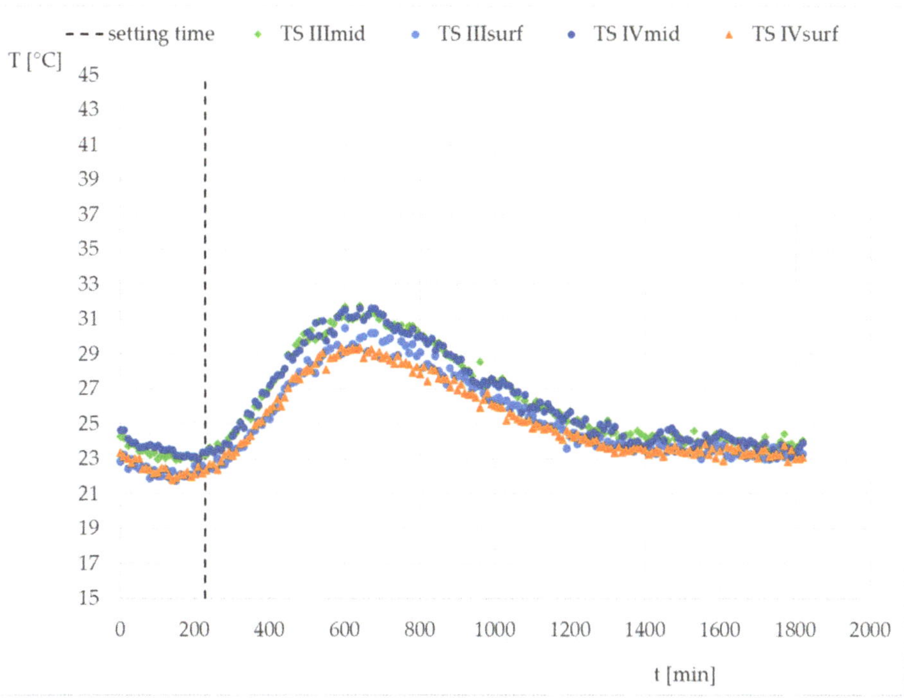

Figure 8. Temperature measurements of the samples PCFP III_{half} and PCFP IV_{half} (using sensors).

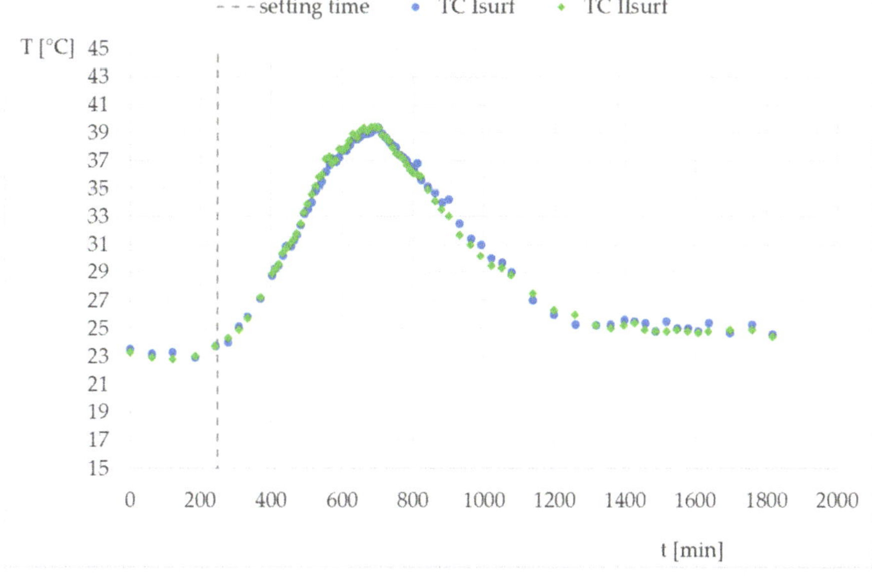

Figure 9. Temperature measurements of the samples I and II (using thermal camera).

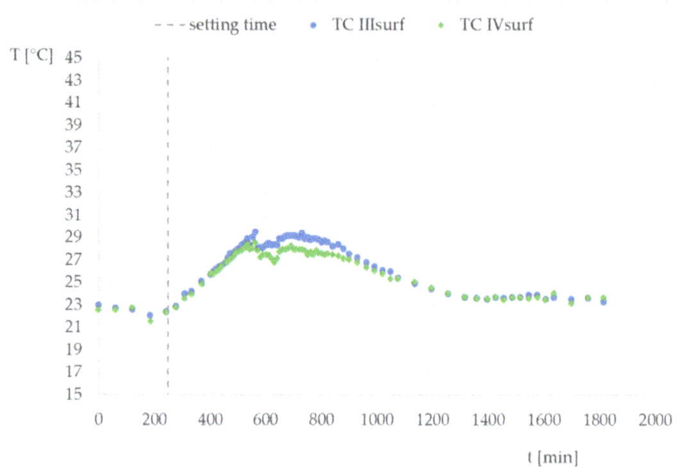

Figure 10. Temperature measurements of the samples III and IV (using thermal camera).

Peak temperatures in the middle of each sample were higher than the peak temperatures measured on the sample surface. The differences were around 4 °C for the full cubic samples, and around 2 °C for the samples with half height. These differences were the consequence of the expected greater development of the temperatures induced by the cement hydration in the core of the samples. The temperatures were reduced towards the surfaces of the samples due to the temperature exchange with the surrounding environment.

Extreme temperatures measured with the thermal camera were 1 to 2 °C lower than the values measured with sensors placed on the surface of the samples. The durations of the temperature increase and final stabilization periods were very similar for all of the samples and measuring methods.

3.2. Modelling

Figures 11 and 12 present the same results, fitted using the asymmetric Gaussian as explained in Section 2. For the purpose of the modelling, the coordinate origin was placed in the moment corresponding to the onset of the temperature increase, while the cutoff was defined as 1200 min. The parameters adopted for the fitting functions are presented in Table 6.

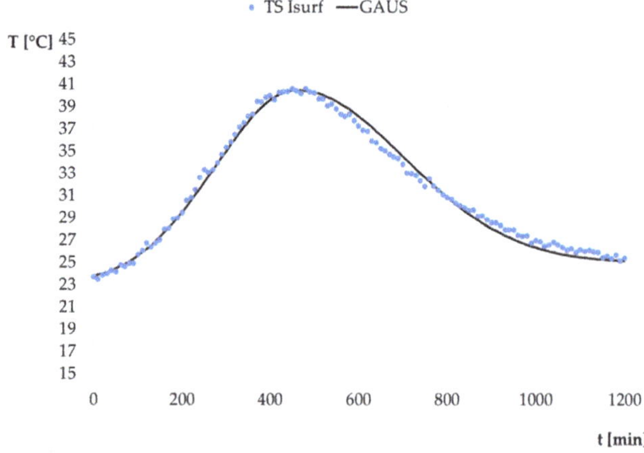

Figure 11. Results of the temperature measurements of sensor TS I_{surf} fitted with asymmetric Gaussian function.

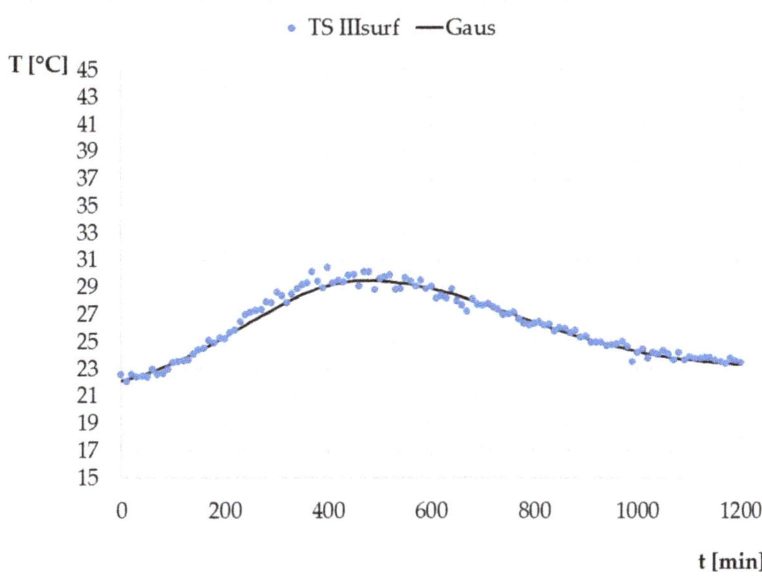

Figure 12. Results of the temperature measurements of sensor TS III$_{surf}$ fitted with asymmetric Gaussian function.

Table 6. Parameters used for the fitting functions.

Coefficient	PCFPI$_{full}$		PCFP II$_{full}$		PCFP III$_{half}$		PCFP IV$_{half}$	
	TS I$_{surf}$	TS I$_{mid}$	TS II$_{surf}$	TS II$_{mid}$	TS III$_{surf}$	TS III$_{mid}$	TS IV$_{surf}$	TS IV$_{mid}$
μ_i	460	440	460	440	470	440	440	440
σ_i	244	236.5	245.5	249	295	303	311	311
r_i	0.758	0.774	0.762	0.791	0.7898	0.686	0.6398	0.686
A_1	15.5	19.5	15.5	19.5	6.5	8.0	6.5	8.0
A_2	17.5	21.5	17.5	21.5	8.5	10.0	8.5	10.0
N_1	25	24	25	24.5	23	23.5	23	23.5
N_2	23	22	23	22.5	21	21.5	21	21.5

The parameter μ_i was defined as the time in minutes needed to reach the maximum temperature in the sample. The same parameter also defines the point of the change in the applied calculation of the two branches in the modeling curve. Parameters σ_i and r_i describe the skewness of the two branches and their mutual relationship. They are the measure of the differences in the rate of temperature change at the onset of the hydration process. Parameters N_1 and N_2 correspond to the temperature of the samples at the beginning (start of the temperature increase)—N_2, and after 1200 min—N_1. Parameters A_1 and A_2 were included in the model in order to adapt the normalized function to the presented temperature measurements. The sum of each of the two parameters $A_1 + N_1$ and $A_2 + N_2$ corresponds to the maximal temperature measured by the sensor in question, while the parameters A_1 and A_2 show the difference in the temperatures when compared to the temperature at the beginning of the test (N_2) and at the end (N_1). The lower values of these two coefficients for sensors TS III$_{surf}$, TS III$_{mid}$, TS IV$_{surf}$. and TS IV$_{mid}$ were derived from the fact that they were placed in samples of lower height, with a smaller volume of paste, which resulted in the development of lower temperatures. These values corresponded to approximately 40% of the A_1 and A_2 coefficient values adopted for larger samples. Figure 13 presents the fitting function for one of the samples measured using the thermal camera. The parameters adopted for the fitting of all four measurements are presented in Table 7.

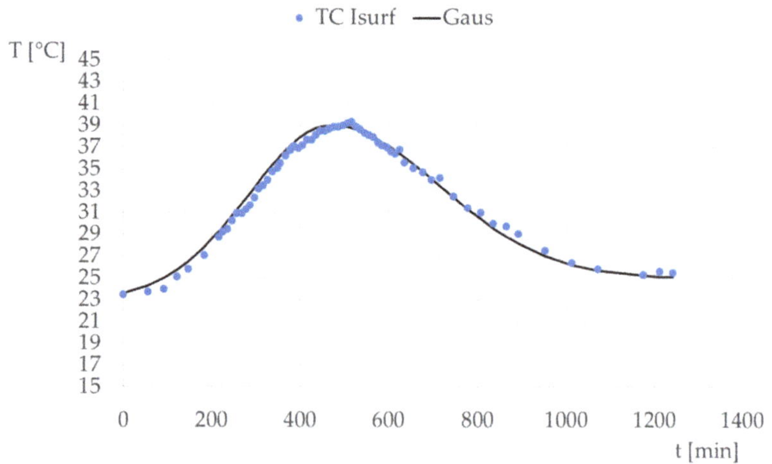

Figure 13. Results of the temperature measurements of the thermal camera for sample I fitted with asymmetric Gaussian function.

Table 7. Parameters used for the creation of the fitting functions for thermal camera measurements.

Coefficient	TC I$_{surf}$	TC II$_{surf}$	TC III$_{surf}$	TC IV$_{surf}$
μ_i	470	470	470	440
σ_i	244	245.5	295	311
r_i	0.758	0.762	0.7898	0.6398
A_1	14	14	6.5	6.5
A_2	16	16	8.5	8.5
N_1	25	25	23	23
N_2	23	23	21	21

The values of the fitting parameters chosen for describing the infrared thermography results were almost identical to the values adopted for the surface sensors, shown in Table 6. The values for samples III and IV were identical. For samples I and II, the maximum temperature was reached with a delay of approximately 10 min, while the highest temperatures were 1–2 °C lower than recorded by the sensors placed on the surface of the samples.

4. Discussion

4.1. The Influence of the Fly Ash Addition to the Cement-Based Composites

The mechanical properties of the mortars containing fly ash were improved, reaching 10.1 MPa for flexural strength and 53.3 MPa for compressive strength after 28 days. Both values were higher than the values measured on the pure cement mortar samples. The beginning and the setting time of the paste containing fly ash were prolonged for one and two hours, respectively, when compared to the pure cement paste.

4.2. Temperature Development Discussion

The maximum temperature measured in the central point of the samples was 44 °C, while the maximum temperature on the surface of the samples was 40 °C. These temperatures were measured for the paste in which 20% of cement was replaced with fly ash. The maximum temperature for the paste containing 40% FA (as measured by Zhang et al. [9]) was 37 °C. The difference was expected due to the larger amount of fly ash used in this study.

If the stages in the temperature development are to be compared to the stages defined in Almattarneh et al. [11], very good compliance can be observed as well. The first stage in the temperature measurements was approximately 200 min (3.3 h), while the second stage lasted until the maximum temperature was reached, lasting for approximately 7.5 h. The third, cooling stage, lasted for approximately 9 h, which was all in accordance with Almattarneh et al. [11].

According to the aforementioned, it can be concluded that the obtained results show very good compliance with the results presented in the literature.

4.3. Comparison between the Results Obtained by Thermal Camera and Thermo-Sensors

The values obtained from the sensor placed in the central part of the samples were expectedly higher than the temperatures measured through sensors placed on the surface or with the thermal camera. The temperatures measured on the surface with two different devices showed very good accordance. Figures 14 and 15 represent the results of thermal camera measurements $\pm 2\,^\circ$C in the central point of the sample and the surface thermal sensors measurements for samples I and III.

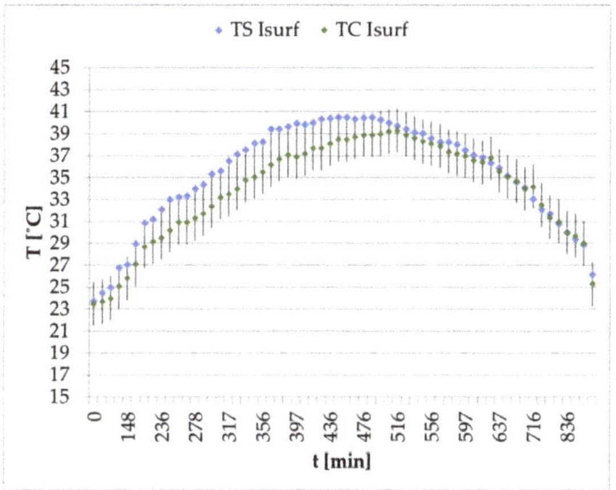

Figure 14. Results of the temperature measurements of the thermal camera for sample I with $\pm 2\,^\circ$C limits together with Sensor TS I$_{surf}$ measurements.

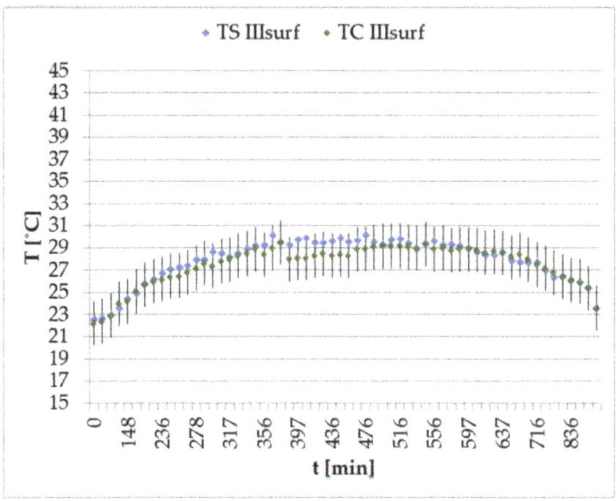

Figure 15. Results of the temperature measurements of the thermal camera for sample III with $\pm 2\,^\circ$C limits together with Sensor TS III$_{surf}$ measurements.

As shown in Figure 14, the highest differences between the measurements occurred in the period of rising temperature and during the temperature peak. Even in this period, the difference was under 2 °C, considered to be the accuracy of the thermal camera measurements. The differences were even smaller for sample III. In conclusion, very good accordance was achieved between the two measuring techniques.

The time when the maximum temperature was reached in the thermal camera recordings slightly differed from the time measured by sensors, but only for the 10 cm high samples.

4.4. Modelling Discussion

All the measurements were modelled through the asymmetric Gaussian function, with very similar coefficients. Small differences between coefficients A_1 and A_2 probably occurred due to the imprecisions in the emissivity measurements, which will be improved through further investigations. The values adopted for the coefficients r_i and σ_i were probably in correlation with the specific heat of the cement paste, which was measured in the literature to be 0.736 J/gK for pastes containing larger amounts of water and using only cement as a binder [13].

For future research, it would be of great importance to perform similar measurements on concrete cubic samples, where the higher heterogeneity of the material could open more interesting questions. Finally, full-scale in situ tests should be performed using the results obtained through the testing of the concrete slabs, bridges, etc.

5. Conclusions

The main focus of the experimental work presented in this paper was the possibility of applying infrared thermography to describe the effects of the temperature development during the early hydration phase of the cement composites. In order to achieve this, a special testing setup was developed and used for temperature measurements in the sample middle point and on the central point of the exposed surface.

The following conclusions were drawn:

The partial replacement of the cement with 20% of fly ash prolonged the setting time of the paste, increased the amount of water necessary for obtaining standard consistency, and increased the flexural and compressive strength of the mortars prepared with these materials.

The results obtained through the system of embedded sensors were very uniform and in accordance with the results found in the literature. As expected, the highest temperatures were measured in the central point of the samples.

Infrared thermography results are comparable with results measured through the embedded thermo-sensors, especially at the surface of the samples.

Although additional confirmatory measurements are necessary, infrared thermography shows the potential to replace embedded sensors on the testing surface of the sample. As noted in the literature, the amount of data collected through the infrared thermography method is greater than by any other method, and its potential is still to be investigated. Further research is necessary to obtain a reliable relation between the inner temperature of the sample and infrared-thermography-based surface measurements.

Author Contributions: Conceptualization, Z.S., A.S. and M.A.; methodology, Z.S.; software, M.S.; hardware, M.S.; validation, Z.S. and A.S.; formal analysis, N.Ž. and M.A.; investigation, N.Ž.; resources, Z.S. and A.S.; data curation, N.Ž. and M.S.; writing—original draft preparation, M.A. and N.Ž.; writing—review and editing, Z.S. and A.S.; visualization, Z.S., A.S., M.S. and M.A.; supervision, Z.S. and A.S. All authors have read and agreed to the published version of the manuscript.

Funding: This research received no external funding.

Data Availability Statement: Not applicable.

Acknowledgments: This research was supported by the Ministry of Education, Science and Technological Development of the Republic of Serbia, contract numbers 451-03-47/2023-01/200131 and 200092.

Conflicts of Interest: The authors declare no conflict of interest.

References

1. *EN 13187:1999*; Thermal Performance of Buildings—Qualitative Detection of Thermal Irregularities in Building Envelopes—Infrared Method. Comite Europeen de Normalisation: Brussels, Belgium, 1999.
2. Lu, X.; Memari, A. Application of infrared thermography for in-situ determination of building envelope thermal properties. *J. Build. Eng.* **2019**, *26*, 100885. [CrossRef]
3. Zheng, Y.; Wang, S.; Zhang, P.; Xu, T.; Zhuo, J. Application of nondestructive testing technology in quality evaluation of plain concrete and RC structures in bridge engineering: A review. *Buildings* **2022**, *12*, 843. [CrossRef]
4. Balaras, C.A.; Argiriou, A.A. Infrared thermography for building diagnostics. *Energy Build.* **2002**, *34*, 171–183. [CrossRef]
5. Bakour, A.; Ftima, M.B. Experimental investigations on the asymptotic fracture energy for large mass concrete specimens using wedge splitting test. *Constr. Build. Mater.* **2021**, *279*, 122405. [CrossRef]
6. Ouyang, J.; Chen, X.; Huangfu, Z.; Lu, C.; Huang, D.; Li, Y. Application of distributed temperature sensing for cracking control of mass concrete. *Constr. Build. Mater.* **2019**, *197*, 778–791. [CrossRef]
7. Savić, A.; Jevtić, D.; Zakić, D.; Radević, A.; Aškrabić, M. The Use of Industrial Byproducts as Fillers in Self-Compacting Concrete. In *Proceedings of the 14th International conference iNDiS 2018, Subconference Eco Build*; Department of Civil Engineering and Geodesy—Faculty of Technical Sciences: Novi Sad, Serbia, 2018; pp. 1267–1273. ISBN 978-86-6022-105-8.
8. Govedarica, O.; Aškrabić, M.; Hadnađev-Kostić, M.; Vulić, T.; Lekić, B.; Rajaković-Ognjanović, V.; Zakić, D. Evaluation of Solidified Wastewater Treatment Sludge as a Potential SCM in Pervious Concrete Pavements. *Materials* **2022**, *15*, 4919. [CrossRef] [PubMed]
9. Zhang, T.; Ma, B.; Jiang, D.; Jiang, Q.; Jin, Z. Comparative research on the effect of various mineral admixtures on the early hydration process of cement. *Constr. Build. Mater.* **2021**, *301*, 124372. [CrossRef]
10. Liu, K.; Ma, Y.; Yuan, Z.; Wu, Y.; Zhang, X.; Zhang, X. Optimisation of early hydration, microstructure, and elevated-temperature resistance of calcium aluminate cement using steel-making slag. *Ceram. Int.* **2022**, *48*, 35328–35339. [CrossRef]
11. Almattarneh, H.; Alwadie, A.; Malkawi, A.; Nuruddin, M.F. A novel method for monitoring hydration process of cement paste material. *Appl. Mechs. Mater.* **2014**, *567*, 333–338. [CrossRef]
12. Bentz, D.; Peltz, M.; Winpigler, J. Early-age properties of cement-based materials: II. Influence of water-to-cement ratio. *J. Mater. Civ. Eng.* **2009**, *21*, 9. [CrossRef]
13. Xu, Y.; Chung, D.L.L. Increasing the specific heat of cement paste by admixture surface treatments. *Cem. Concr. Res.* **1999**, *29*, 1117–1121. [CrossRef]
14. Azenha, M.; Faria, R.; Figueiras, H. Thermography as a technique for monitoring early age temperatures of hardening concrete. *Constr. Build. Mat.* **2011**, *25*, 4232–4240. [CrossRef]
15. *NEN 5970:2001 nl*; Determination of the Compressive Strength Development of Young Concrete Based on the Weighted Maturity. Royal Netherlands Standardization Institute: Delft, The Netherlands, 2001.
16. *EN 196-3:2016*; Methods of Testing Cement—Part 3: Determination of Setting Times and Soundness. Comite Europeen de Normalisation (CEN): Brussels, Belgium, 2016.
17. *EN 196-1:2016*; Methods of Testing Cement—Part 1: Determination of Strength. Comite Europeen de Normalisation (CEN): Brussels, Belgium, 2016.
18. LM35 Precision Centigrade Temperature Sensors, Texas Instruments. 2017. Available online: https://www.ti.com/product/LM35/datasheet (accessed on 20 October 2022).
19. NI USB-6008/6009, National Instruments, USA. 2015. Available online: http://www.ni.com/pdf/manuals/371303n.pdf (accessed on 20 October 2022).
20. NI LabVIEW Documentation. Available online: https://www.ni.com/docs/en-US/category/labview-docs (accessed on 21 October 2022).
21. Stević, Z.; Stević, M.; Radovanović, I.; Stolić, P.; Radivojević, M.; Petronić, S. PC and LabVIEW based voltage and current source for electrochemical investigations. In Proceedings of the International Scientific-Practical Conference, Modern Information and Electronic Technologies, Odessa, Ukraine, 24–28 May 2021; pp. 46–49.
22. Kato, T.; Omachi, S.; Aso, H. Asymmetric Gaussian and Its Application to Pattern Recognition. In *Structural, Syntactic, and Statistical Pattern Recognition, Proceedings of the SSPR /SPR 2002, Windsor, ON, Canada, 6–9 August 2002*; Lecture Notes in Computer Science; Caelli, T., Amin, A., Duin, R.P.W., de Ridder, D., Kamel, M., Eds.; Springer: Berlin, Germany, 2002; Volume 2396, p. 2396. [CrossRef]

 *buildings*

Article

Experimental and Numerical Assessment of the Thermal Bridging Effect in a Reinforced Concrete Corner Pillar

Gianpiero Evola * and Antonio Gagliano

Department of Electric, Electronic and Computer Engineering (DIEEI), University of Catania, Via S. Sofia 64, 95123 Catania, Italy; antonio.gagliano@unict.it
* Correspondence: gevola@unict.it

Abstract: This paper discusses experimental and simulated data regarding the thermal bridging effect in a reinforced concrete corner pillar, which belongs to a building dating back to the 1980s and located in Southern Italy. The thermal field determined by the concrete pillar corner has been evaluated, introducing an experimental procedure based on both direct measurements and indirect observations of the inner superficial temperature by means of thermal imaging techniques and surface temperature probes. Moreover, indoor and outdoor air temperature and relative humidity were measured to provide suitable boundary conditions in the numerical simulations, performed with a commercial software tool widely used in Italy based on 2D finite element techniques. The experimental measurements show that, at more than 50 cm from the corner, the surface temperatures become almost constant, meaning that the thermal bridging effect becomes less evident. However, the surface temperature in the corner is around 1.5 °C lower than in the undisturbed flanking walls. In terms of local heat flux, the discrepancy between simulations and measurements is below 3%. Finally, this paper verifies the effectiveness of External Thermal Insulation Composite System (ETICS) renovation in reducing the thermal bridging effect of the corner pillar. The results also include the calculation of the linear thermal transmittance with a series of relations available in well-known atlases for thermal bridges and show that these relations are more reliable in the case of uninsulated pillar than for the insulated one.

Keywords: thermal bridge; reinforced concrete; corner pillar; heat flux; linear thermal transmittance; 2D numerical simulations; experimental measurements

Citation: Evola, G.; Gagliano, A. Experimental and Numerical Assessment of the Thermal Bridging Effect in a Reinforced Concrete Corner Pillar. *Buildings* **2024**, *14*, 378. https://doi.org/10.3390/buildings14020378

Academic Editor: Grzegorz Ludwik Golewski

Received: 11 January 2024
Revised: 25 January 2024
Accepted: 29 January 2024
Published: 1 February 2024

1. Introduction

According to a recent estimation by the Odyssee-Mure project, buildings are responsible for about 39% of the final energy demand in the European Union [1], with 2/3 of this share coming from households. Such a significant energy consumption by the existing building stock can be mostly attributed to the poor thermal performance of their envelope, especially for those buildings built before 1990, when current EU regulations addressing energy efficiency in buildings were not in force. Thus, it is possible to estimate that around 75% of the current EU building stock is not energy efficient [2].

A non-negligible contribution to heat losses in buildings comes from thermal bridges. Thermal bridges are those parts of the building envelope where the otherwise uniform thermal resistance is modified either by a local change in the geometry or by penetration of the building envelope by materials with different thermal conductivity [3]. Common examples of thermal bridges are balconies, floor-to-wall connections, corners, and window sills/reveals. The impact of thermal bridges on the energy balance is particularly relevant in buildings with reinforced concrete (RC) structures. Indeed, the thermal conductivity of RC ranges from 2.0 to 2.5 $W \cdot m^{-1} \cdot K^{-1}$, depending on the quantity of reinforcing steel bars, while other common building materials (e.g., cement screed, mortar, bricks, non-reinforced concrete) have a much lower thermal conductivity (namely, between 0.4 and

$1.0\ \mathrm{W\cdot m^{-1}\cdot K^{-1}}$). Hence, RC beams, pillars, and balconies locally modify the temperature field, increase heat transfer, and determine "cold points" where mould growth can occur [4].

While the addition of thermal insulation to the building envelope is an effective way to drastically reduce the transmission heat losses through walls and slabs, only an accurate correction of the thermal bridging details may lead to a significant reduction in their impact [5]; thus, thermal bridges may become particularly important also in well-insulated buildings where the construction detailing has not been designed with good accuracy [6]. For instance, Ilomets et al. underlined that neglecting thermal bridges can lead to underestimating by 23% the transmission heat losses in an uninsulated building in Northern Europe, but ETICS renovation is likely to even increase this share up to 34% in case of "bad practice" correction, especially because of window lintels and balconies. Instead, "good practice" ETICS insulation reduces thermal bridges to around 10% of the transmission heat losses [7].

Some studies analysed thermal bridges in buildings with lightweight concrete or clay bricks and RC structures and demonstrated that in these buildings, the most important thermal bridges were due to balconies, pillars, and the contact between the wall and window [5,8,9]. Curto et al. provided a detailed analysis of thermal bridges in a building with tuff blocks and RC structures and drew up a list of "good practice" precautions to insulate the building and reduce the transmission heat losses in the thermal bridges to 9% of the total [10]. Thermal breakers emerge as an interesting, innovative solution to correct the thermal bridge between exterior walls and concrete slabs, both with and without a balcony, leading to a reduction in the space heating demand ranging from 5% to 25% according to the climate and the building geometry [8,11–14].

Moreover, it is interesting to underline that in Southern Europe, "good practice" insulation with corrected thermal bridges can reduce the space heating demand of a building by around 30% in the winter, but the effects of thermal bridge correction in the summer is practically negligible, because of the limited temperature gradient between indoors and outdoors [15–17]. However, in very hot climates such as in the Gulf area, the intense solar radiation hitting the structural elements, especially in the west façades, makes heat gains through thermal bridges not negligible in the summer; covering the exposed structural elements with a layer of insulation of no more than 100 mm can prevent such a negative effect [18]. Finally, the mitigation of thermal bridges is also of crucial importance in the refurbishment of traditional buildings, as pointed out by Cirami et al. [19].

Now, given the importance of thermal bridges and their non-negligible impact on the overall heat losses in buildings, their evaluation must be accurately carried out with reliable tools, such as via numerical finite element analysis or by means of thermal bridge atlases compliant with the recently updated Standard EN ISO 14683:2017 [20]. Previously adopted approximate methods, e.g., based either on a fixed percentage increase applied to the transmission heat losses or on a fixed surcharge to the U-value, regardless of the building's features, are no longer admitted [21].

The current thermal bridge calculation tools aim at assessing the linear thermal transmittance (ψ): this parameter measures the increase in the transmission heat losses caused by a thermal bridge per unit length if compared to a thermally undisturbed envelope with parallel isothermal lines. Only a few years ago, commercial finite element numerical tools were not widespread; thus, the linear thermal transmittance of thermal bridges was mainly determined through atlases and abacuses, including a large variety of common cases.

For instance, the CENED abacus, published in 2011 under the initiative of the Lombardy Region and based on the research carried out by Politecnico of Milan [22], includes more than one hundred thermal bridges with equations to calculate their linear thermal transmittance (ψ) as a function of technical parameters like the wall thickness, the U-value, and the thermal conductivity of the insulating material (if any). One more atlas available in Italy is edited by Edilclima [23]; this includes a collection of common thermal bridges, and their linear thermal transmittance is reported in tabular form. However, no analytic relations are available. The Standard EN ISO 14683:2017 also includes default ψ-values for a limited range of commonly occurring types of two-dimensional thermal bridges, which can be used when there is no detailed information about the specific building node [20]; other available catalogues are the Swiss atlas distributed by the Office Federal de l'Energie OFEN [24] and the Passive House Institute atlas [25].

All these catalogues consider recurring construction techniques and are sufficiently reliable (accuracy $\pm 20\%$) if the dimensions and the thermal properties of the included thermal bridges are similar—or less favourable—to those of the real detail; otherwise, catalogues may become very inaccurate and usually underestimate the heat losses through the thermal bridge [26,27].

On the contrary, the finite element analysis has the advantage of avoiding any geometric limitations and allows for an accurate calculation of the depicted construction detail according to the Standard EN ISO 10211:2017, with an error below $\pm 5\%$ [3]; furthermore, it provides both the heat flux and the temperature distribution in the building component, thus giving information on the possible occurrence of mould growth [10]. The use of 2D finite element numerical tools to determine the linear thermal transmittance has recently become very common, even in non-scientific applications such as for the release of Energy Performance Certificates. Indeed, many commercial software tools now include suitable plug-ins to support this.

The present paper shows experimental and simulated analyses regarding the thermal field in a reinforced concrete protruding corner pillar, which belongs to a building with uninsulated lightweight concrete walls. The building dates back to the 1980s; it is located in Southern Italy and displays a very common building solution in Italy, at least until the first comprehensive energy-saving regulation was issued in 1991. The proposed investigation aims to define an experimental procedure to evaluate the thermal bridging effect using both direct measurements and indirect observations of the inner surface temperature around the reinforced concrete protruding corner pillar. In particular, the internal surface temperatures of both the corner pillar and the adjacent building elements were measured by using several Pt 1000 probes and a thermal imaging camera. Indeed, infrared thermography has been recently recognized as a useful, non-invasive method to provide both qualitative and quantitative measures of the actual thermal bridging performance [28,29]. The infrared thermography allows for a precise visualization of the temperature field around the corner pillar, which is here enforced by the direct measurements of the surface temperatures in a grid of points within the investigated area. The thermal bridge was also studied numerically through a commercial software tool widely used in Italy, based on 2D finite element techniques. Moreover, the effectiveness of ETICS renovation in reducing the thermal bridging effect of the corner pillar has been verified.

The findings of this paper demonstrate that the proposed experimental practice can provide a reliable assessment of the thermal field around corner pillars, highlighting the extension of the coldest area, as well as the increased heat flux transmitted through the thermal bridge. One of the strengths of the proposed procedure is its simplicity, both in terms of timing and equipment. Thus, the proposed approach could be followed by technicians to quantify the thermal bridging effect in existing buildings with reinforced concrete structures as well as to verify the efficacy of the retrofit interventions for the mitigation of this kind of thermal bridge. Moreover, this paper casts light on the ability of catalogues and atlases to reliably calculate the so-called "linear thermal transmittance" for this type of thermal bridge, both in the case of insulated and uninsulated walls, thus providing useful methodological information for a reliable assessment of corner pillars in a very high number of buildings.

2. Materials and Methods

The case study here addressed a corner pillar made with reinforced concrete that belonged to a detached house with uninsulated hollow concrete blocks built in the 1980s. The building is in Ragalna (province of Catania), a municipality located on the Southern slope of Mount Etna, at 830 m above sea level (Lat: $37°38'$; Long: $14°56'$). According to the

Italian Presidential Decree DPR 412/93 [30], the Heating Degree Days attributed to this municipality are HDD = 1879 °C·day (climate zone D). The observed mean minimum daily air temperature in the winter is 3 °C, whereas the mean maximum daily temperature in the summer is 28 °C.

2.1. A 2D Numerical Modelling

The selected thermal bridge has been numerically simulated with the software tool "IRIS", provided by ANIT (National Association for Thermal and Acoustic Insulation) [31]. The two-dimensional model built in IRIS is shown in Figure 1; as suggested by the Standard EN ISO 10211:2017 [3], the length of the two walls that constitute the thermal bridge (flanking elements) is at least equal to the greater number between 100 cm and is three times the thickness of the flanking elements (here, 34 cm). The thermal field in the modelled detail is supposed not to vary along the third dimension and is determined in steady-state conditions.

The thermal properties of the materials set in the simulations are reported in Table 1. The hollow concrete blocks are described as a "full" layer with equivalent thermal conductivity and density; this means that the equivalent layer has, for the same given thickness (30 cm), the same thermal resistance and surface mass as the real hollow blocks, including 12-mm mortar joints between the blocks. The reported values are taken from an Italian Standard [32] and refer to hollow blocks with 50% void area, typically used in the 1980s in Southern Italy. The internal and external overall surface thermal resistance values have been set to $R_{si} = 0.13$ m^2·K·W^{-1} and $R_{se} = 0.04$ m^2·K·W^{-1}, respectively [33]. The resulting wall thermal transmittance is $U_{WALL} = 1.26$ W·m^{-2}·K^{-1}.

Regarding the boundary conditions used in the simulations, indoor and outdoor temperatures are given the same values observed during the measurement campaign, as explained in Section 2.2. As a result, the numerical simulation provides the temperature values in all points of the building detail and the transmitted heat flux (q) per unit length. The tool is validated against the cases provided in Appendix C of the Standard EN ISO 10211:2017, showing an error below 0.1 °C on the minimum temperature.

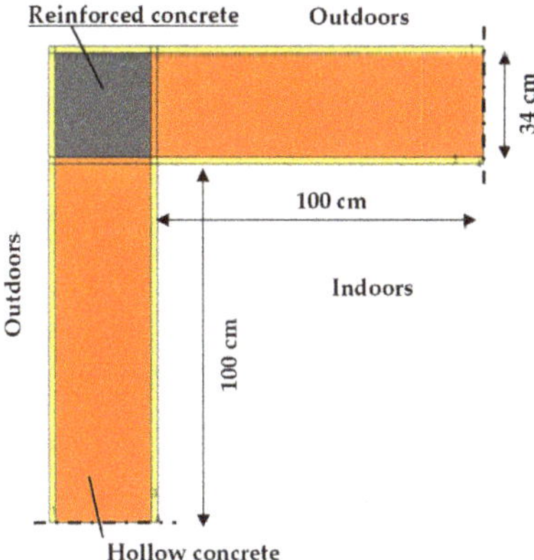

Figure 1. Cross section of the simulated 2D building detail.

Table 1. Thermal properties of the selected building materials.

Material	Thickness [cm]	Conductivity [W·m^{-1}·K^{-1}]	Density [kg·m^{-3}]
Inner cement plaster	2	0.70	1300
Hollow concrete blocks [32]	30	0.55	770
Reinforced concrete	30	2.50	2300
Outer gypsum plaster	2	0.40	1800

2.2. Experimental Measurements

Several devices were used in this study to measure the internal surface temperatures, as well as the indoor and outdoor conditions. The measurement campaign was carried out on the 22nd of February 2022, in the morning; the two walls are oriented north and west, but they are constantly shaded by a porch and are not hit by direct solar radiation. The room was preliminarily warmed for around two hours with a fan heater until a sufficient temperature difference was observed between the indoor and outdoor environment, and the walls could show a reasonably fully developed and steady thermal field. Once these conditions were achieved, all measured parameters were collected for ten minutes with an acquisition step of one minute, which was deemed sufficient to identify and describe an almost steady behaviour. More in detail, indoor and outdoor conditions were acquired using a Testo 645 high-precision humidity/temperature measuring instrument equipped with a Pt100 probe (Figure 2), with the following features:

- Temperature range: from $-200\ °C$ to $+200\ °C$;
- Temperature resolution: $\pm 0.1\ °C$;
- Temperature accuracy: $\pm 0.2\ °C$;
- Relative Humidity range: from 0% to +100%;
- Relative Humidity resolution: $\pm 0.1\%$.

The mean air temperature and relative humidity over the acquisition period are as follows:

- Indoors: $T_I = 17.0\ °C$ and $RH_I = 52.1\%$;
- Outdoors: $T_E = 10.7\ °C$ and $RH_E = 58.4\%$.

Instead, two ThermoZig dataloggers, manufactured by Carlesi Strumenti (Bologna, Italy), were used to measure the internal wall surface temperature at several different points: while one datalogger was equipped with two Pt1000 (Class A) temperature probes, the second one had a Pt1000 temperature probe plus a heat flux sensor (Figure 2), which was, however, only used to collect a surface temperature value. The features of the probes are as follows:

- Temperature range: from $-50\ °C$ to $+150\ °C$;
- Temperature resolution: $\pm 0.01\ °C$;
- Temperature accuracy: $\pm 0.15\ °C$.

Figure 2. ThermoZig datalogger (**left**) and Testo 645 digital humidity/temperature measurement device (**right**).

Figures 3 and 4 show the position of the probes: two of them were placed at 25 cm from the corner on both sides in order to verify possible asymmetries; two more probes were placed at 50 cm and 75 cm from the corner, on the same side. All probes were positioned at the height of 1.5 m, that is to say, half the room height; no probe was placed in the corner.

Moreover, when dealing with thermal bridges, using a thermal imaging camera allows for a more comprehensive view of the thermal bridging effect and provides a further reference for the assessment of the local heat transfer. Thus, the thermal field on the internal wall surface was investigated through a thermal imaging camera (Thermacam B4) manufactured by Teledyne Flir LCC (Wilsonville, OR, USA) with the following performance parameters:

- Resolution: 320×240 pixels;
- Thermal sensitivity: 0.08 °C at 30 °C;
- Spectral range: 7.5 to 13 µm;
- Temperature range: from −20 °C to +55 °C;
- Temperature accuracy: ±2 °C.

The thermal emissivity was set at $\varepsilon = 0.93$. Using a thermal imaging camera is appropriate when the structures are preliminarily warmed up and when no solar radiation influences the thermal field observed with the camera. This justifies the choice of both the measurement time (early morning) and the position of the investigated walls (north and west fronts).

In this paper, the measured surface temperature values coming from the two different instruments will be compared to the simulated values, and the consequences of the possible differences in the assessment of the heat losses will be discussed.

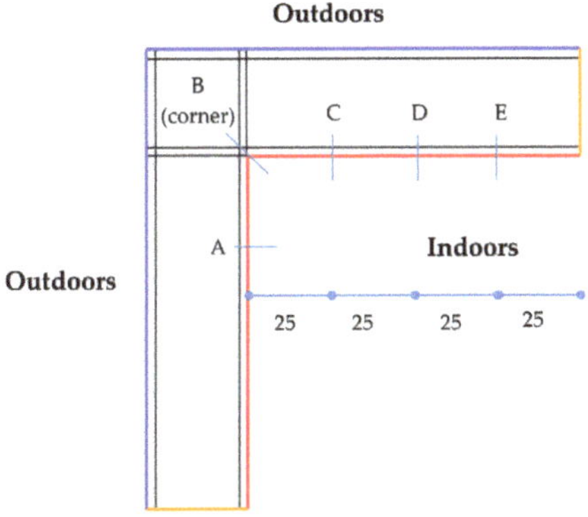

Figure 3. Position and labelling of the points where the inner surface temperatures are measured (dimensions in cm).

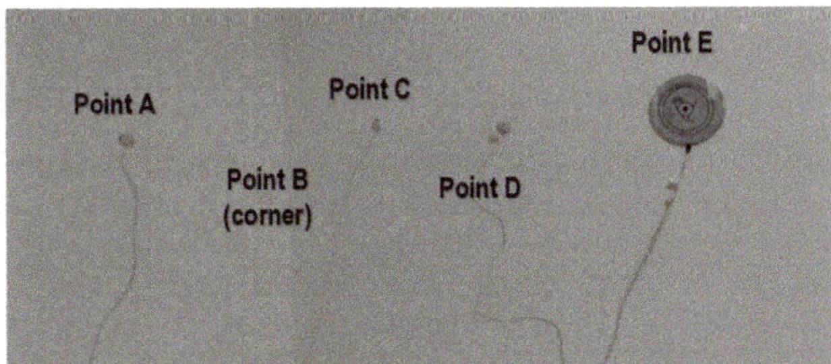

Figure 4. Picture of the probes installed on the wall surface.

2.3. Determination of the Linear Thermal Transmittance

The first way to determine the linear thermal transmittance consists of elaborating on the results of the numerical simulations described in Section 2.1. Indeed, once the transmitted heat flux per unit length (q) is calculated, the linear thermal transmittance can be determined as follows [3]:

$$\psi = \frac{q}{(T_I - T_E)} - U \cdot A \tag{1}$$

In Equation (1), (A) is the area of the flanking elements per unit length that can be measured between the finished external or internal faces, giving rise to two different values of the linear thermal transmittance, respectively identified as ψ_E and ψ_I.

Moreover, the CENED abacus [22] proposes the relations reported in Equations (2) and (3). The thermal bridge is identified via the code ASP.004 and is described by the parameters shown in Figure 5. The "non-dimensional" thermal transmittance U* is the ratio of two U-values, respectively referred to as the diagonal of the pillar (U_{PIL}) and to the wall (U_{WALL}), whereas the "equivalent thermal conductivity" (λ_{eq}) refers to the wall, but it does not include any insulation layers, if available.

$$\psi_E = 0.075 + 0.025 \cdot U^* - 1.056 \cdot \lambda_{eq} \tag{2}$$

$$\psi_I = 0.350 - 0.003 \cdot U^* + 0.103 \cdot \lambda_{eq} \tag{3}$$

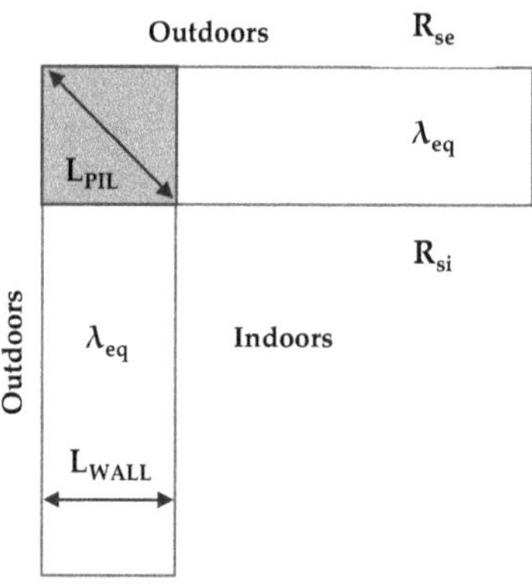

Figure 5. Horizontal section of the investigated thermal bridge (ASP.004: uninsulated protruding corner pillar and uninsulated wall) [22].

On the other hand, in the Edilclima atlas [23], the case which is closest to the investigated thermal bridge is C15 ("Joint between two walls with distributed thermal insulation and uninsulated protruding pillar"); the linear thermal transmittance is assigned for nine different combinations of wall thickness (25 cm, 30 cm and 40 cm) and U-value (0.32, 0.73 and 1.3 $\mathrm{W \cdot m^{-2} \cdot K^{-1}}$). It is worth highlighting that this atlas only provides the "outer" linear thermal transmittance ψ_E; furthermore, the thermal conductivity of reinforced concrete elements is set by default to λ_{RC} = 2.0 $\mathrm{W \cdot m^{-1} \cdot K^{-1}}$ and cannot be modified.

A further possible reference to calculate the linear thermal transmittance of a corner pillar comes from an outdated French regulation [34]. Here, the linear thermal transmittance of the investigated thermal bridge—actually, with hollow bricks in place of the lightweight concrete blocks—is provided as a function of the wall thickness. Other atlases available in the literature—including the Standard EN ISO 14683:2017—as well as the more recent versions of the French regulations, do not consider the case of a reinforced concrete corner pillar.

3. Results and Discussion

3.1. Surface Temperature

This section discusses the comparison among the surface temperature values determined through the various techniques described in Section 2, both experimentally and numerically. Starting from the thermal imaging survey, Figure 6 shows the thermal image taken in the middle of the measuring campaign five minutes after the first data acquisition. The thermal image allows for visualizing the thermal bridging effect, with the coldest point in the corner (Point B) reaching 14.2 °C along the horizontal plane where the ThermoZig probes belong. The thermal image also shows a slight asymmetry, with the left-hand side (Point A) being only 0.2 °C warmer than the corresponding point on the right-hand side (Point C). The surface temperature tends to become stable after point D, farther than 50 cm from the corner: the maximum observed difference in terms of surface temperature (from Point E to Point B, in the corner) amounts to 1.4 °C. Finally, a slight gradient in the surface temperature can be observed along the vertical direction, possibly due to buoyancy effects.

On the other hand, Figure 7 shows the mean values of the surface temperatures measured by the four ThermoZig probes during the ten-minute experiment (no probe has been installed in Point B in the corner). Figure 7 confirms that Point A and Point C, both at 25 cm from the corner, show almost the same temperature: indeed, Point A on the left-hand side keeps at slightly higher values (+0.06 °C on average). These two points are 0.3 °C colder than Point D, placed 25 cm farther. Then, Point D and Point E have practically the same temperature, thus suggesting that the thermal bridging effect becomes negligible beyond 50 cm from the corner.

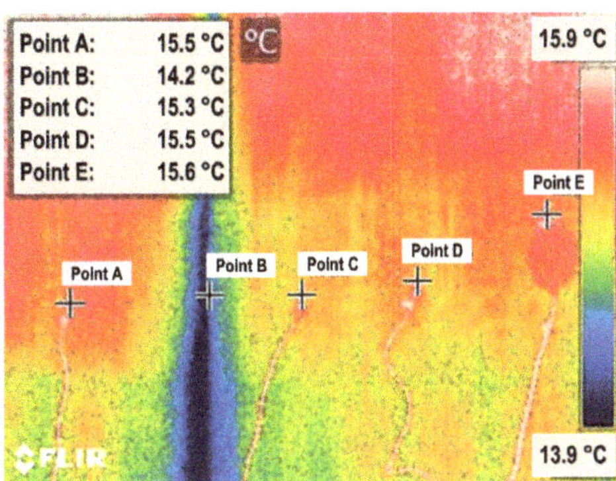

Figure 6. Thermal image and corresponding surface temperatures in the five selected points.

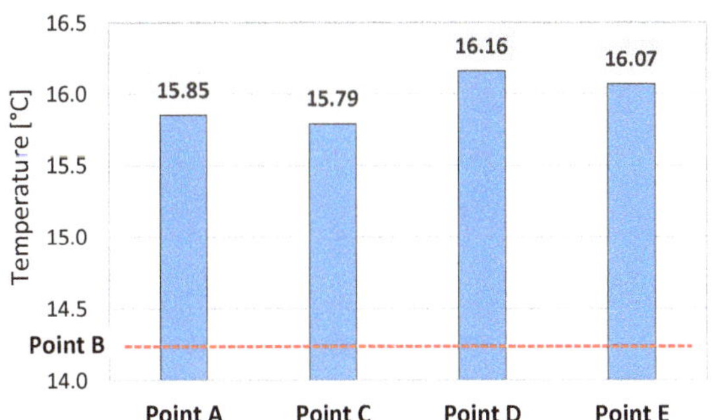

Figure 7. Mean surface temperatures measured by the ThermoZig probes.

Furthermore, Figure 8 shows the temperature distribution inside the corner pillar and the flanking elements resulting from the 2D numerical simulation. The simulation relies on a subdivision in 645 elements, which provided a variation of less than 0.05% on the total heat flux compared with a simulation based on 384 elements. Here, it is evident that not only does the surface temperature become stable farther than 50 cm from the corner, but that this is the distance after which the thermal bridging effect vanishes, and all isothermal lines become parallel. The RC pillar is signify/cantly colder than the adjacent walls; indeed, almost the entire pillar keeps below 12 °C, while a local temperature increase is observed close to the corner, with a minimum surface temperature corresponding to 14.3 °C. Finally,

Figure 9 proposes a comparison between measured and simulated results. There is a very good agreement between numerical simulations and ThermoZig measurements, with a discrepancy in the order of 0.1 °C, while the thermal imaging tends to underestimate the surface temperatures, with a difference of around 0.5 °C from the simulations emerging in Point D and Point E. These results must be interpreted considering the temperature probes' accuracy: on the one hand, this is 0.15 °C for the Pt1000 ThermoZig probes (Carlesi Strumenti, Bologna, Italy), which is the same order of magnitude as the discrepancy between measurements and numerical simulations. On the other hand, the accuracy of the thermal imaging camera is 2 °C, which may explain the higher discrepancy obtained using this kind of instrument.

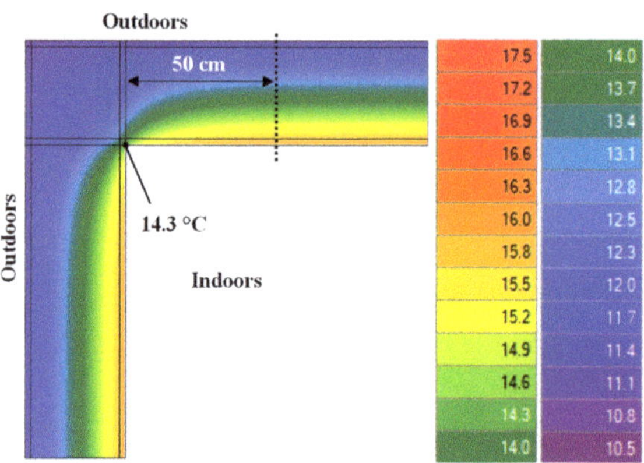

Figure 8. Temperature distribution inside the uninsulated walls and corner pillar, resulting from 2D numerical simulations with IRIS.

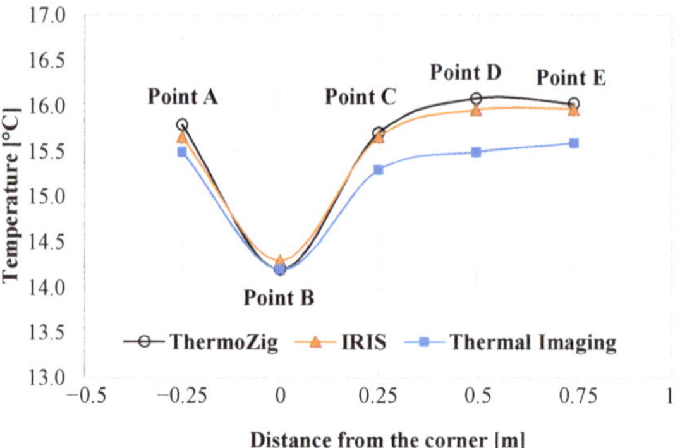

Figure 9. Comparison between measured and simulated temperatures (ThermoZig values correspond to the time average).

3.2. Transmitted Heat Flux

Based on the internal surface temperature values collected by the ThermoZig dataloggers, one can also estimate the heat flux (q) locally exchanged by the surface of the investigated wall per unit length in steady-state conditions and compare it to the corresponding result of the numerical simulation. Indeed, given the symmetry of the component, which is confirmed by the very close temperature values in Point A and Point C, the following set of equations holds. The ratio behind these equations is that the two symmetrical adjacent walls are divided into four equal elements (25 cm each); the surface temperature attributed to each element is the average of the values measured at its extreme points.

$$q = 2 \cdot (q_1 + q_2 + q_3 + q_4) \tag{4}$$

$$q_1 = \frac{0.25}{R_{si}} \cdot \left[T_I - \frac{(T_{sB} + T_{sC})}{2} \right] = 3.8 \, \frac{W}{m} \tag{5}$$

$$q_2 = \frac{0.25}{R_{si}} \cdot \left[T_I - \frac{(T_{sC} + T_{sD})}{2} \right] = 2.2 \, \frac{W}{m} \tag{6}$$

$$q_3 = \frac{0.25}{R_{si}} \cdot \left[T_I - \frac{(T_{sD} + T_{sE})}{2} \right] = 1.8 \, \frac{W}{m} \tag{7}$$

$$q_4 = \frac{0.25}{R_{si}} \cdot [T_I - T_{sE}] = 1.8 \, \frac{W}{m} \tag{8}$$

Since the temperature in Point B was not measured by the ThermoZig device, in Equation (5), the thermal imaging value is used as T_B. Moreover, Equation (8) implies that the temperature in Point E holds until the far end of the flanking element, which is reasonable due to the stabilization of the temperature values beyond 50 cm from the corner. According to this approach, which relies on experimentally measured temperatures, the estimated heat flux transferred by the thermal bridge with its flanking walls amounts to 19.2 W·m^{-1}. Considering the accuracy in the measurement of the indoor air temperature T_I (0.2 °C) and the surface temperatures in the various points (0.15 °C), the uncertainty in the estimated heat flux according to Equations (4)–(8) is 1.4 W·m^{-1}. On the other hand, the value provided by the numerical simulation in IRIS is 18.6 W·m^{-1}, with a discrepancy of only 2.7% from the estimated central value.

3.3. Linear Thermal Transmittance

This section compares the linear thermal transmittance of the uninsulated corner pillar, determined through the various methods discussed in Section 2.3, including the numerical simulations in IRIS based on the approach reported in the Standard EN ISO 10211:2017. Table 2 compares the various values.

Table 2. Linear thermal transmittance of the investigated thermal bridge, according to different sources.

Source	ψ_I (W·m^{-1}·K^{-1})	ψ_E (W·m^{-1}·K^{-1})
Numerical simulation (IRIS)	0.385	−0.471
CENED abacus	0.401	−0.456
Edilclima atlas	-	−0.380
French regulation (Th-K 77)	0.15	-

On the one hand, the CENED abacus implies the calculation of the "equivalent thermal conductivity" (λ_{eq}) and the "non-dimensional thermal transmittance" (U*). Here, the above-mentioned parameters assume the values calculated using Equations (9)–(11), where the wall thickness is L_{WALL} = 0.34 m:

$$\lambda_{eq} = \frac{L_{WALL}}{\left(\frac{1}{U_{WALL}} - R_{si} - R_{se} \right)} = 0.545 \, \frac{W}{m \cdot K} \tag{9}$$

$$U_{PIL} = \left(R_{si} + \sum_{j=1}^{3} \frac{L_j}{\lambda_j} + R_{se} \right)^{-1} = \left(0.13 + \frac{0.028}{0.7} + \frac{0.424}{2.5} + \frac{0.028}{0.4} + 0.04 \right)^{-1} \tag{10}$$

$$= 2.22 \, \frac{W}{m^2 \cdot K}$$

$$U^* = \frac{U_{PIL}}{U_{WALL}} = 1.76 \tag{11}$$

Under these circumstances, Equations (2) and (3) can be used since they hold if $0.23 \leq \lambda_{eq} \leq 0.81$ and $1.5 \leq U^* \leq 4.5$. On the other hand, the ψ-value derived from the Edilclima atlas refers to 30 cm thick walls with $U = 1.3$ W·m^{-2}·K^{-1}, i.e., very close to the investigated configuration. Finally, in the French regulation, a thickness between 30 and 34 cm provides $\psi_I = 0.15$ W·m^{-1}·K^{-1}, but no data about ψ_E are available.

Table 2 comparison proves that the formulation available in the CENED abacus is reliable as it deviates from the detailed 2D numerical simulations by only 3%. On the other hand, the Edilclima atlas overestimates the heat losses in the corner pillar by around 20%. Finally, the outdated French standard cannot be considered reliable since it underestimates the linear thermal transmittance by more than 60%. The main reason is that these sources have some underlying hypotheses that do not always correspond to the current case study (for instance, the Edilclima atlas considers $\lambda_{RC} = 2.0$ W·m^{-1}·K^{-1} for reinforced concrete, and this value cannot be modified). In all cases, the linear thermal transmittance is lower when using the external building dimensions (ψ_E); indeed, the heat loss surface is already overestimated, and a minor additional heat loss must be accounted for.

One further interesting result consists of defining the increase in the heat flux caused by the corner pillar if compared to a couple of real undisturbed uninsulated walls. Indeed, if one considers two real walls with an internal size $A_I = 10.8$ m^2 each (height $H = 2.7$ m, width $W = 4$ m), the heat losses in the absence of the thermal bridging effect (Q_{noTB}) would be determined as in Equation (12). Instead, taking the thermal bridge into account, one should adopt Equation (13):

$$Q_{noTB} = U_{WALL} \cdot (T_I - T_E) \cdot 2 \cdot A_I = 171.5 \text{ W} \tag{12}$$

$$Q_{TB} = q_{noTB} + \psi_I \cdot (T_I - T_E) \cdot H = 178.0 \text{ W} \tag{13}$$

This result suggests that the RC corner pillar causes an increase of 3.8% in the local heat losses, thus confirming that thermal bridges associated with RC corner pillars in uninsulated walls have a relatively low impact on the heat losses while determining possible mould growth due to the low internal surface temperatures.

4. Effect of ETICS Renovation on Thermal Bridges

This section explores the effectiveness of ETICS insulation in reducing the thermal bridging effect in the RC corner pillar. To this aim, a layer of extruded polystyrene (XPS) with $\rho = 40$ kg·m^3 and $\lambda = 0.035$ W·m^{-1}·K^{-1} is added on the external side of the wall; several different thickness values are tested, from 6 cm to 14 cm. The thermal performance of the insulated building detail is evaluated numerically by means of IRI, in terms of minimum internal surface temperature and linear thermal transmittance; for the sake of comparison, the latter is also assessed using the CENED abacus and the Edilclima atlas.

In particular, the CENED abacus considers the following set of equations to calculate the linear thermal transmittance of a protruding corner pillar with externally insulated walls, identified by the code ASP.005:

$$\psi_E = -0.281 + 0.147 \cdot U^* + 0.143 \cdot L_{WALL} \tag{14}$$

$$\psi_I = 0.385 - 0.116 \cdot U^* - 0.198 \cdot L_{WALL} \tag{15}$$

Here, the wall thickness L_{WALL} and the "non-dimensional" thermal transmittance U^*, calculated as in Equation (11), take the insulating layer into account. Equations (14) and (15) hold if 0.30 m $\leq L_{WALL} \leq 0.65$ m and $0.76 \leq U^* \leq 1.18$, which is the case of the investigated thermal bridge ($U^* = 0.78$ to 0.87, $L_{WALL} = 0.40$ m to 0.48 m).

Coming to the Edilclima atlas, the relevant case is C14 ("Joint between two walls with continuous external insulation and insulated protruding pillar"). Table 3 reports the suggested values for the "outer" linear thermal transmittance ψ_E holding if the uninsulated portion of the wall has an equivalent thermal conductivity $\lambda_{eq} = 0.5$ W·m^{-1}·K^{-1}, i.e., only 10% lower than the investigated wall. Table 4 resumes the various results; those pertaining to the Edilclima atlas were interpolated from Table 3 by considering a thickness of 34 cm and the U-values reported in the second row of Table 4 itself.

Table 3. Linear thermal transmittance values (ψ_E) according to Edilclima atlas [23]—thermal bridge C14 ("Joint between two walls with continuous external insulation and insulated protruding pillar").

| | Linear Thermal Transmittance ψ_E | |
U-Value	Thickness = 30 cm	Thickness = 40 cm
0.70 W·m^{-2}·K^{-1}	−0.16 W·m^{-1}·K^{-1}	−0.21 W·m^{-1}·K^{-1}
0.60 W·m^{-2}·K^{-1}	−0.13 W·m^{-1}·K^{-1}	−0.17 W·m^{-1}·K^{-1}
0.50 W·m^{-2}·K^{-1}	−0.11 W·m^{-1}·K^{-1}	−0.14 W·m^{-1}·K^{-1}
0.40 W·m^{-2}·K^{-1}	−0.09 W·m^{-1}·K^{-1}	−0.11 W·m^{-1}·K^{-1}
0.30 W·m^{-2}·K^{-1}	−0.07 W·m^{-1}·K^{-1}	−0.09 W·m^{-1}·K^{-1}
0.20 W·m^{-2}·K^{-1}	−0.06 W·m^{-1}·K^{-1}	−0.07 W·m^{-1}·K^{-1}
0.10 W·m^{-2}·K^{-1}	−0.06 W·m^{-1}·K^{-1}	−0.06 W·m^{-1}·K^{-1}

The thickness refers to the wall without insulation, with $\lambda_{eq} = 0.5$ W·m^{-1}·K^{-1}.

Table 4. Main parameters describing the thermal performance of the insulated corner pillar (the acronym NA indicates non-available data).

Insulation Thickness	6 cm	8 cm	10 cm	12 cm	14 cm
U_{WALL} [W·m^{-2}·K^{-1}]	0.399	0.325	0.274	0.237	0.209
Minimum T_{SI} (°C)	16.0	16.1	16.2	16.3	16.4
Linear thermal transmittance—external ψ_E (W·m^{-1}·K^{-1})					
Numerical simulation (IRIS)	−0.099	−0.083	−0.073	−0.067	−0.063
CENED abacus (case ASP.005)	−0.096	−0.098	−0.099	−0.098	−0.097
EDILCLIMA atlas (case C14)	−0.100	−0.085	−0.075	−0.070	−0.065
Linear thermal transmittance—internal ψ_I (W·m^{-1}·K^{-1})					
Numerical simulation (IRIS)	0.220	0.190	0.168	0.151	0.137
CENED abacus (case ASP.005)	0.205	0.205	0.204	0.201	0.199
EDILCLIMA atlas (case C14)	NA	NA	NA	NA	NA

The results suggest that increasing the insulation thickness from 6 cm to 14 cm has a minor effect on the internal surface temperature, which changes from 16.0 °C to 16.4 °C; in any case, the ETICS allows for increasing this temperature by around 2.0 °C compared to the uninsulated corner pillar (point B in Figure 9), and this avoids the risk of mould growth on the internal surface. Moreover, higher insulation thickness means lower internal linear thermal transmittance (ψ_I): this effect is particularly evident in the numerical simulations (IRIS), while the results of the CENED abacus are not sensitive to the insulation thickness and show a non-negligible discrepancy from the detiled IRIS simulations, ranging from 7% underestimation (6 cm of insulation) to 50% overestimation (14 cm of insulation). On the contrary, the Edilclima atlas reflects very well the numerical predictions in this case. Previous studies highlighted that thermal bridge catalogues are not always very reliable: in some cases, their use allows for calculating the overall heat losses through thermal bridges with an error below 10%, while in other cases, very high discrepancies were observed [35].

Further interesting information comes from Figure 10, which shows the temperature distribution inside the insulated walls, resulting from IRIS simulations. Here, please consider that the size of the flanking elements is not the same as in the uninsulated thermal bridge (Figure 8): indeed, the flanking element must be at least three times as long as the wall thickness [3], leading to a length of 1.2 m and 1.44 m, respectively, with 6 cm and 14 cm of thermal insulation. The two diagrams show that in both cases, the isothermal lines become parallel at more than 50 cm, precisely at around 60 cm in the case of medium-thermal insulation (6 cm) and 70 cm in the case of high-thermal insulation (14 cm). Interestingly, Hallik et al. demonstrated that, in the case of a well-insulated building envelope with RC structures, the sufficient length of the flanking element to adequately calculate the linear thermal transmittance is equal to approximately the wall's thickness [36]. In our study, this minimum distance approximates 1.5 times the wall thickness; by the way, it is confirmed that the minimum distance imposed by the Standard EN ISO 10211:2017 is, in many cases, excessive and that it could be reduced by around 50% to minimize computational time.

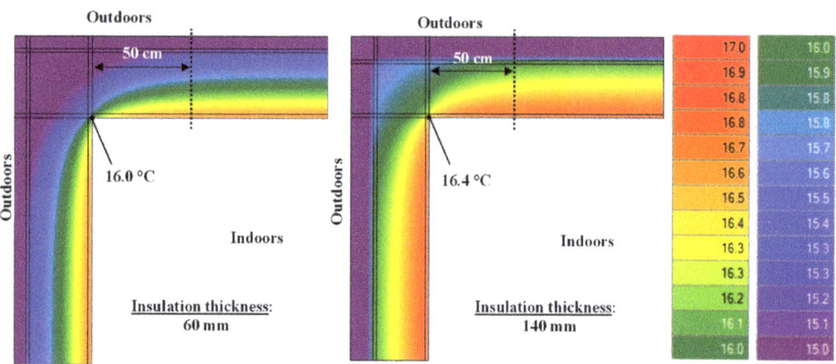

Figure 10. Temperature distribution inside the insulated walls and corner pillar, resulting from 2D numerical simulations with IRIS (left: 6 cm insulation; right: 14 cm insulation).

Finally, it is interesting to repeat the same exercise discussed at the end of Section 3.3, aimed at quantifying the local increase in the heat losses due to the thermal bridge between two real walls with an internal size $A_I = 10.8$ m^2 each. In the case of the ETICS application, the heat flux transferred with no thermal bridging—see Equation (12)—would range from 28.4 W (insulation: 14 cm) to 54.3 W (insulation: 6 cm). Instead, taking the thermal bridge into account, Equation (13) would provide heat losses ranging from 30.7 W (insulation: 14 cm) to 58.0 W (insulation: 6 cm). Hence, the insulated RC corner pillar locally increases the heat losses by 7% to 8%, which is twice as high as in the case of the uninsulated corner pillar; this is coherent with the findings of Figure 10, showing that the thermal bridging effect penetrates deeper into the flanking elements in case of insulated corner pillar. In conclusion, insulating an RC corner pillar with an ETICS solution obviously drastically reduces the transmission heat losses, but the role of the thermal bridge becomes more important—in percentage—than in the uninsulated corner.

5. Conclusions

This research has studied, both experimentally and numerically, the thermal field in a thermal bridge consisting of a reinforced concrete corner pillar, which belongs to a building dating back to the 1980s and located in Southern Italy. The inner surface temperatures near the corner pillar have been measured using Pt 1000 temperature probes and thermal imaging techniques. Moreover, 2D finite element simulations were performed based on the same boundary conditions as those measured in the experimental campaign. This study has allowed us to verify that a 30 cm thick RC corner pillar generates, in a building made with lightweight concrete blocks ($U_{WALL} = 1.26$ W·m^{-2}·K^{-1}), a local increase of around 4% in the heat flux, if compared to undisturbed one-dimensional heat transfer through the flanking elements. The thermal bridging effect vanishes farther than 50 cm from the corner, where the isothermal lines inside the walls become parallel; however, in only 50 cm, the measured inner surface temperature drops from 15.5 °C to 14.2 °C (in the corner), which also has significant effects in terms of increased radiant heat transfer, increased thermal discomfort, and higher risk of mould formation.

The first message resulting from the 2D numerical simulations conducted with the software tool IRIS and based on the same average boundary conditions as in the measurement campaign is that this tool provides very reliable results since the discrepancy with the measured inner surface temperatures is in the order of 0.1 °C; moreover, the estimated heat flux in the two cases differs by less than 3%. Furthermore, the adoption of thermal imaging techniques is very useful for obtaining visual information about the presence and the extension of the thermal bridging effect, but it proves less accurate for quantifying it, with a discrepancy of even 0.5 °C in terms of surface temperature. The results also suggest that the relations proposed by the CENED abacus can be reliably used to calculate the linear thermal transmittance of this uninsulated thermal bridge since their outcome deviates by around 3% from accurate numerical simulations. Other sources, such as the Edilclima atlas and the French regulation Th-K 77, provide less accurate values, and their use is not fully recommended in this case.

Moreover, numerical simulations were used to study the thermal performance of the same building detail in the case of the ETICS insulation by adding from 6 cm to 14 cm of XPS to the external side of the wall. The ETICS allows for increasing the minimum internal surface temperature by around 2.0 °C compared to the uninsulated corner pillar, thus reducing the risk of mould growth. However, the relative impact of the thermal bridge on the local transmission heat losses becomes higher than in the uninsulated building, doubling from 4% to around 8%. This paper also shows that the relations proposed by the CENED abacus are less reliable to describe the insulated corner pillar and can even lead to an overestimation by almost 50% of the local transmission heat losses.

Finally, the proposed experimental approach provides a reliable assessment of the steady thermal field around corner pillars: it consists of installing a limited number of Pt1000 probes on both sides of the corner in order to verify possible asymmetries and the extensions of the thermal bridging effect. The knowledge of the surface temperature values plus the indoor and outdoor temperature allows for estimating the transmitted heat flux as in Equations (4)–(8); the linear thermal transmittance can then be assessed by dividing the heat flux by the temperature difference between indoors and outdoors. The measurements must be carried out in the absence of direct solar radiation hitting the investigated structures and only after the indoor temperature has been warmed up.

The reliability of this procedure has been backed up by the numerical simulations and the thermal imaging camera, which makes it suitable for technicians to experimentally quantify the thermal bridging effect in existing buildings with reinforced concrete structures, as well as to verify the effect of renovation actions for the mitigation of this kind of thermal bridge

Author Contributions: Conceptualization, G.E. and A.G.; methodology, G.E. and A.G.; software, G.E.; formal analysis, G.E. and A.G.; investigation, G.E. and A.G.; resources, A.G.; data curation, G.E. and A.G; writing—original draft preparation, G.E.; writing—review and editing, G.E. and A.G.; supervision, A.G.; funding acquisition, G.E. and A.G. All authors have read and agreed to the published version of the manuscript.

Funding: This research was funded by the University of Catania in the framework of the Research Projects SIS-RENEW and MENTOR (Piano di Incentivi per la Ricerca 2020–2022).

Data Availability Statement: The data presented in this study are available on request from the corresponding author. The data are not publicly available due to privacy.

Conflicts of Interest: The authors declare no conflicts of interest.

Nomenclature

Symbol	Quantity		Unit
A	Area per unit length		$m^2 \cdot m^{-1}$
H	Height		m
HDD	Heating Degree Days		$^{\circ}C \cdot day$
L	Thickness		m
q	Heat flux per unit length		$W \cdot m^{-1}$
R	Thermal resistance		$m^2 \cdot K \cdot W^{-1}$
RH	Relative Humidity		%
T	Temperature		$^{\circ}C$
U	Thermal transmittance		$W \cdot m^{-2} \cdot K^{-1}$
W	Width		m
ε	Thermal emissivity		-
λ	Thermal conductivity		$W \cdot m^{-1} \cdot K^{-1}$
ρ	Density		$kg \cdot m^{-3}$
ψ	Linear thermal transmittance		$W \cdot m^{-1} \cdot K^{-1}$
Subscript	Meaning		
eq	Equivalent		
PIL	Pillar		
RC	Reinforced Concrete		
se	External surface		
si	Internal surface		
Acronyms	Meaning		
ETICS	External Thermal Insulation Composite System		
NA	Not Available		
XPS	Extruded Polystyrene		

References

1. Odyssee-Mure Project Website, Sectoral Profile—Overview, Final Energy Consumption by Sector in EU. Available online: https://www.odyssee-mure.eu/publications/efficiency-by-sector/overview/final-energy-consumption-by-sector.html (accessed on 4 January 2024).
2. European Commission, Department of Energy. In Focus: Energy Efficiency in Buildings. Brussels. 17 February 2020. Available online: https://ec.europa.eu/info/news/focus-energy-efficiency-buildings-2020-lut-17_en (accessed on 23 December 2023).
3. *EN ISO 10211*; CEN, European Committee for Standardisation. Thermal Bridges in Building Construction—Heat Flows and Surface Temperatures—Detailed Calculations. CEN: Brussels, Belgium, 2017.
4. Ilomets, S.; Kalamees, T. Evaluation of the criticality of thermal bridges. *J. Build. Pathol. Rehabil.* **2016**, *1*, 11. [CrossRef]
5. Alhawari, A.; Mukhopadhyaya, P. Thermal bridges in building envelopes—An overview of impacts and solutions. *Int. Rev. Appl. Sci. Eng.* **2018**, *9*, 31–40. [CrossRef]
6. Janssen, A.; Van Londersele, E.; Vandermarcke, B.; Roels, S.; Standaert, P.; Wouters, P. Development of limits for the linear thermal transmittance of thermal bridges in buildings. In Proceedings of the X International Conference on Thermal Performance of the Exterior Envelopes of Whole Buildings—ASHRAE, Clearwater Beach, FL, USA, 1–5 December 2013.
7. Ilomets, S.; Kuusk, K.; Paap, L.; Arumägi, E.; Kalamees, T. Impact of linear thermal bridges on thermal transmittance of renovated apartment buildings. *J. Civ. Eng. Manag.* **2017**, *23*, 96–104. [CrossRef]
8. Zhang, X.; Jung, G.-J.; Rhee, K.-N. Performance evaluation of thermal bridge reduction method for balcony in apartment buildings. *Buildings* **2022**, *12*, 63. [CrossRef]
9. Evola, G.; Costanzo, V.; Urso, A.; Tardo, C.; Margani, G. Energy performance of a prefabricated timber-based retrofit solution applied to a pilot building in Southern Europe. *Build. Environ.* **2022**, *222*, 109442. [CrossRef]
10. Curto, D.; Franzitta, V.; Guercio, A.; Martorana, P. FEM analysis: A review of the most common thermal bridges and their mitigation. *Energies* **2022**, *15*, 2318. [CrossRef]
11. Kim, M.-Y.; Kim, H.-G.; Kim, J.-S.; Hong, G. Investigation of thermal and energy performance of the thermal bridge breaker for reinforced concrete residential buildings. *Energies* **2022**, *15*, 2854. [CrossRef]
12. Ge, H.; McClung, V.R.; Zhang, S. Impact of balcony thermal bridges on the overall thermal performance of multi-unit residential buildings: A case study. *Energy Build.* **2013**, *60*, 163–173. [CrossRef]
13. Baba, F.; Ge, H. Dynamic effect of balcony thermal bridges on the energy performance of a high-rise residential building in Canada. *Energy Build.* **2016**, *116*, 78–88. [CrossRef]
14. Aghasizadeh, S.; Kari, B.M.; Fayaz, R. Thermal performance of balcony thermal bridge solutions in reinforced concrete and steel frame structures. *J. Build. Eng.* **2022**, *48*, 103984. [CrossRef]
15. Theodosiou, T.G.; Papadopoulos, A.M. The impact of thermal bridges on the energy demand of buildings with double brick wall constructions. *Energy Build.* **2008**, *40*, 2083–2089. [CrossRef]

16. Evola, G.; Margani, G.; Marletta, L. Energy and cost evaluation of thermal bridge correction in Mediterranean climate. *Energy Build.* **2011**, *43*, 2385–2393. [CrossRef]
17. Kotti, S.; Teli, D.; James, P.A.B. Quantifying thermal bridge effects and assessing retrofit solutions in a Greek residential building. *Procedia Environ. Sci.* **2017**, *38*, 306–313. [CrossRef]
18. Alayed, E.; O'Hegarty, R.; Kinnane, O. Solutions for exposed structural concrete bridged elements for a more sustainable concrete construction in hot climates. *Buildings* **2022**, *12*, 176. [CrossRef]
19. Cirami, S.; Evola, G.; Gagliano, A.; Margani, G. Thermal and economic analysis of renovation strategies of a historic building in Mediterranean area. *Buildings* **2017**, *7*, 60. [CrossRef]
20. *EN ISO 14683*; CEN, European Committee for Standardisation. Thermal Bridges in Building Construction—Linear Thermal Transmittance—Simplified Methods and Default Values. CEN: Brussels, Belgium, 2017.
21. Berggren, B.; Wall, M. Calculation of thermal bridges in (Nordic) building envelopes—Risk of performance failure due to inconsistent use of methodology. *Energy Build.* **2013**, *65*, 331–339. [CrossRef]
22. CENED, Atlante dei Ponti Termici. 2011. Available online: https://www.cened.it/download (accessed on 25 March 2023). (In Italian)
23. Capozzoli, A.; Corrado, V.; Gorrino, A.; Soma, P. *Atlante Nazionale dei Ponti Termici: Conforme alle UNI EN ISO 14683 e UNI EN ISO 10211*; Edizioni Edilclima: Novara, Italy, 2011. (In Italian)
24. Office Fédéral de L'ÉNergie (OFEN). *Catalogue des Ponts Thermiques*; Office Fédéral de L'ÉNergie (OFEN): Bern, Switzerland, 2003.
25. Passive House Institute (PHI). *Thermal Bridge Catalogue*; Passive House Institute: Darmstadt, Germany, 2019.
26. Basiricò, T.; Cottone, A.; Enea, D. Analytical mathematical modeling of the thermal bridge between reinforced concrete wall and inter-floor slab. *Sustainability* **2020**, *12*, 9964. [CrossRef]
27. Theodosiou, T.; Tsikaloudaki, K.; Kontoleon, K.; Giarma, C. Assessing the accuracy of predictive thermal bridge heat flow methodologies. *Renew. Sustain. Energy Rev.* **2021**, *136*, 110437. [CrossRef]
28. O'Grady, M.; Lechowska, A.A.; Harte, A.M. Infrared thermography technique as an in-situ method of assessing heat loss through thermal bridging. *Energy Build.* **2017**, *135*, 20–32. [CrossRef]
29. Baldinelli, G.; Bianchi, F.; Rotili, A.; Costarelli, D.; Seracini, M.; Vinti, G.; Asdrubali, F.; Evangelisti, L. A model for the improvement of thermal bridges quantitative assessment by infrared thermography. *Appl. Energy* **2018**, *211*, 854–864. [CrossRef]
30. Presidential Decree 412/93. Regulation Containing the Laws for Design, Installation, Operation and Maintenance of Buildings Heating Systems in order to Reduce Energy Consumptions, for the implementation of the item 4, paragraph 4, of the Law 9th January 1991, n. 10. Available online: http://www.normattiva.it/atto/caricaDettaglioAtto?atto.dataPubblicazioneGazzetta=1993 -10-14&atto.codiceRedazionale=093G0451 (accessed on 29 January 2024). (In Italian)
31. IRIS v.5.1, Software ANIT. Available online: https://www.anit.it/iris (accessed on 12 April 2023).
32. *UNI 10355*; UNI, Ente Nazionale di Normazione. Murature e Solai. Valori Della Resistenza Termica e Metodo di Calcolo. UNI: Rome, Italy, 1994.
33. *EN ISO 6946*; CEN, European Committee for Standardisation. Building Components and Building Elements—Thermal Resistance and Thermal Transmittance—Calculation Methods. CEN: Brussels, Belgium, 2017.
34. *Règles Th-K 77: Règles de Calcul des Caractéristiques Thermiques Utiles des Parois de Construction, Mises à Jour au 1er Janvier 1990*; Centre Scientifique et Technique du Bâtiment, CSTB: Marne-la-Vallée, France, 1990.
35. Bergero, S.; Chiari, A. The influence of thermal bridge calculation method on the building energy need: A case study. *Energy Procedia* **2018**, *148*, 1042–1049. [CrossRef]
36. Hallik, J.; Kalamees, T. The effect of flanking element length in thermal bridge calculation and possible simplifications to account for combined thermal bridges in well insulated building envelopes. *Energy Build.* **2021**, *252*, 111397. [CrossRef]

 *buildings*

Article

Experimental and Numerical Performance Evaluation of Bio-Based and Recycled Thermal Break Strips in LSF Partition Walls

Paulo Santos *, David Abrantes, Paulo Lopes and Diogo Mateus

ISISE, Department of Civil Engineering, University of Coimbra, 3030-788 Coimbra, Portugal
* Correspondence: pfsantos@dec.uc.pt

Abstract: The thermal performance of Lightweight Steel Framed (LSF) walls could be strongly compromised due to steel's high thermal conductivity and their related thermal bridges. In this paper, the performance of bio-based (pine wood) and recycled (rubber–cork composite) Thermal Break Strip (TBS) materials, to mitigate the thermal bridge effect originated by steel profiles in LSF partition walls, is evaluated. This assessment was achieved by measurements under controlled laboratory conditions and by predictions using some numerical simulation models. Regarding the measurements, two climatic chambers (cold and hot) were used to impose a nearly constant temperature difference (around 35 °C), between the LSF partition test samples' surfaces. To measure the overall surface-to-surface thermal resistance (R-value) of the evaluated LSF wall configurations, the Heat Flow Meter (HFM) method was used. Moreover, the measured values were compared with the calculations by 2D (THERM models) and 3D (ANSYS models) numerical simulations, exhibiting an excellent agreement (less than ±2% difference). Three TBS locations and three materials are evaluated, with their thermal performance improvement compared with a reference interior partition LSF wall, having no TBS. The top performance was accomplished by the aerogel super-insulating TBS material. The bio-based material (pine wood) and the recycled rubber–cork composite present quite similar results, with a slight advantage for the pine wood TBSs, given their higher thickness. Considering the TBS location, the inner and outer side present comparable performances. When using TBSs on both sides of steel profile flanges, there is a relevant thermal performance improvement, as expected. The thickness of the TBS also presents a noteworthy influence on the LSF partition thermal resistance.

Keywords: lightweight steel frame; thermal break strips; partition walls; bio-based material; measurements; numerical simulations; thermal performance

Citation: Santos, P.; Abrantes, D.; Lopes, P.; Mateus, D. Experimental and Numerical Performance Evaluation of Bio-Based and Recycled Thermal Break Strips in LSF Partition Walls. *Buildings* **2022**, *12*, 1237. https://doi.org/10.3390/buildings12081237

Academic Editor: Cinzia Buratti

Received: 23 July 2022
Accepted: 11 August 2022
Published: 14 August 2022

Publisher's Note: MDPI stays neutral with regard to jurisdictional claims in published maps and institutional affiliations.

1. Introduction

Climate change, global warming, as well as shortages and high price of fossil fuels, have led to a growing concern and demand regarding energy efficiency, particularly in buildings. This sector consumes nearly forty percent of the total energy in Europe and accounts for approximately 36% of the CO_2 emissions [1]. Improvements of buildings' energy efficiency leads to reduced energy consumption, reduced CO_2 emissions, and also decreasing costs during the operational phase. Energy efficiency and thermal comfort in buildings could be mostly influenced by the physical properties (e.g., thermal resistance) of the building envelope. To reduce CO_2 emissions, the use of biomaterials and recycled materials should be prioritized. The construction sector has been experiencing a change in constructive technology, turning to more lightweight and industrialized alternatives.

The Light Steel Framing (LSF) constructive system is one of these technologies, that has been proliferating due to its advantages compared to the heavyweight reinforced concrete and brick masonry systems, such as: light weight, facilitating transportation and handling; high mechanical strength; easy prefabrication ensuring better quality controlled building components, as well as permitting modular construction; faster construction resulting in time savings; a more sustainable construction due to its potential of recyclability and reuse, which can exceed 95%; in case of humidity, excellent stability of shape; decrease in waste; reduced water consumption; and less need for heavy machinery and an intensive workforce [2–4].

Regarding Lightweight Steel Framed (LSF) walls, the high thermal conductivity of steel may originate important thermal bridges, that should be predicted in the building design phase and treated at the construction stage [5]. These thermal bridges can reduce the building energy efficiency by more than 30%. Thermal Break Strips (TBSs), an Exterior Thermal Insulation Composite System (ETICS), and slotted steel studs are some of the thermal mitigation strategies usually used in LSF buildings.

Usually, an LSF wall is made of three main constituents: (1) cold formed steel internal frame; (2) sheathing panels on internal side of walls and ceilings (e.g., gypsum plasterboard) and on external side (e.g., Oriented Strand Board—OSB); and (3) insulation materials (air cavity insulation, e.g., mineral wool, and/or Exterior Thermal Insulation Composite System—ETICS) [6,7]. The mineral wool placed inside the air cavity, in addition to the thermal insulation role, causes an important acoustic performance improvement as well, increasing the noise insulation [8,9]. Moreover, the thermal insulation effectiveness is significantly related to the position in the LSF walls [10,11] and the LSF construction type [12].

Indeed, the type of LSF construction system is defined by the presence of the insulation layer and its location on the wall, that can be categorized in three typologies: (1) warm, (2) cold, and (3) hybrid steel frame. In warm frame construction, all insulation is continuous, being positioned on the exterior side of the wall (e.g., ETICS). In cold frame construction, all insulation is located inside the air gap (between the metallic profiles), with their thickness limited to the steel profile depth. Given its usual best thermal performance, the hybrid steel frame type is the most commonly used [8]. In this LSF system, both kinds of insulation are used, namely external continuous, as well as discontinuous batt insulation placed between steel studs [13].

There are three main methodologies to assess the thermal transmittance (U-value) and/or thermal resistance (R-value) of building elements: (1) analytical formula computations, (2) numerical simulations, and (3) experimental measurements [14].

Concerning the analytical formulas, they are the simplest method of those mentioned above, being very useful and effortless to be used at the early design stage [15,16]. However, these analytical formulas can only be used, and are available, for very simplified wall arrangements, so their use very restricted. Additionally, these formulations commonly assume a basic one-dimensional (1D) steady-state heat flow and neglect the materials' heat storage effect and/or the variation in thermal properties (e.g., thermal conductivity change with humidity and temperature) [17].

Regarding numerical simulations, the fastest and easiest ones are provided by two-dimensional (2D) models [10,12,18], while the more time-consuming and complete are the three-dimensional (3D) models [19–21]. Notice that this numerical simulation approach has the benefit of permitting a fast comparison among several building element configurations, saving time and money, in comparison with an experimental approach. However, there is a need for a specific and adequate calculation tool (software). To verify their algorithms' reliability, the computed results must be verified using, for instance, the test cases prescribed in standard ISO 10211 [22]. Moreover, to validate the implemented numerical model, the calculated results should be verified by comparison with experimental measurements. These measurements could be carried out in laboratory-controlled conditions or in situ [23,24]. Some often used measurement methods are: Infrared Thermography (IRT); the Hot Box (HB) method, that can be Guarded HB (GHB) or Calibrated HB (CHB); the Heat Flow Meter (HFM) method; and the Guarded Hot Plate (GHP) method [23].

To improve the thermal performance of LSF buildings, several thermal bridge mitigation techniques could be used, such as: flange stud indentation [25,26]; slotted thermal steel studs [19,27,28]; Thermal Break Strips (TBSs) along the steel flanges [12,28,29]; external continuous thermal insulation (e.g., ETICS) [14,30–32]. Another approach to increase the performance of an LSF element, when there is a cavity inside the wall, is by decreasing the radiation heat transfer. With this purpose, a reflective aluminum foil or a paint having low emissivity inside the LSF wall air gap could be used [33–36]. Regarding the use of TBSs in LSF walls, no study was found in the literature using a bio-based material with this purpose.

In this research work, the thermal performance of bio-based (pine wood) and recycled (rubber–cork composite) materials, used as Thermal Break Strips (TBSs) in LSF partition walls, is evaluated using experimental and numerical approaches. Thus, the surface-to-surface *R*-value of LSF partition walls, using different configurations, was assessed under controlled laboratory conditions. The main advantage of using these conductive *R*-values is that they do not depend on the surrounding environmental conditions. Therefore, any surface thermal resistance value could be added to obtain the total thermal resistance of the LSF wall.

These laboratory measurements were completed using a set of two climatic chambers (one cold and the other hot). The experimental tests were performed using the HFM method [37]. Moreover, three tests were carried out for each wall, placing the sensors at different locations (bottom, middle, and top), on the surfaces of the LSF wall prototype, corresponding to a total of 30 laboratory tests. Besides the previous bio-based and eco-friendly recycled TBS materials, a high-performance state-of-the-art insulation material (aerogel) was also evaluated. These TBS materials were used on three different LSF wall positions: outer flange, inner flange, as well as on the two steel flanges. Moreover, the measured values were matched with the predictions obtained by 2D (THERM models [38]) and 3D (ANSYS models [39]) numerical simulations, exhibiting an excellent agreement (up to ±2% difference).

This manuscript is organized as follows. After this brief introductory section, the material and methods are described, including the evaluated LSF partition walls, the laboratory measurements, and the numerical simulations. Next, the obtained results are presented and discussed, namely the conductive thermal resistances, the Infrared (IR) images, and the heat flux predictions. Finally, some concluding remarks about this study are provided.

2. Materials and Methods

2.1. Description of LSF Partitions

The description of the dimensions, geometry, materials, and thermophysical properties of the LSF reference wall and the Thermal Break Strips (TBSs) is presented in this section. The reference LSF wall corresponds to an LSF wall configuration used as an internal partition. As displayed in Figure 1, this reference LSF partition wall is 140 mm thick, where the inner and outer sheathing surfaces are made of 2 Gypsum Plaster Boards (GPBs) (12.5 mm thick) on each lateral side of the vertical metallic studs (C90 × 37 × 5 × 0.6 mm) spaced 400 mm apart, with the air gap entirely occupied with Mineral Wool (MW) thermal insulation (90 mm thick). Two sheathing drywall layers were adopted to improve the mechanical resistance, as well as to slightly decrease the steel frame thermal bridge effect. In Table 1 are presented the thermal conductivity and thickness values of the materials that make up the reference partition LSF wall.

The TB strips analyzed have a nominal thickness equal to 10 mm and are 50 mm wide. Figure 2 illustrates the materials used, namely Pine Wood (PW), Aerogel (AG), recycled Rubber and Cork (RC) composite. Notice that the thickness of PW strips is slightly higher (13 mm), than the nominal thickness (10 mm). In Table 2 are displayed the thermal conductivities for the TBS materials, changing from 130 mW/(m·K) for PW, down to 15 mW/(m·K) for AG. As shown in Figure 3, the TBSs are placed in three different configuration positions along both steel stud flanges; the inner flange; and the outer flange.

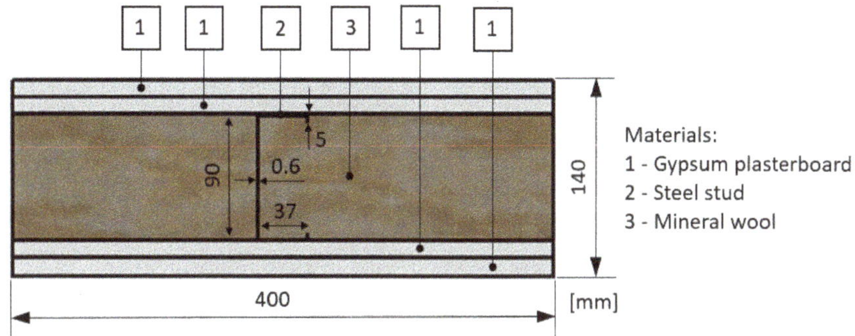

Figure 1. Materials, dimensions, and geometry of the reference LSF partition cross-section.

Table 1. Thicknesses (d) and thermal conductivities (λ) of materials used in the reference LSF partition wall.

Constitutive Materials	d [mm]	λ [W/(m·K)]	Ref.
Gypsum Plaster Board (2 × 12.5 mm)	25	0.175	[40]
Mineral Wool (MW)	90	0.035	[41]
Steel Studs (C90 × 37 × 5 × 0.6 mm)	—	50.000	[42]
Gypsum Plaster Board (2 × 12.5 mm)	25	0.175	[40]
Global Thickness	**140**	—	—

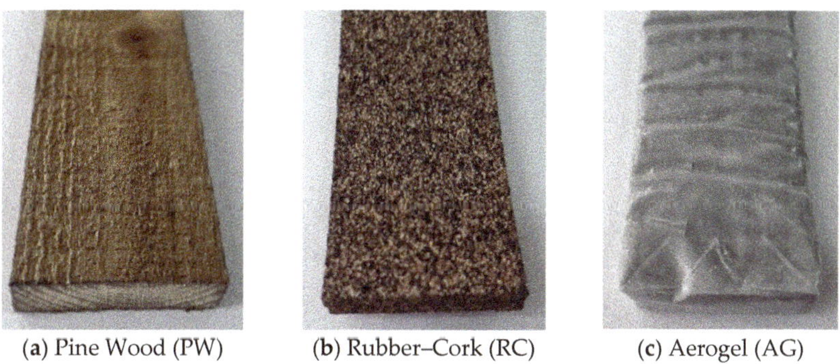

(**a**) Pine Wood (PW)   (**b**) Rubber–Cork (RC)   (**c**) Aerogel (AG)

Figure 2. Thermal break strip materials assembled on the LSF partitions.

Table 2. Thermal conductivities (λ) of the thermal break strips.

Materials (Abbreviation)	λ [mW/(m·K)]	Ref.
Pine Wood (PW)	130	[42]
Rubber and Cork composite (RC)	88	[43]
Aerogel (AG)	15	[44]

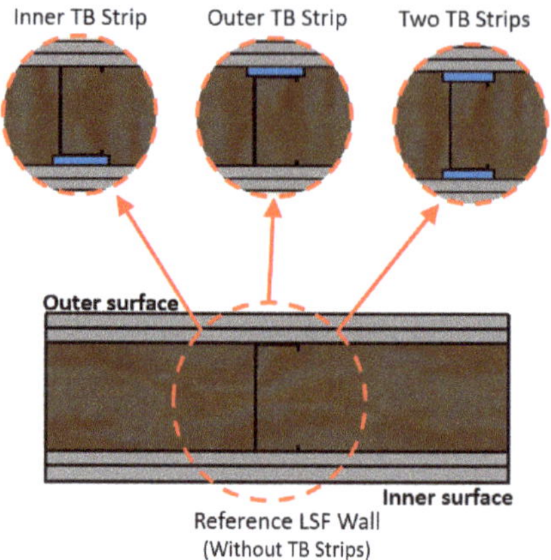

Figure 3. Positionings of the Thermal Break Strips (TBSs).

2.2. Lab Measurements

The lab measurements were achieved with the aid of a mini hot box apparatus, where the wall prototype is set in the middle of two climatic chambers (cold and hot), as shown in Figure 4a. Notice that, even though not being illustrated in Figure 4a, the perimeter of the wall sample was insulated using a 80 mm layer of polyurethane foam, having a thermal conductivity equal to 36 mW/(m·K), mitigating the heat losses through the LSF wall test sample perimeter. The cold chamber was cooled by a refrigerator and the hot chamber was warmed up by an electrical resistance of 70 watts. The tested wall samples are 1060 mm wide and 1030 mm high, with a structure composed of three vertical steel profiles, which were set 400 mm apart, where the central one is in the middle.

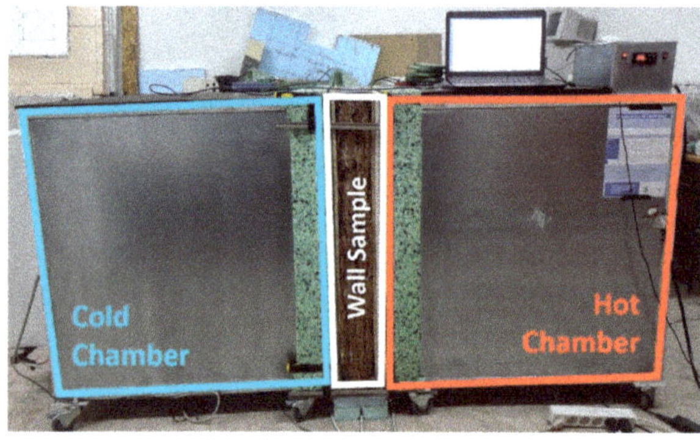
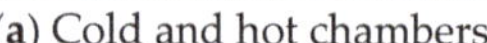

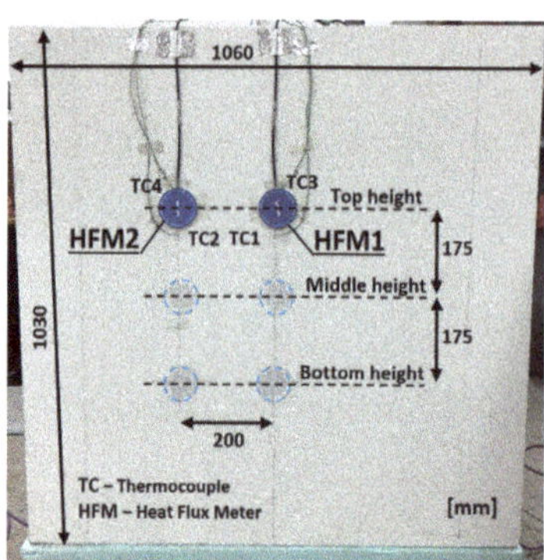

(**a**) Cold and hot chambers   (**b**) Wall surface sensor location

Figure 4. Experimental lab test illustrations.

In this work, the performance of the LSF wall test samples was obtained by making use of the Heat Flow Meter (HFM) method [37], improved for two HFM sensors, one on each wall surface [23]. Four Hukseflux sensors (model HFP01), with a precision of ±3%, were used to quantify the heat flux passing across the test sample walls, two on the cold side and the other two on the hot side, reducing the duration needed for the tests and improving the precision of the laboratorial measurements, as recommended by Rasooli and Itard [27]. On both sides (cold and hot), one HFM was placed in the middle of the insulation cavity, while the another one was placed over the centered vertical steel stud, at 3 distinct heights, as displayed in Figure 4b, to assess the thermal behavior in 2 different zones.

The temperature measurements were obtained using 12 Type K (1/0.315) PFA insulated Thermocouples (TCs), having class 1 precision certification. Half of them were placed on the hot side, while the other half on the cold side. Moreover, two TCs were used to measure the air temperature inside both chambers (hot and cold), another two TCs were used to measure the wall surface temperatures near the HFMs, and the remaining two were used to measure the air temperature near the wall surface, as displayed in Figure 4b for the cold partition surface.

The cold and hot chambers were set to maintain an average temperature of 5 °C and 40 °C, respectively. These two climatic chambers were well insulated to minimize surrounding heat losses and to ensure that the lab measurements were taken in a condition of near steady-state heat transmission. Notice that in a real building application context, the walls are exposed to transient conditions with variable temperature differences and are exposed to wind and solar radiation effects. The main advantage of using conductive (or surface-to-surface) thermal resistances to quantify the performance of the LSF walls is that it is possible to add any intended surface thermal resistance to calculate the total thermal resistance (or transmittance) depending on the surrounding environmental conditions.

The temperatures and heat flux quantified during the tests were recorded with the aid of one Pico TC-08® data-logger, with an accuracy of ±0.5 °C, on each side of the LSF partition wall. The management of this acquired data was performed with the software PicoLog®, version 6.1.10, on a laptop computer connected to the data-loggers.

2.3. Numerical Simulations

The adopted Finite Element Method (FEM) software, for the thermal bidimensional numerical simulations, was THERM® (version 7.6.1). These simulations only consider a representative cross-section of the LSF walls with a steel stud spacing of 400 mm, as shown before in Figure 1 for the reference LSF partition. The hygrothermal properties for the different materials used in these numerical models are displayed in Tables 1 and 2. Furthermore, the error on the FEM results was limited to 2%, for the numerical models evaluated in this research. The boundary conditions regarding environment air temperatures and the surface film coefficients were defined for each simulation.

The air temperatures for the cold and hot environments were equal to 5 °C and 40 °C, respectively. These are average values measured inside the cold and hot chambers. The surface coefficients of heat transfer or film coefficients, h [W/m^2·K], of the LSF wall surfaces were set equal to the average values obtained for each test, considering the difference between surface and air temperatures, ΔT [K], as well as the superficial heat flux, q [W/m^2], as indicated in the following expression:

$$h = \frac{q}{\Delta T} \quad \left(= \frac{1}{R_\mathrm{s}}\right). \tag{1}$$

The thermal resistances of the internal and external surfaces, R_si and R_se, ranged within the conventional values predefined in the standard ISO 6946 [45] for the horizontal heat flux, [0.04, 0.13] m^2·K/W.

The verification of the software THERM® 2D models' accuracy was also confirmed by comparing with 3D models, which were developed using the software ANSYS® (version 19.1). With this purpose, two distinct reference wall models were compared, i.e., with and without steel studs, as displayed in Figure 5. The numerical simulations' boundary conditions of the models are the ones mentioned in the previous paragraph. Figure 5 displays the colored temperature distribution for the reference partition wall models configured in the following software: (a) THERM and (b) ANSYS. As illustrated, the simulated colored temperature distributions of the LSF partition wall are very similar. Additionally, the obtained conductive R-values are analogous, with a nearly zero percentage difference (i.e., −0.8%).

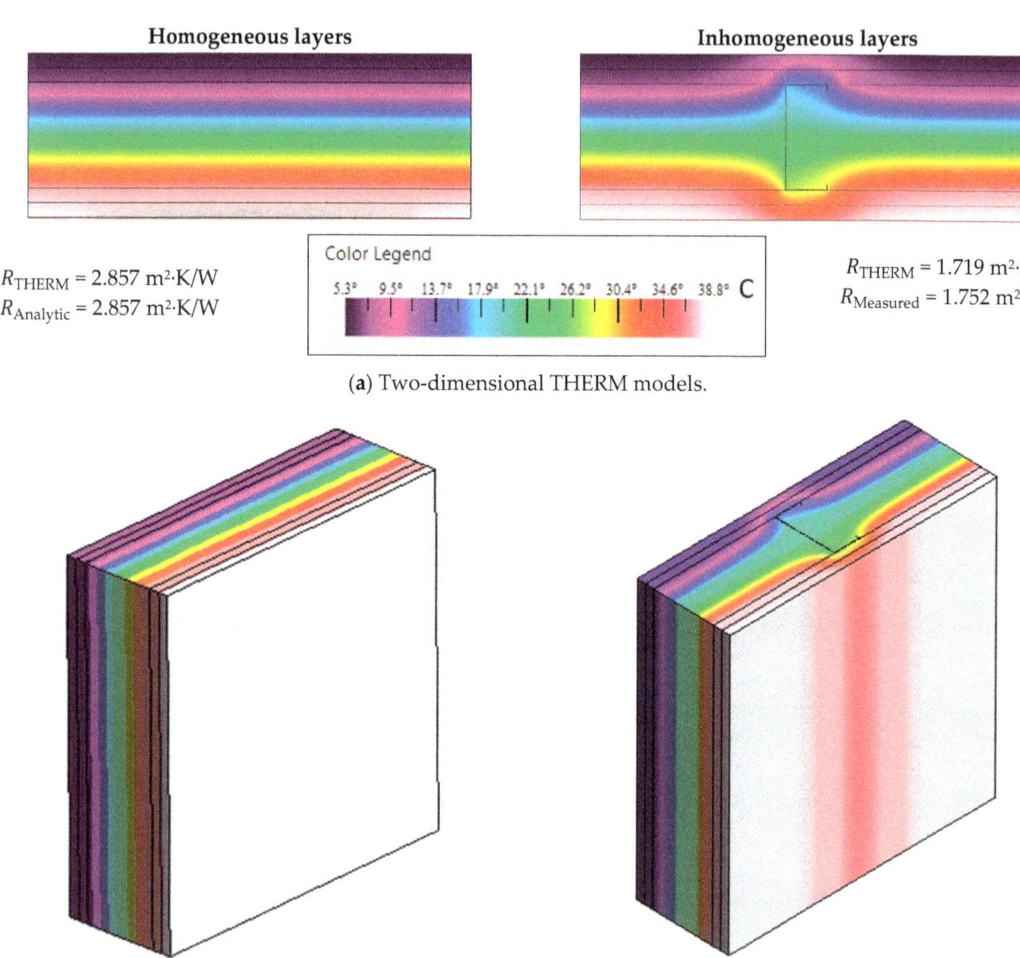

(**a**) Two-dimensional THERM models.

(**b**) Three-dimensional ANSYS model.

Figure 5. THERM models' accuracy: conductive thermal resistances and temperature distribution.

3. Results and Discussion

3.1. Conductive Thermal Resistances

In Table 3 are displayed the measured laboratory values and the corresponding computed values by 2D Finite Element Method (FEM) models using the software THERM for the surface-to-surface R-values of the studied LSF partitions, the percentage and absolute differences among them. The results are divided into four sets: (1) reference LSF partition wall (Ref.); (2) LSF partition walls with an inner TBS (PW_{in}, RC_{in}, and AG_{in}); (3) LSF partition walls with an outer TBS (PW_{out}, RC_{out}, and AG_{out}); and (4) LSF partition walls with a TBS on two sides, outer and inner (PW_{x2}, RC_{x2}, and AG_{x2}).

The laboratorial measurements and the numerical predicted R-values are quite similar, with the biggest differences in percentages being ±2%. Thus, the accuracy of both computed and measured R-values is ensured. Moreover, the TBSs can mitigate the heat losses originated by the steel frame thermal bridges, increasing the R-values of the LSF partitions. This improvement mainly depends on the number of TBSs, their thickness, and material thermal conductivity.

Table 3. Numerical (THERM) and experimental (measured) thermal resistances (conductive *R*-values).

Wall Code Wall Description	R-Value		Difference	
	THERM $[(m^2 \cdot K)/W]$	Measured $[(m^2 \cdot K)/W]$	Absolute $[(m^2 \cdot K)/W]$	Percentage [%]
Ref. Reference LSF Partition Wall	1.719	1.752	+0.033	+2%
$\mathbf{PW_{in}}$ Inner Pine Wood TBS	1.976	1.931	−0.045	−2%
$\mathbf{RC_{in}}$ Inner Rubber–Cork TBS	2.006	2.006	+0.000	0%
$\mathbf{AG_{in}}$ Inner Aerogel TBS	2.359	2.404	+0.045	+2%
$\mathbf{PW_{out}}$ Outer Pine Wood TBS	1.981	1.976	−0.005	0%
$\mathbf{RC_{out}}$ Outer Rubber–Cork TBS	1.975	1.965	−0.010	−1%
$\mathbf{AG_{out}}$ Outer Aerogel TBS	2.358	2.414	+0.056	+2%
$\mathbf{PW_{x2}}$ Double Pine Wood TBSs	2.254	2.304	+0.050	+2%
$\mathbf{RC_{x2}}$ Double Rubber–Cork TBSs	2.236	2.202	−0.034	−2%
$\mathbf{AG_{x2}}$ Double Aerogel TBSs	2.892	2.885	−0.007	0%

TBS—Thermal Break Strip.

In Figure 6, the measured *R*-values are graphically displayed, for an easier visualization and comparison. Since the Mineral Wool (MW) batt insulation is expansible, the *R*-value rise due to a homogeneous MW layer increment of 10 and 20 mm, equivalent to the thickness of one and two TBSs, respectively, is also displayed.

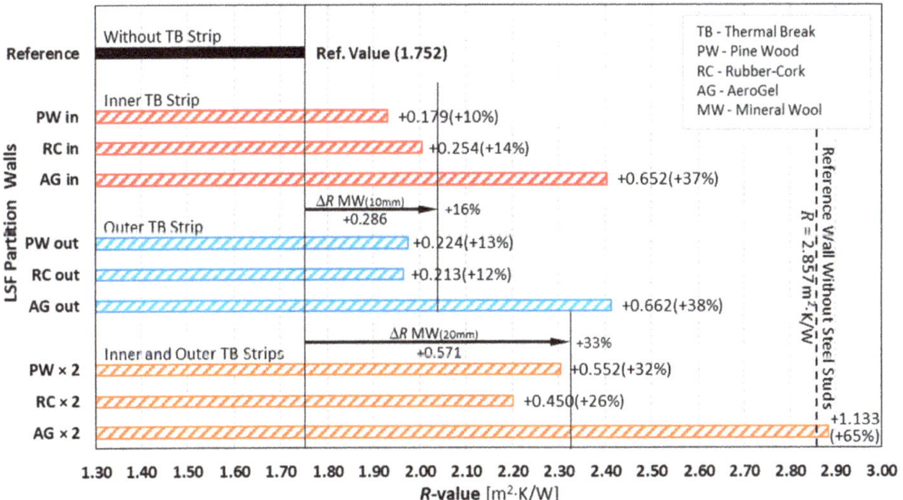

Figure 6. Measured conductive thermal resistances of LSF partition walls.

The thermal conductivity of MW batt insulation (0.035 W/m·K) is lower than Rubber–Cork (RC) composite (0.088 W/m·K) and Pine Wood (PW) (0.130 W/m·K), being higher in relation to Aerogel (AG) (0.015 W/m·K)). So, as expected, when using TBS materials with lower thermal conductivities, the thermal performance improvement is lower than the one expected for a homogeneous MW layer. This *R*-value increase for a single TBS ranges from +10% up to +14% and for double TBSs from +26% to +32%. However, the *R*-values for the LSF partition walls having aerogel TBSs are higher than the expected ones for the homogeneous MW increased layer, with the thermal performance enhancement for these TBSs being equal to: +37% (inner); +38% (outer); and +65% (double TBSs).

Notice that the configuration with aerogel TBSs on both sides of the metallic stud presents a conductive *R*-value of 2.885 m²·K/W, which means that it fully mitigates the thermal bridge effect created by the steel frame, since it reaches the *R*-value of a homogeneous layered wall without any steel stud, which is 2.857 m^2·K/W, as graphically displayed in Figure 6 as a vertical dashed line.

Another interesting and quite surprising feature in Figure 6 is that for the TBSs on the outer flange and with TBSs on both sides of the steel studs, the *R*-values measured when using Pine Wood (PW) are higher than when using rubber–cork (RC) composite, even when PW has a higher thermal conductivity (0.130 W/(m·K)). This happens since pine wood TBSs, instead of having the nominal 10 mm thickness, are 13 mm thick, which also originates a bigger MW expansion of 3 and 6 mm, for single and double TBSs, respectively.

3.2. Thermographic Images

In Figure 7 are illustrated the Infrared (IR) images taken for the tested LSF partition wall cold surface, for the reference wall, and for the LSF partition walls with aerogel TBSs. The purpose was to assess the linear thermal bridge mitigation effect originated by the steel studs' high thermal conductivity. Additionally, in Figure 8 are displayed the profiles for the recorded surface temperatures along the lines (L1 to L4), as presented in Figure 7.

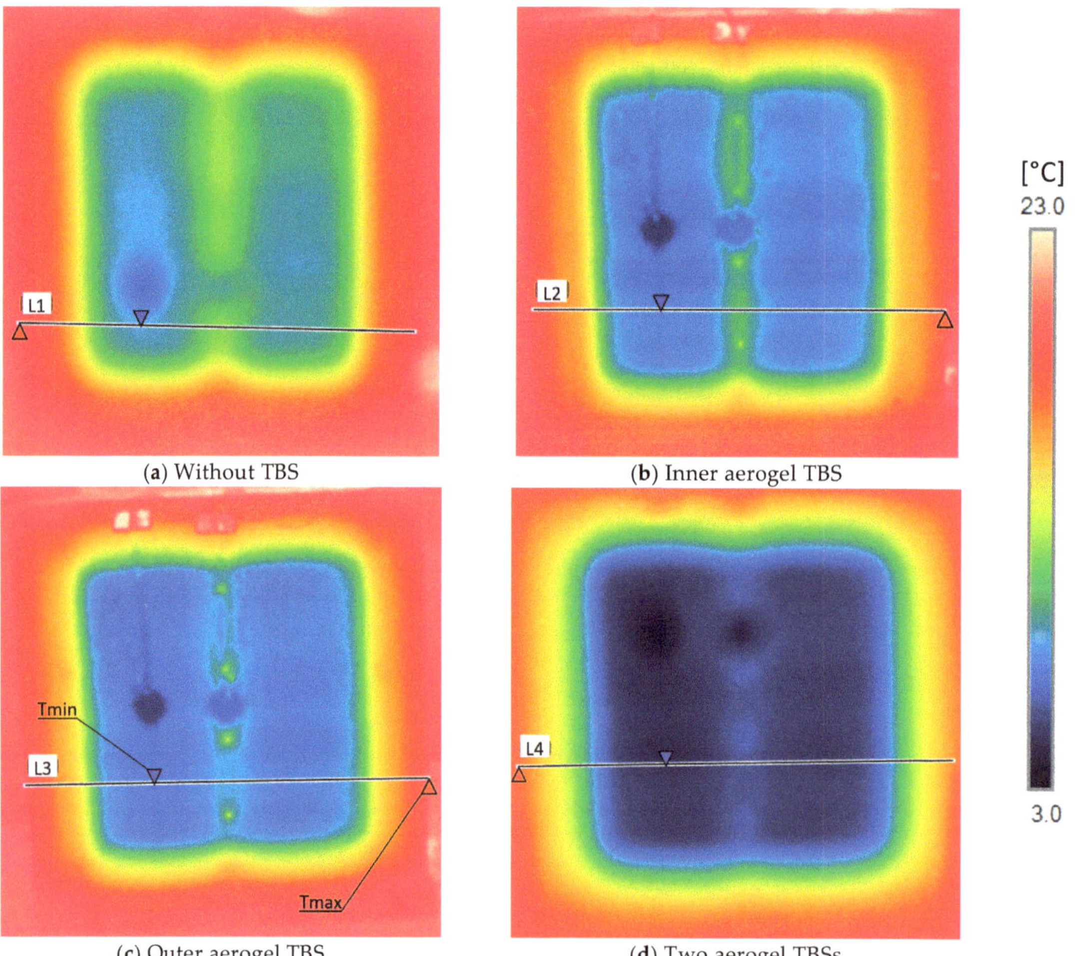

(**a**) Without TBS

(**b**) Inner aerogel TBS

(**c**) Outer aerogel TBS

(**d**) Two aerogel TBSs

Figure 7. Infrared thermography photos of the LSF partition walls: cold side.

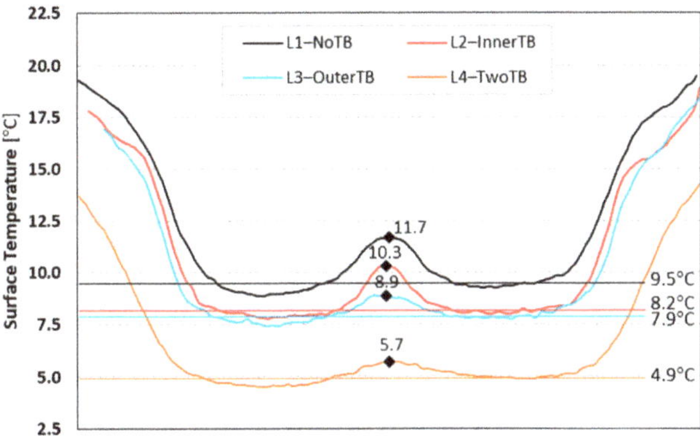

Figure 8. Surface temperatures in the four horizontal lines, L1–L4, previously identified in Figure 7.

Looking at the IR images, the vertical steel stud in the center can easily be detected in Figure 7a (LSF partition wall without TBSs), due to the increased localized heat transfer, originating a higher surface temperature on the cold surface. In comparison, the central vertical steel stud in Figure 7d is the least pronounced, since the heat transfer for the LSF partition wall with aerogel TBSs on both sides of the steel studs is mitigated the most, reducing very significantly the related thermal bridge effect.

Figure 8 also illustrates these features well, where the maximum temperatures are in the central steel profile. In comparison with the average temperature of the wall surface, the differences could be ordered, from higher to lower, as follows: without TBSs (L1); with TBSs in the interior side flange (L2); with TBSs in the outer flange (L3); with TBSs in both flanges (L4).

3.3. Heat Flux Predictions

A similar assessment was performed using 2D FEM models, computed in THERM, as illustrated in Figure 9, where the predicted heat flux distribution on the cross-section of the four LSF partitions previously illustrated in Figure 7 is displayed.

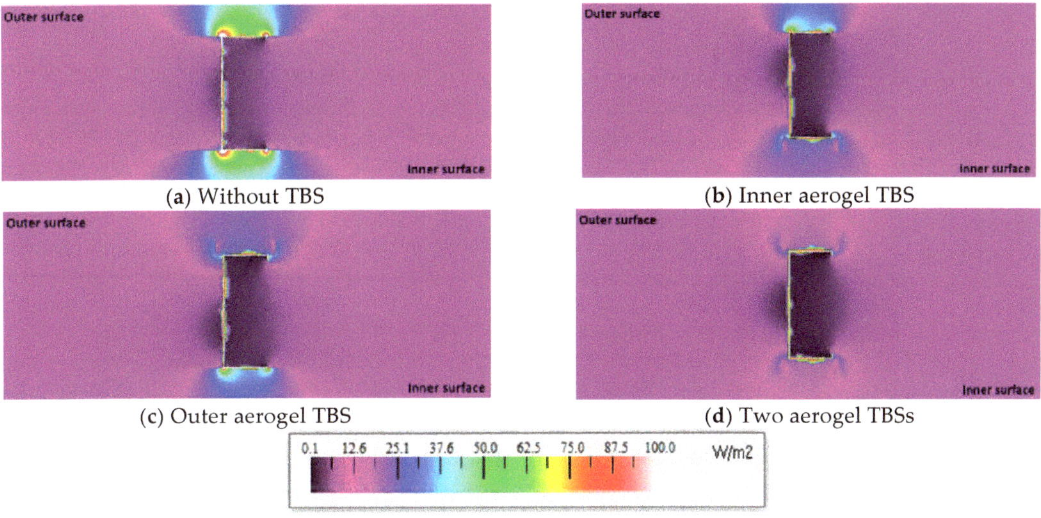

Figure 9. Heat flux distribution predicted for the LSF wall cross-sections.

In Figure 9a, there is a higher heat flux near the steel stud flanges, due to the thermal bridge effect. This increased heat flux is related to the higher heat conduction along the steel profile, being more observable near the two sheathing layers on the two sides of the wall, mainly in the vicinity of the metallic flanges, which spreads heat to the gypsum plaster boards. Notice that in the web of the steel stud there is also a significant heat flux, which is not easily visible in this scale. When an aerogel TBS is positioned on the inner (Figure 9b) or outer (Figure 9c) steel flange, the heat flux through the steel flange shows a significant decrease on the TBS side. However, in all the remaining parts the reduction in the heat flux is not as pronounced.

Finally, in Figure 9d the heat flux distribution when a TBS is used on both flanges of the steel profile is illustrated. As expected, the heat flux on both sides of these flanges is highly mitigated, leaving the major heat flux values mainly within the web steel stud. This last illustration allows us to visualize how these two aerogel TBSs can completely attenuate the metallic studs' thermal bridge effect.

4. Conclusions

The performance of Thermal Break Strips (TBSs) on Light Steel Framed (LSF) partitions were measured under quasi-steady-state laboratory conditions. The measured surface-to-surface thermal resistances were compared with the computations provided by Finite Element Method (FEM) simulation models. Several types of TBS materials were tested: (1) a bio-based material (pine wood); (2) an eco-friendly material (recycled rubber–cork composite); and (3) a super-insulation material (aerogel). Moreover, several arrangements regarding the location of the TBS were considered (outer, inner, and on both flanges of the steel studs). For each of the aforementioned configurations, three types of results were analyzed: (1) the measured and the predicted conductive R-values, for all TBS materials; (2) the recorded temperatures at the cold surface through infrared images, using the super-insulating aerogel TBS; and (3) the THERM predicted heat flux for the same aerogel TBS LSF wall configurations.

Next, we summarize the main achievements of this research work:

- A very good agreement between the R-value measurements and the numerical simulation predictions was achieved, having differences smaller than $\pm 2\%$.
- The thermal performance was quite analogous when a single TBS was utilized on outer or inner steel stud flanges, given the LSF wall symmetry.
- The use of TBSs on both metallic stud flanges significantly enhances the thermal resistance, when compared to the use of only one TBS and without any TBS.
- The best thermal performance was achieved by the aerogel TBS material, given their very reduced thermal conductivity, when compared with the remaining materials.
- The use of the high-performance aerogel TBSs on both steel profile flanges was the unique configuration able to fully mitigate the thermal bridge effect due to the steel frame, reaching the thermal resistance provided by the reference wall having homogeneous layers, i.e., without steel profiles.
- The bio-based pine wood TBS exhibited a better thermal performance when compared to the recycled rubber–cork composite TBS, for some wall configurations (outer and two TBSs).
- The TBSs' thickness also has a significant influence on the wall thermal resistance, not only because it mitigates the local thermal bridge effect on the steel stud, but also given the mineral wool expansion, which increases the wall cavity thermal resistance (between metallic studs).

Author Contributions: Conceptualization, P.S. and D.M.; Formal analysis, D.M.; Funding acquisition, P.S.; Investigation, P.S., D.A. and P.L.; Methodology, P.S.; Project administration, P.S.; Resources, P.S. and D.M.; Software, D.A. and P.L.; Supervision, P.S.; Validation, D.A. and P.L.; Visualization, D.A. and P.L.; Writing—original draft, D.A. and P.L.; Writing—review & editing, P.S. and D.M. All authors have read and agreed to the published version of the manuscript.

Funding: This research was funded by FEDER funds through the Competitivity Factors Operational Programme—COMPETE—and by national funds through the Foundation for Science and Technology FCT—within the scope of the project POCI-01-0145-FEDER 032061.

Cofinanciado por: POCI-01-0145-FEDER-032061

Acknowledgments: The authors also want to thank the support provided by the following companies: Pertecno, Gyptec Ibéria, Volcalis, Sotinco, Kronospan, Hulkseflux, Hilti, and Metabo.

Conflicts of Interest: The authors declare no conflict of interest.

References

1. EU. Directive (EU) 2018/844 of the European parliament and of the council on the energy performance of buildings and on energy efficiency. *Off. J. Eur. Union* **2018**, *156*, 75–91.
2. Zhan, Q.; Xiao, Y.; Musso, F.; Zhang, L. Assessing the hygrothermal performance of typical lightweight steel-framed wall assemblies in hot-humid climate regions by monitoring and numerical analysis. *Build. Environ.* **2021**, *188*, 107512. [CrossRef]
3. Soares, N.; Santos, P.; Gervásio, H.; Costa, J.J.; Simões da Silva, L. Energy efficiency and thermal performance of lightweight steel-framed (LSF) construction: A review. *Renew. Sustain. Energy Rev.* **2017**, *78*, 194–209. [CrossRef]
4. Perera, D.; Upasiri, I.; Poologanathan, K.; Gatheeshgar, P.; Sherlock, P.; Hewavitharana, T.; Suntharalingam, T. Energy performance of fire rated LSF walls under UK climate conditions. *J. Build. Eng.* **2021**, *44*, 03293. [CrossRef]
5. Ribeiro, T.; Santos, P.; Mateus, D. Performance of Thermal Break Strips in Lightweight Steel Framed Walls. In Proceedings of the 1st International Conference on Water Energy Food and Sustainability (ICoWEFS 2021), Leiria, Portugal, 10–12 May 2021; Springer International Publishing: Cham, Switzerland, 2021; pp. 382–389. [CrossRef]
6. Santos, P. Chapter 3-Energy Efficiency of Lightweight Steel-Framed Buildings. In *Energy Efficient Buildings*; Yap, E.H., Ed.; InTech: London, UK, 2017; pp. 35–60. [CrossRef]
7. Silvestre, N.; Pires, J.; Santos, A. *Manual de Conceção de Estruturas e Edifícios em LSF-Light Steel Framing, 1a*; CMM: Coimbra, Portugal, 2013.
8. Roque, E.; Santos, P.; Pereira, A. Thermal and sound insulation of lightweight steel-framed façade walls. *Sci. Technol. Built Environ.* **2019**, *25*, 156–176. [CrossRef]
9. Wang, C.; Mynors, D. Acoustic performance of cold-formed steel buildings. In *Recent Trends in Cold-Formed Steel Construction*; Woodhead Publishing: Sawston, UK, 2016; pp. 173–182. [CrossRef]
10. Roque, E.; Santos, P. The effectiveness of thermal insulation in lightweight steel-framedwalls with respect to its position. *Buildings* **2017**, *7*, 13. [CrossRef]
11. Atsonios, I.A.; Mandilaras, I.D.; Kontogeorgos, D.A.; Founti, M.A. Two new methods for the in-situ measurement of the overall thermal transmittance of cold frame lightweight steel-framed walls. *Energy Build.* **2018**, *170*, 183–194. [CrossRef]
12. Santos, P.; Lemes, G.; Mateus, D. Thermal transmittance of internal partition and external facade LSF walls: A parametric study. *Energies* **2019**, *12*, 2671. [CrossRef]
13. Santos, P.; Simões da Silva, L.; Ungureanu, V. *Energy Efficiency of Light-Weight Steel-Framed Buildings*, 1st ed.; Technical Committee 14-Sustainability & Eco-Efficiency of Steel Construction; European Convention for Constructional Steelwork (ECCS): Brussels, Belgium, 2012; ISBN 978-92-9147-105-8.
14. Santos, P.; Gonçalves, M.; Martins, C.; Soares, N.; Costa, J.J. Thermal transmittance of lightweight steel framed walls: Experimental versus numerical and analytical approaches. *J. Build. Eng.* **2019**, *25*, 100776. [CrossRef]
15. Gorgolewski, M. Developing a simplified method of calculating U-values in light steel framing. *Build. Environ.* **2007**, *42*, 230–236. [CrossRef]
16. Kosny, J.; Christian, J.E.; Barbour, J.; Goodrow, J. *Thermal Performance of Steel Framed Walls*; ORNL Report; Oak Ridge National Laboratory: Oak Ridge, TN, USA, 1994.
17. ASHRAE. *Handbook of Fundamentals*, SI ed.; ASHRAE-American Society of Heating, Refrigerating and Air-Conditioning Engineers: Atlanta, GA, USA, 2017.
18. Muzzi, T.A.; Souza, H.A.; Gomes, A.P. Heat transfer analysis of the vertical closing system in light steel framing using the isothermal planes method and finite element method. *REM-Int. Eng. J.* **2021**, *74*, 425–431. [CrossRef]
19. Martins, C.; Santos, P.; Simoesdasilva, L. Lightweight steel-framed thermal bridges mitigation strategies: A parametric study. *J. Build. Phys.* **2016**, *39*, 342–372. [CrossRef]
20. Santos, P.; Martins, C.; Simoesdasilva, L.; Bragança, L. Thermal performance of lightweight steel framed wall: The importance of flanking thermal losses. *J. Build. Phys.* **2014**, *38*, 81–98. [CrossRef]
21. Manzan, M.; De Zorzi, E.Z.; Lorenzi, W. Numerical simulation and sensitivity analysis of a steel framed internal insulation system. *Energy Build.* **2018**, *158*, 1703–1710. [CrossRef]

22. *ISO 10211*; Thermal Bridges in Building Construction-Heat Flows and Surface Temperatures-Detailed Calculations. ISO-International Organization for Standardization: Geneva, Switzerland, 2017.
23. Soares, N.; Martins, C.; Gonçalves, M.; Santos, P.; da Silva, L.S.; Costa, J.J. Laboratory and in-situ non-destructive methods to evaluate the thermal transmittance and behavior of walls, windows, and construction elements with innovative materials: A review. *Energy Build.* **2019**, *182*, 88–110. [CrossRef]
24. Gori, V.; Marincioni, V.; Biddulph, P.; Elwell, C.A. Inferring the thermal resistance and effective thermal mass distribution of a wall from in situ measurements to characterise heat transfer at both the interior and exterior surfaces. *Energy Build.* **2017**, *135*, 398–409. [CrossRef]
25. Santos, P.; Poologanathan, K. The importance of stud flanges size and shape on the thermal performance of lightweight steel framed walls. *Sustainability* **2021**, *13*, 3970. [CrossRef]
26. ThermaChannel. Website 'ThermaChannel-Energy-Efficient Steel Framing'. 2018. Available online: www.thermachannel.com (accessed on 5 February 2019).
27. Lupan, L.M.; Manea, D.L.; Moga, L.M. Improving Thermal Performance of the Wall Panels Using Slotted Steel Stud Framing. *Procedia Technol.* **2016**, *22*, 351–357. [CrossRef]
28. Váradi, J.; Toth, E. Thermal Improvement of Lightweight Façades containing Slotted Steel Girders. In Proceedings of the Twelfth International Conference on Civil, Structural and Environmental Engineering Computing, Funchal, Portugal, 1–4 September 2009.
29. Santos, P.; Mateus, D. Experimental assessment of thermal break strips performance in load-bearing and non-load-bearing LSF walls. *J. Build. Eng.* **2020**, *32*, 101693. [CrossRef]
30. Kapoor, D.R.; Peterman, K.D. Quantification and prediction of the thermal performance of cold-formed steel wall assemblies. *Structures* **2021**, *30*, 305–315. [CrossRef]
31. Kempton, L.; Kokogiannakis, G.; Green, A.; Cooper, P. Evaluation of thermal bridging mitigation techniques and impact of calculation methods for lightweight steel frame external wall systems. *J. Build. Eng.* **2021**, *43*, 102893. [CrossRef]
32. Kosny, J.; Christian, J.E. Thermal evaluation of several configurations of insulation and structural materials for some metal stud walls. *Energy Build.* **1995**, *22*, 157–163. [CrossRef]
33. Santos, P.; Ribeiro, T. Thermal performance improvement of double-pane lightweight steel framed walls using thermal break strips and reflective foils. *Energies* **2021**, *14*, 6927. [CrossRef]
34. Bruno, R.; Bevilacqua, P.; Ferraro, V.; Arcuri, N. Reflective thermal insulation in non-ventilated air-gaps: Experimental and theoretical evaluations on the global heat transfer coefficient. *Energy Build.* **2021**, *236*, 110769. [CrossRef]
35. Jelle, B.P.; Kalnæs, S.E.; Gao, T. Low-emissivity materials for building applications: A state-of-the-art review and future research perspectives. *Energy Build.* **2015**, *96*, 329–356. [CrossRef]
36. Fantucci, S.; Serra, V. Investigating the performance of reflective insulation and low emissivity paints for the energy retrofit of roof attics. *Energy Build.* **2019**, *182*, 300–310. [CrossRef]
37. *ISO 9869-1*; Part 1: Heat Flow Meter Method. Thermal Insulation-Building Elements-In-Situ Measurement of Thermal Resistance and Thermal Transmittance. ISO-International Organization for Standardization: Geneva, Switzerland, 2014.
38. THERM. Software Version 7.6.1. Lawrence Berkeley National Laboratory: Berkeley, CA, USA. 2017. Available online: https://windows.lbl.gov/software/therm (accessed on 14 February 2019).
39. ANSYS Workbench. Software Version 19.1. ANSYS, Inc.: Canonsburg, PA, USA. 2018. Available online: www.ansys.com/products/ (accessed on 8 June 2020).
40. Gyptec, I. Technical Sheet: Standard Gypsum Plasterboard. 2022. Available online: https://www.gyptec.eu/documentos/Ficha_Tecnica_Gyptec_A.pdf (accessed on 10 February 2022).
41. Volcalis. Technical Sheet: Alpha Mineral Wool. 2022. Available online: https://www.volcalis.pt/categoria_file_docs/fichatecnica_volcalis_alpharollo-386.pdf (accessed on 10 February 2022).
42. Santos, C.; Matias, L. *ITE50-Coeficientes de Transmissão Térmica de Elementos da Envolvente dos Edifícios*; LNEC-Laboratório Nacional de Engenharia Civil: Lisboa, Portugal, 2006. (In Portuguese)
43. MS-R0 Test Report. Test Report Ref. 5015 PE 1977/08-Determination of Thermal Conductivity (Acousticork MS-R0). In *Departamento de Engenharia Têxtil*; Universidade do Minho: Braga, Portugal, 2008.
44. Proctor Group. Spacetherm® CBS Specification. 2018. Available online: https://www.proctorgroup.com/assets/PerformanceSpecs/SpacethermCBSPerformanceSpecification.pdf (accessed on 10 February 2022).
45. *ISO 6946*; Building Components and Building Elements—Thermal Resistance and Thermal Transmittance—Calculation Methods. ISO-International Organization for Standardization: Geneva, Switzerland, 2017.

 *buildings*

Article

Evaluation of Building Energy Savings Achievable with an Attached Bioclimatic Greenhouse: Parametric Analysis and Solar Gain Control Techniques

Dimitrios Kaliakatsos, Francesco Nicoletti *, Francesca Paradisi, Piero Bevilacqua and Natale Arcuri

Mechanical, Energy and Management Engineering Department, University of Calabria, 87036 Rende, CS, Italy
* Correspondence: francesco.nicoletti@unical.it

Abstract: Bioclimatic solar greenhouses are passive solar systems of relevant interest in the building sector, as they allow the reduction of energy needs related to air-conditioning. The aim of this work is to analyze the thermal behavior of a bioclimatic solar greenhouse attached to a residential building. It is equipped with photovoltaic solar blinds (SPBs) to manage solar inputs and produce electricity. Automated control systems are implemented to activate the vents and SPBs. The parametric performance analysis conducted using the dynamic simulation software EnergyPlus allowed the evaluation of the influence of glass type, thermal mass, size, ventilation and location. The results show how the automation of the vents allows the maximization of heat exchange throughout the year, leading to a reduction in consumption even during the summer period. Analyses conducted for some cities in the Mediterranean area show that the maximum energy saving obtained is greater than 13%; in addition, photovoltaic solar shading contributes to the production of more than 1000 kWh/year of electricity.

Keywords: energy savings; solar greenhouse; solar PV blinds; thermal mass; vents opening

Citation: Kaliakatsos, D.; Nicoletti, F.; Paradisi, F.; Bevilacqua, P.; Arcuri, N. Evaluation of Building Energy Savings Achievable with an Attached Bioclimatic Greenhouse: Parametric Analysis and Solar Gain Control Techniques. *Buildings* **2022**, *12*, 2186. https://doi.org/10.3390/buildings12122186

Academic Editor: Xing Jin

Received: 17 November 2022
Accepted: 7 December 2022
Published: 9 December 2022

Publisher's Note: MDPI stays neutral with regard to jurisdictional claims in published maps and institutional affiliations.

1. Introduction

In developed countries, the residential sector is responsible for about 40% of total energy consumption, of which more than 50% is used for the air-conditioning of buildings, contributing to the release of 30% of total CO_2 emissions [1]. Over the years, the need to reduce CO_2 emissions in the building sector has led to the emergence of the concept of bioclimatic architecture. The benefits are linked to site and building geometry, with the aim of exploiting solar radiation as much as possible to reduce energy consumption [2,3]. The main systems used for this purpose are passive solar systems, which can help maintain indoor comfort without the aid of active devices powered by external energy sources. Passive solar systems are based on capturing solar energy and storing and/or distributing heat inside the building by means of thermal air circulation [4,5].

Bioclimatic solar greenhouses, consisting of a glassed-in environment adjacent to the building, are used to reduce the energy demand of adjacent air-conditioned rooms during the winter period [6,7]. The high temperature achieved in the greenhouse allows the air-conditioned room to be heated [8] by directly supplying warm air through vents [9]. The attached solar greenhouses, which are an open thermodynamic system with energy and mass exchange, receive heat from the sun's rays passing through the glass envelope [10]. The installation of a solar greenhouse also provides additional benefits, not related to energy savings, such as the renovation and weather protection of the building's exterior façade [11].

The attached greenhouse is the most common type; however, the large number of dispersing surfaces reduces the energy gain and also increases the risk of overheating [12]. These drawbacks are mitigated in the built-in greenhouse, which is characterized by a smaller external surface area [13]. The benefits of using attached greenhouses are also closely dependent on parameters such as thermal storage capacity, type of glazing, size, and night shading. A careful analysis is needed to assess the influence of these factors in the design phase.

Various studies have been conducted to evaluate their performance in different climate zones and latitudes as the operation of solar greenhouses is dependent on outdoor conditions. Mihalakakou and Ferrante [9] simulated the thermal behavior of a south-facing attached greenhouse in four European cities (Milan, Dublin, Athens and Florence), concluding that in all the cities analyzed, the greenhouse led to a reduction in heating demand in the winter period, despite the occurrence of more or less pronounced over-heating phenomena in the summer. Aelenei et al. [13] evaluated energy savings following the installation of an attached greenhouse in six climate zones in Portugal. They obtained energy savings during the winter period, against higher consumption during the summer period, with a total energy saving of between 15 and 55%. However, the role played by the greenhouse's construction characteristics and thermal storage mass was not emphasized in these studies. Chiesa et al. [14] evaluated the impact on energy consumption of an attached solar greenhouse at 50 locations in central and southern Europe. In contrast to a significant reduction in winter consumption, they observed an increase in summer consumption in colder climates if no action was taken to prevent overheating.

Mihalakakou [15] demonstrated how it is possible to reduce the internal greenhouse temperature and limit the effect of overheating by adopting different passive cooling systems, such as night ventilation, solar shading systems and the use of underground ducts. Bataineh et al. [10] achieved a consumption reduction of 42% with an attached solar greenhouse in Jordan, minimizing overheating problems by adopting passive cooling systems. In the present work, considering the problems found in these articles, control logic will be implemented to move the blinds so as to optimize the inputs in winter and minimize them in summer to avoid excessively high temperatures.

Ulpiani et al. [8] found that in Italy, a greenhouse with convective heat exchange with double glazing and reduced depth reduced daily consumption more than 27%, highlighting how the use of a greenhouse with reduced depth leads to greater savings. Grudzinska [11] achieved energy demand savings of 33% by using a double-glazed envelope with selective coating in Poland. These articles do not investigate the performance using low-e double glazing.

The design of a solar greenhouse must consider the geometric configuration: the rectangular shape allows for a larger area of south-facing glazing and the contact surface between the greenhouse and the indoor environment, increasing solar gains and heat exchange with the building [15].

Numerous studies have shown that a southern exposure maximizes the amount of incident solar radiation in the winter period and reduces it in the summer period [9,16,17]. Attention must be paid to the glazed surfaces constituting the sides and roof, as they may cause overheating in summer and a reduction in benefits during winter [8,16]. The roof or south façade of the greenhouse could be tilted to maximize winter gain by reducing the angle of incidence [18].

The optical properties of the glazing greatly influence the behavior of the greenhouse; for proper functioning, the glass chosen should allow an adequate level of insulation without an excessive penalization of solar gain [11]. In sites with cold climates or harsh winters, the use of double glazing is advisable, as opposed to sites with mild winters, where single glazing is sufficient [13,19]. Regardless of the climate, good results can also be achieved using single glazing if the building is not insulated, while otherwise the use of high-performance glass is necessary [14].

The presence of thermal mass brings the following benefits [20]: increased average and minimum temperatures, reduced daily temperature fluctuations and reduced maximum air temperature inside the greenhouse. Rempel et al. [21] found that the dimensioning of the thermal mass depends on the time of day in which the stored heat is to be made available. If heating is required in the daytime, the mass must be minimal. If it is required in the evening, it must be greater. Moreover, thermal mass can provide cooling benefits and can prevent overheating [22]. Chiesa et al. [14] found that thermal mass has little influence on the performance of an attached solar greenhouse. They argue that this system requires further parametric analysis depending on the type of glass adopted.

The purpose of solar shading is to reduce consumption in the summer and avoid overheating, ensuring thermal and visual comfort for occupants [23]. Shading systems are subdivided according to type (fixed or movable) and positioning with respect to the glazing (external, internal, integrated). The main types of fixed shading systems are: overhanging, horizontal slat, vertical slat and egg-crate [24]. Their effectiveness depends on the geometric characteristics of the shading elements, the climate and the latitude of the installation site [25]. Mobile shading systems, which can be placed either externally or internally, can be adjusted according to the user's needs. The main types are Venetian blinds, vertical slat blinds and roller blinds [24]. Venetian blinds are the most widely used shading devices due to their versatility, and their effectiveness depends on the width of the slats, the ratio of the length to the distance between the slats, the angle of inclination and the place of installation [26]. The setup of mobile solar shading can be adjusted using three types of control: manual, motorized and automated. Various studies have shown that the use of automated solar shading contributes to improved indoor environmental conditions and increased energy savings [26,27]. In particular, Nicoletti et al. [26] evaluated the energy savings resulting from the implementation of a strategy to control the slat angle of Venetian blinds, obtaining 15% higher energy savings than Venetian blinds with fixed slat orientation. The inclination angle of solar shading and the reflection coefficient of their external surfaces influence energy savings in the summer [28]. There are no studies that consider the use of automated blinds to manage the performance of a solar greenhouse.

Building-integrated photovoltaic (BIPV) systems consist of photovoltaic cells integrated in the building that not only produce electricity but also form the building envelope [29]. Based on their location, BIPV systems are divided into three groups [30]: roof-integrated systems, façade-integrated systems and systems integrated into external architectural elements. Solar photovoltaic blinds (SPB) are part of façade-integrated BIPV systems, the function of which is the shielding of solar radiation and the production of photovoltaic energy [31]. For the estimation of producibility, it is necessary to consider the mutual shading between the louvers, the reflection coefficient of the back of the louvers and the type of photovoltaic cells used [32,33].

Indirect-gain solar greenhouses with an insulated separation wall and heat exchange by means of thermal air circulation constitute a configuration that has been poorly analyzed in the literature. Furthermore, the installation of photovoltaic solar shading moved by intelligent logic in a solar greenhouse has not been contemplated in any of the cited studies.

The influence of all these characteristics (type of glazing, thermal mass, size, shading and ventilation systems, location) are investigated for this paper. The aim of the work is to provide a general overview of the characteristics that a solar greenhouse attached to a building located in the Mediterranean area should have. In fact, this area is characterized by a hot summer climate and, therefore, attention must be paid to the temperatures obtained. An attached solar greenhouse is studied to evaluate the influence of all parameters on the reduction of energy consumption for a residential building. In the analyzed case, the separating wall is insulated and photovoltaic blinds are placed to screen the greenhouse glazing. The opening of the vents placed on the separation wall and the angle of inclination of the SPBs are controlled by appropriate control systems based on IoT (internet of things) technology. The adoption of this technology makes it possible to extend the savings achieved to the entire year, not limiting it to the winter period.

The study is conducted using the EnergyPlus simulation software, where a reference building with an attached solar greenhouse is modeled, implementing the algorithms related to the control techniques. The evaluation of the impact of the solar greenhouse on the building's consumption is conducted with parametric analyses, studying the savings achieved as the following parameters varied: depth, type of glass constituting the envelope, thickness of the thermal mass, use of solar shading, ventilation during the summer period, location. Finally, the overall electrical energy produced by photovoltaic solar shading is quantified.

2. Materials and Methods

This section describes the reference building analyzed and how the attached sunspace and control techniques are modeled for the numerical calculations.

2.1. Reference Building

The study is conducted considering Rome (Italy) as the reference locality. Climate data from the Gianni De Giorgio database (IGDG) are used to perform the dynamic simulations. These data are recognized as reliable by the scientific community and are recorded in the period 1951–1970 by the weather stations of the Meteorological Service of the Italian Air Force. Monthly average values of diffuse solar radiation on the horizontal plane, of direct normal solar radiation and outdoor air temperature are shown in Figure 1. The annual average outdoor temperature, the annual average amount of diffuse solar radiation and direct normal radiation are 15.3 °C, 625.3 kWh/m^2 e 969.0 kWh/m^2, respectively.

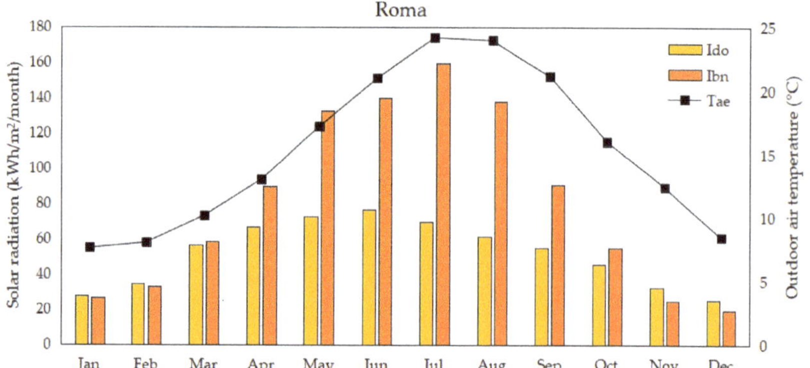

Figure 1. Monthly average climate data for Rome: diffuse irradiance on horizontal plane, direct normal irradiance and outdoor air temperature.

The building is a semi-detached house consisting of a basement, ground floor raised from ground level and attic. The basement consists of a cellar, garage and tavern, while the ground floor consists of an entrance hall, kitchen, living room, three bedrooms and bathroom. The air-conditioned floor area of the building is 110 m^2. The following rooms are not air-conditioned: cellar, cellar, stairwell, attic, bathrooms. The height of the rooms is 2.7 m. The 3D model of the reference building and the layout of the rooms are illustrated in Figure 2.

The composition of vertical structures and the relative transmittance values are shown in Table 1. The internal walls consist of a 0.08 m thick layer of perforated brick, covered on both sides with 0.01 m of plaster. The composition of horizontal structures is shown in Table 2. In addition, values of the linear transmission coefficients characterizing the thermal bridges of the building envelope have been included for precautionary purposes.

The internal gains related to occupancy are obtained by setting the occupancy density, the metabolic rate according to the activity carried out in each zone (Table 3) and the hourly occupancy schedule. Figure 3 shows, for each hour of the day, the percentage of time the area is occupied.

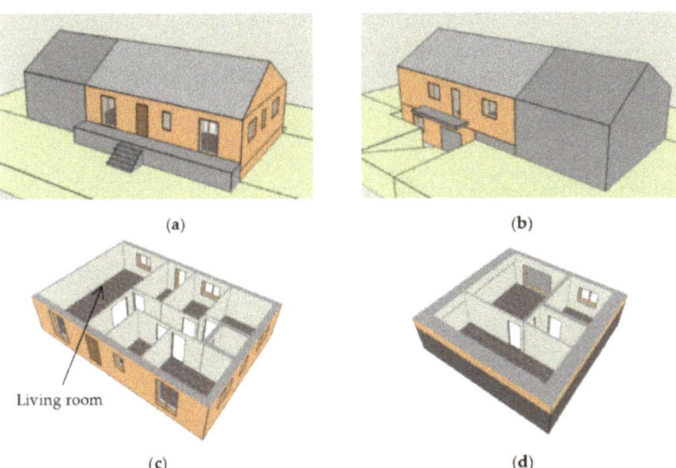

Figure 2. Reference building: (**a**) Front view—south side; (**b**) rear view—north side; (**c**) ground floor section; (**d**) basement section.

Room	1	2	3	4	5	6	7	8	9	10	11	12	13	14	15	16	17	18	19	20	21	22	23	24
Living room - rumpus room																	X	X	X	X	X	X	X	
Kitchen								X	X	X										X	X	X	X	
Entrance - corridor								X	X	X								X	X	X	X	X	X	
Bedroom	X	X	X	X	X	X	X	X															X	X
Bathroom								X	X	X										X	X	X	X	

Legend (percentages relative to the maximum value): 1, 0.75, 0.67, 0.5, 0.25, 0.2

Figure 3. Daily schedule of occupancy and percentages relative to the maximum value.

Table 1. Properties of the wall layers.

	Material	Thickness (m)	Conductivity (W/m K)	Spec. Heat (J/kg K)	Density (kg/m^3)
Insulated external wall ($U = 0.281$ W/m^{2}K)	External plaster	0.005	1.4	1000	2000
	EPS insulation	0.1	0.0385	1200	30
	Hollow bricks	0.3	0.39	840	866.67
	Internal plaster	0.015	0.7	1000	1400
Insulated basement wall ($U = 0.268$ W/m^{2}K)	Pebbles	0.4	1.2	840	1700
	Synthetic material sheets	0.01	0.23	900	1100
	EPS insulation	0.1	0.0418	1200	30
	Hollow bricks	0.3	0.39	840	866.67
	Internal plaster	0.02	0.7	1000	1400
Insulated internal wall ($U = 0.274$ W/m^{2}K)	Internal plaster	0.01	0.7	1000	1400
	Masonry (hollow bricks)	0.08	0.48	840	2000
	EPS insulation	0.1	0.0385	1200	30
	Masonry (hollow bricks)	0.08	0.48	840	2000
	Internal plaster	0.01	0.7	1000	1400
Adjacent units partition wall ($U = 0.736$ W/m^{2}K)	Internal plaster	0.02	0.7	1000	1400
	Soundproofing bricks	0.3	0.265	1000	1200
	Internal plaster	0.02	0.7	1000	1400

Table 2. Properties of floors and roof layers.

	Material	Thickness (m)	Conductivity (W/m K)	Spec. Heat (J/kg K)	Density (kg/m^3)
Insulated floor slab (U = 0.248 W/m^{2}K)	Ceramic tiles	0.01	1.3	840	2300
	Concrete mortar screed	0.06	1.06	1000	2000
	EPS insulation	0.14	0.0418	1200	30
	Slab blocks	0.26	0.67	840	842.31
	Internal plaster	0.01	0,7	1000	1400
Roof cover (U = 0.257 W/m^{2}K)	Stainless steel	0.002	17	460	7900
	EPS insulation	0.05	0.0418	1200	30
	Steel	0.002	50	450	7800
	Concrete	0.03	1.162	1000	2000
	EPS insulation	0.09	0.0418	1200	30
	Slab blocks	0.26	0.7429	840	1146.15
	Internal plaster	0.02	0.7	1000	1400
Ground floor (U = 0.270 W/m^{2}K)	Ceramic tiles	0.01	1.3	840	2300
	Concrete mortar screed	0.08	1.08	1000	1600
	EPS insulation	0.09	0.034	1200	50
	Reinforced concrete	0.315	1.91	1000	2400
	Synthetic material sheets	0.005	0.23	900	1100
	Pebbles and crushed stones	0.4	0.7	840	1500

Table 3. Occupancy density and metabolic rate.

Zone	Occupancy Density (People per m^2)	Metabolic Rate (W per Person)
Living/rumpus room	0.0188	110
Kitchen	0.0237	160
Entrance—corridor	0.0196	180
Bedrooms	0.0229	90
Bathroom	0.0187	120

The internal loads due to the equipment and lighting system are assumed to be 4.1 W/m^2, constant throughout the day. In each air-conditioned zone, 20, 26 and 24 °C are set as set-point temperatures for heating, cooling and ventilation, respectively. In the non-air-conditioned rooms, internal loads and heating and cooling systems are absent. The building's heating system is active from 1 November to 15 April, as regulated by Italian regulations, while the cooling system is active from May to September.

The exchange of air in the rooms is by means of natural ventilation in order to activate the free-cooling regime in summer. In particular, natural ventilation is used if the air temperature inside the zone is higher than both the outside air temperature and the set point temperature for ventilation.

The windows are composed of a 70 mm thick softwood frame and a low-emission double glazing unit with argon in the cavity, with the thickness of the glass panes and the cavity being 3 and 13 mm, respectively. The frame has a transmittance of 1.258 W/m^{2}K, while the double glazing has a solar gain of 0.649 and a transmittance of 1.512 W/m^{2}K. During the summer (June–September), windows are protected by Venetian blinds placed outside the glazing. The solar shading consists of 0.025 m deep slats with a fixed 80° inclination, spaced 0.01875 m and characterized by a reflection coefficient of 0.8. The external entrance door on the ground floor of the building and the insulated internal doors separating the air-conditioned and non-air-conditioned areas have transmittance values of 1.551 W/m^{2}K and 1.761 W/m^{2}K, respectively.

2.2. Modelling of the Attached Solar Greenhouse

The solar greenhouse under study is attached on the south façade of the building near the living room (Figure 4). The greenhouse has the following gross dimensions: height 2.7 m and width 4.5 m. Since the separating wall with the living room is insulated and has a glazed opening, the greenhouse will be an insulated, direct-gain hybrid system. Heat exchange between the greenhouse and the interior of the building will take place due to the presence of four openings along the top and bottom of the partition, in order to ensure the thermo-circulation of air. The greenhouse has fully glazed east, south and west walls and roof, supported by a wooden frame. The thermo-circulation of air is ensured by four rectangular air vents with dimensions of 0.8 m × 0.2 m.

Figure 4. Reference building model with the attached sunspace.

The thermal storage mass inside the greenhouse consists of concrete blocks with the characteristics shown in Table 4. The thickness of the thermal mass varied parametrically and it is distributed on both the separating wall and the floor. The floor has a 5 cm thick layer of insulation material at its base, with the aim of preventing the loss of accumulated heat to the ground [20,34]. The surface of the storage material is covered with a layer of plaster, which is dark-colored with an absorption coefficient of 0.9.

Table 4. Properties of thermal storage material.

	Conductivity (W/m K)	Specific Heat (J/kg K)	Density (kg/m^3)
Concrete block	1.63	1000	2300

The choice of making the absorption coefficient of the wall higher than that of the floor is justified by the monthly variability of the total absorption coefficient of the greenhouse, with a minimum value in June and a maximum value in December. This monthly variability is exploited to increase the absorption of solar radiation as much as possible during the winter period and reduce it in the summer, thus limiting overheating problems. The glass walls of the greenhouse are protected externally by automated photovoltaic blinds, characterized by the same geometry as the screens used to protect the building's window elements. The roof is instead protected by a solar blind with a high reflection coefficient in the period between June and September. Natural ventilation of the greenhouse during the summer period is achieved by opening the east and west windows.

2.3. Control Techniques

One of the objectives of this study concerns the integration of IoT technology to improve greenhouse performance. Dynamic control systems are used to open the vents and manage solar radiation. The latter is controlled by the inclination of the SPBs. To this end, a method for opening the vents is proposed and a control method taken from the literature [26] for the Venetian blinds is implemented. The use of photovoltaic blinds on an attached solar greenhouse has never been analyzed in the literature. These control processes are modeled through the EMS (energy management system) functionality of EnergyPlus, in which the Erl programming language is used (EnergyPlus Runtime Language).

The algorithm for opening and closing the vents requirs two air temperature sensors: for the living room and for the solar greenhouse. The control system exploited the daily temperature excursion inside the greenhouse to enable summer night ventilation. The vents are opened if:

- The air temperature in the greenhouse is lower than in the living room and the air temperature in the living room is higher than 25 °C (cooling mode);
- The air temperature in the greenhouse is higher than in the living room and the air temperature in the living room is below 21 °C (heating mode).

With this configuration, the greenhouse is used not only for the reduction of consumption related to heating, but also for the reduction of consumption related to the cooling of the building (free cooling). The actuators are electric servomotors that allow the slats of the vents to be moved. The information would be managed by a local network in which the sensors and actuator are connected to a control unit that processes the simple algorithm. For energy simulations conducted in EnergyPlus, the algorithm is implemented in the EMS section in a dynamic manner.

The strategy used to control SPBs is the one developed by Nicoletti et al. [26], in which the angle of inclination of the lamellae ω is adjusted with the aim of maximizing solar gains while avoiding overheating phenomena. Furthermore, the slats are adjusted so that the occupants are not subject to glare phenomena, thus guaranteeing visual comfort. The latter condition is not considered in this case, since the solar greenhouse is conceived as a non-air-conditioned environment not subject to occupants.

The sensors detect the air temperature inside the living room T, where thermal comfort conditions must be maintained, and the global solar radiation G incident on the walls of the greenhouse (using a cheap PV cell which provides a voltage signal proportional to the incident radiation). Figure 5 shows the algorithm implemented to control the SPBs. The angle of inclination depends on the solar altitude α and solar azimuth evaluated with respect to the wall orientation $(\gamma - \gamma_w)$. The sun horizontal angle profile β also appears in the functional diagram and is defined as:

$$\beta = \arctan\left(\frac{\tan\alpha}{\cos(\gamma - \gamma_w)}\right) \qquad (1)$$

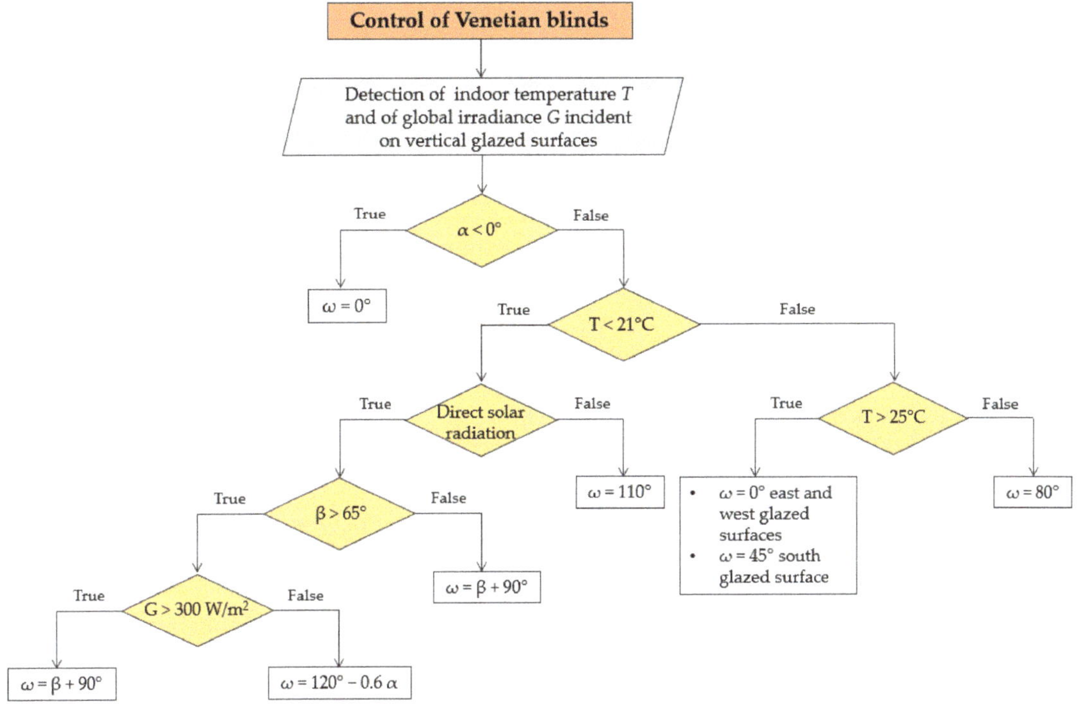

Figure 5. Control algorithm adopted for the regulation of SPBs.

Photovoltaic blinds consists of organic cells with an electrical efficiency of 10%. The electrical producibility of PV blinds is estimated in accordance with procedure shown in [35].

2.4. Parametric Analysis

The reference case study concerns a solar greenhouse located in Rome with the following characteristics: a depth of 1.5 m, low emissivity double glazing with argon in the cavity, a floor thickness of 5 cm, use of SPBs and ventilation in summer with east and west windows open 50% at night. The optimal greenhouse configuration is obtained through a series of parametric analyses and the results are compared to the reference building without sunspace, evaluating the energy savings achieved. In each analysis, the individual parameter of interest is varied, leaving the other characteristics belonging to the base case unchanged.

The parameters analyzed are listed below:

- Greenhouse depth:
  1. 1.5 m;
  2. 2 m;
  3. 2.5 m.
- Type of glass constituting the greenhouse envelope (values for solar gain, direct solar transmission and transmittance (W/m²K) are given in brackets):
  1. 6 mm single-glazing (0.819, 0.775, 5.778);
  2. 3/13/3 mm double glazing with air in the cavity (0.764, 0.705, 2.716);
  3. 3/13/3 mm double glazing with argon in the cavity (0.764, 0.705, 2.556);
  4. 3/13/3 mm low-emission double glazing with argon in the cavity (0.649, 0.538, 1.512).

- Thermal storage mass inside the greenhouse:
  1. 5 cm floor thickness, no thermal mass on the wall;
  2. 10 cm floor thickness, no thermal mass on the wall;
  3. 20 cm floor thickness, 10 cm floor thickness;
  4. 30 cm floor thickness, 20 cm floor thickness.
- Use of SPBs:
  1. With SPBs;
  2. Without SPBs.
- Greenhouse ventilation methods in summer:
  1. East and west windows open at 50% at night;
  2. East and west windows open at 100% at night;
  3. East and west windows open at 50% all day;
  4. East and west windows open at 100% all day.
- Locality:
  1. Genoa, Italy (latitude $44°23'$);
  2. Rome, Italy (latitude $41°54'$);
  3. Capo Palinuro, Italy (latitude $40°1'$).

The climatic characteristics of the locations chosen for comparison are shown in the results section. The sites are chosen to be representative of northern, central and southern Italy. They all belong to the same climatic zone with the Italian classification so that the building transmittances do not need to be changed. Therefore, a comparison with the same building envelope is possible.

3. Results and Discussion

This section discusses the results obtained, considering the period of one year. The results for the reference building (without greenhouse) show that the annual thermal energy demand for heating was 1718 kWh and for cooling was 954.2 kWh.

3.1. Parametric Performance Analyses with Attached Solar Greenhouse

3.1.1. Influence of Sunspace Depth

Heating consumption increases as the depth of the greenhouse increases, in contrast to summer consumption, as shown in Table 5: a depth of 1.5 m results in greater savings in the winter period (8.6%), while a depth of 2.5 m results in greater savings in the summer period (18.2%).

Table 5. Building's annual energy needs and percentage change relative to the reference case varying sun space's depth.

Depth (m)	Energy Needs (kWh)		Variation from Reference Case	
	Heating	Cooling	Heating	Cooling
1.5	1570.0	805.7	−8.6%	−15.6%
2.0	1580.3	790.9	−8.0%	−17.1%
2.5	1591.8	780.1	−7.3%	−18.2%

The results are influenced by the low thermal resistance characterizing glass walls; consequently, heat losses to the outside increase as the envelope surface area increases. Figure 6a shows the average monthly heat gains per unit of glazed area, net of losses to the outside. In the winter months, due to lower insolation, the glazing is more susceptible to heat losses to the outside and the reduction of the dispersing surfaces increases the gains. In the intermediate periods, the greater amount of solar radiation incident on the glazing balances out the losses. This leads to an increase in gains as the glazed area increases. During the summer months, the gains tend to even out due to the activation of the solar shading; however, the absence of incident solar radiation favors heat losses as the depth of the greenhouse increases. Thermal gains influence the air temperature inside the greenhouse (Figure 6b). Consequently, they influence the heat exchanges with the living room (Figure 6c,d) with the thermo-circulation of air through the vents on the separation wall.

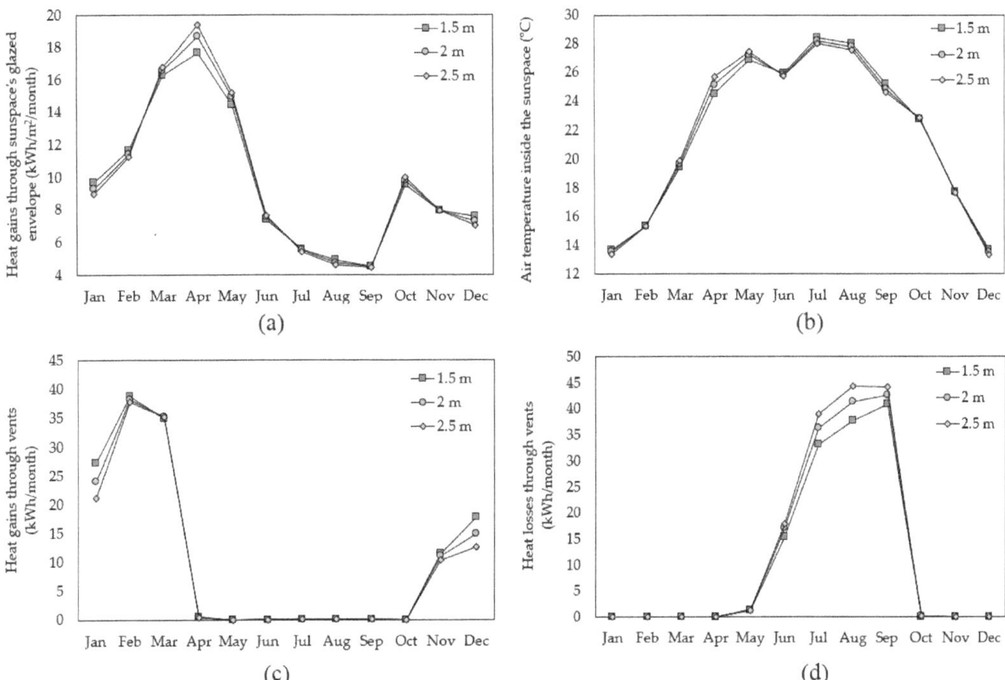

Figure 6. Depth analysis: (**a**) Monthly average heat gains through sun space's envelope per unit of glazed surface; (**b**) monthly average air temperature inside the sun space; (**c**) monthly average living room heat gains through vents; (**d**) monthly average living room heat losses through vents.

3.1.2. Influence of Glass Type

The results in Table 6 show that the use of low-e double glazing is the optimal solution during the winter period. It leads to a saving of 8.6% compared to 4.3% achieved with single glazing. However, in the summer period the situation is reversed: single glazing leads to a greater reduction in consumption (20.6%) than low-e glass (15.6%). The performance obtained in the case of double glazing is intermediate with respect to those discussed; moreover, the presence of argon in the cavity leads to negligible energy savings compared to the case where there is air.

Table 6. Building's annual energy needs and percentage change relative to the reference case varying the glazing type constituting the envelope of the sun space.

Glazing Type	Energy Needs (kWh)		Variation from Reference Case	
	Heating	Cooling	Heating	Cooling
Single	1643.6	757.7	−4.3%	−20.6%
Double—Air	1608.8	776.7	−6.4%	−18.6%
Double—Argon	1604.5	778.7	−6.6%	−18.4%
Double LoE—Argon	1570.0	805.7	−8.6%	−15.6%

The use of low-emissivity double glazing allows higher temperatures inside the greenhouse (Figure 7a), although the amount of solar radiation transmitted is penalized (Figure 7b). This behavior is caused by the low-emissivity coating being opaque to the high wavelength radiation emitted by the surfaces. The glass is able to retain heat inside the greenhouse and the low transmittance value results in less heat loss [36]. The opposite behavior is observed in the case of single-glazing: despite the greater amount of solar radiation transmitted into the greenhouse, the high transmittance characterizing the glass leads to greater dispersion with a consequent reduction in gains. By analyzing the temperatures inside the greenhouse, it is possible to evaluate the effects on heat gains and losses in the living room through the vents (Figure 7c,d).

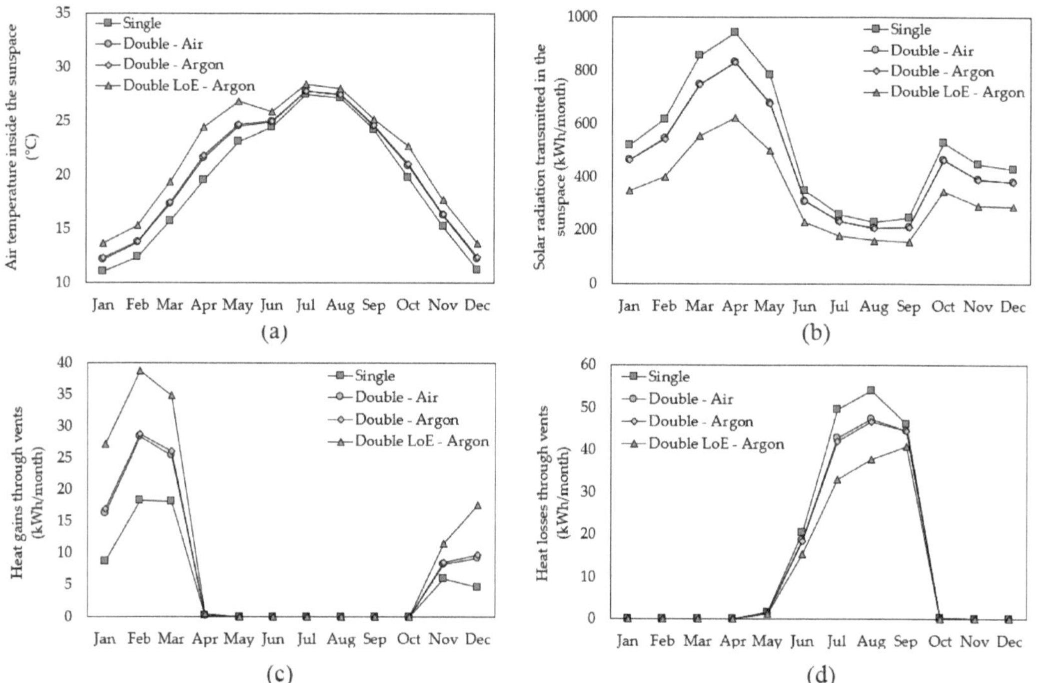

Figure 7. Glazing type analysis: (**a**) Monthly average air temperature inside the sun space; (**b**) monthly average solar radiation transmitted in the sun space; (**c**) monthly average living room heat gains through vents; (**d**) monthly average living room heat losses through vents.

3.1.3. Influence of Thermal Mass

From the values shown in Table 7, it is evident how the presence of storage was counterproductive: as the thickness increases, the consumption for both heating and cooling increases, resulting in a reduction of energy savings (from 8.6% to 4.0% for heating and from 15.6% to 11.0% for cooling).

Table 7. Building's annual energy needs and percentage change relative to the reference case varying the heat storage mass.

Thermal Mass	Energy Needs (kWh)		Variation from Reference Case	
	Heating	Cooling	Heating	Cooling
Floor 5 cm	1570.0	805.7	−8.6%	−15.6%
Floor 10 cm	1581.8	817.2	−7.9%	−14.4%
Floor 20 cm, wall 10 cm	1630.7	847.1	−5.1%	−11.2%
Floor 30 cm, wall 20 cm	1649.9	849.2	−4.0%	−11.0%

The greenhouse analyzed is predominantly an insulated gain system, as the separating wall is insulated. Therefore, the temperatures of the thermal mass do not directly influence the conditions inside the building. The greenhouse only acts as a heat collector and not as a thermal buffer between the interior and exterior environment. In particular, the main gain is the thermal exchange through the vents when the air temperature inside the greenhouse is between 21 °C and 25 °C.

The presence of the accumulation mitigats daily temperature fluctuations, which is a counterproductive effect, as the lower temperature peaks during the winter period reduce heat exchange with the living room. The same is true for the summer period when, in order to have a greater heat exchange with the interior, night temperatures must be as low as possible. The days of 5 January and 5 July were chosen as representative for the heating and cooling periods, respectively, whose daily temperature trends inside the greenhouse are shown in Figure 8a,b. Figure 8c,d shows the daily trends of thermal gains (5 January) and losses (5 July) in the living room.

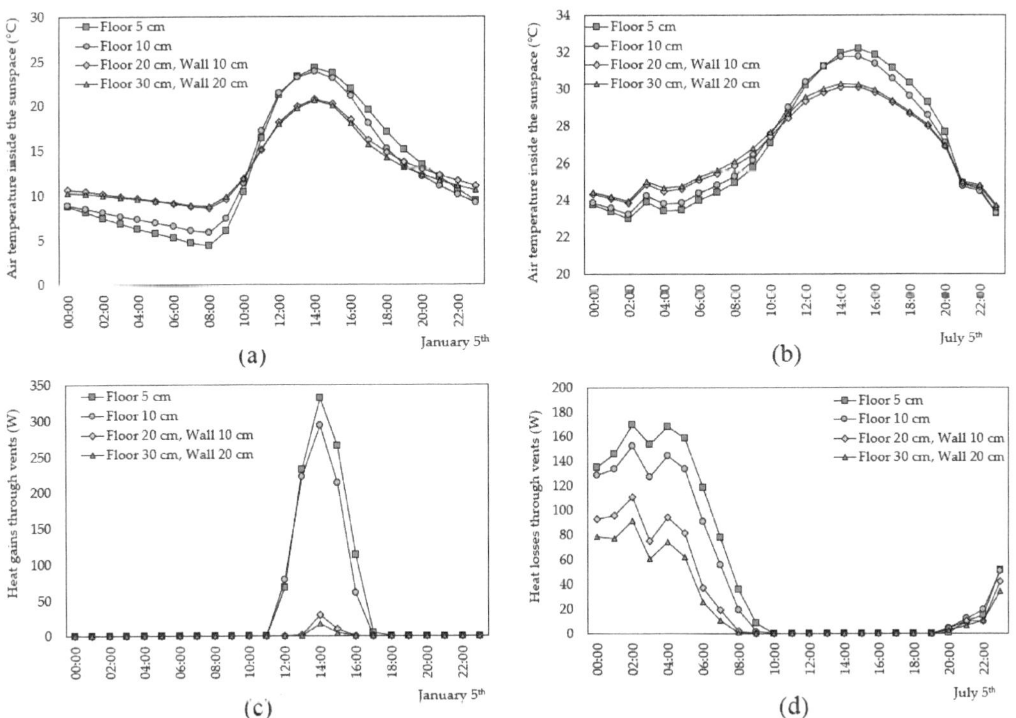

Figure 8. Heat storage mass analysis: (**a**) Air temperature inside the sun space (5 January); (**b**) air temperature inside the sunspace (5 July); (**c**) monthly average living room heat gains through vents (5 January); (**d**) monthly average living room heat losses through vents (5 July).

3.1.4. Influence of Solar Photovoltaic Blinds

The energy requirements with and without solar shading to protect the greenhouse are shown in Table 8.

Table 8. Building's annual energy needs and percentage change relative to the reference case with and without SPBs.

Presence of SPBs	Energy Needs (kWh)		Variation from Reference Case	
	Heating	Cooling	Heating	Cooling
With SPBs	1570.0	805.7	−8.6%	−15.6%
Without SPBs	1510.0	1071.7	−12.1%	+12.3%

In the winter period, the absence of shading results in an increase in energy savings from 8.6% to 12.1%. Although SPBs are controlled in such a way as to let in as much solar radiation as possible, their design causes the glazing to be partially hidden from the sun's rays. This effect can be seen in Figure 9a, which shows the monthly average trends of solar radiation transmitted into the greenhouse. Shading, on the other hand, is essential during the summer period. Their absence leads to a 12.3% increase in consumption compared to the building without a greenhouse. This phenomenon is caused by the overheating of the greenhouse (Figure 9b), which is in turn favored by the presence of low-emissivity glazing.

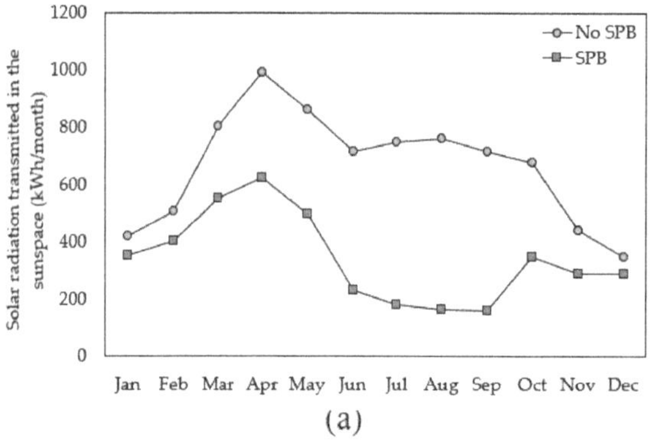

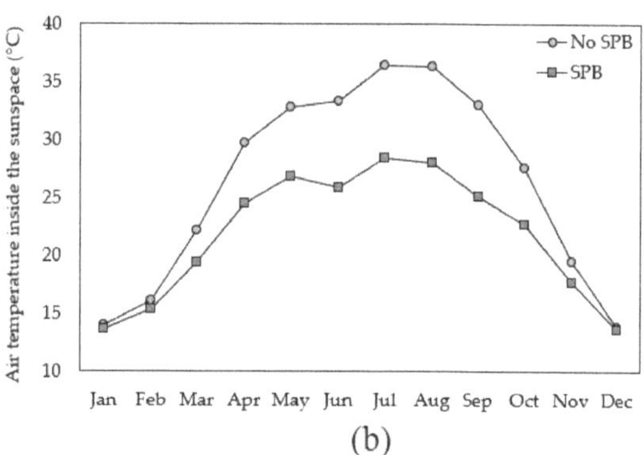

Figure 9. Analysis with and without SPBs: (**a**) Monthly average solar radiation transmitted in the sun space; (**b**) monthly average air temperature inside the sun space.

3.1.5. Influence of Greenhouse Summer Natural Ventilation

Table 9 shows the results with the different opening modes of the side windows of the solar space. Consumption during the cooling period decreased both as the percentage of window opening increased and as the duration of the ventilation period increased. The maximum saving during the cooling period was 22.6%. In the winter period, of course, consumption remained unchanged.

Table 9. Building's annual energy needs and percentage change relative to the reference case varying ventilation strategy.

Ventilation	Energy Needs (kWh)		Variation from Reference Case	
	Heating	Cooling	Heating	Cooling
50%—only night	1570.0	805.7	−8.6%	−15.6%
100%—only night	1570.0	775.7	−8.6%	−18.7%
50%—all day	1570.0	775.5	−8.6%	−18.7%
100%—all day	1570.0	738.8	−8.6%	−22.6%

The reduction in consumption is caused by the lower temperatures inside the greenhouse (Figure 10a), which promote the removal of heat from the living room through increased thermo-circulation of air (Figure 10b).

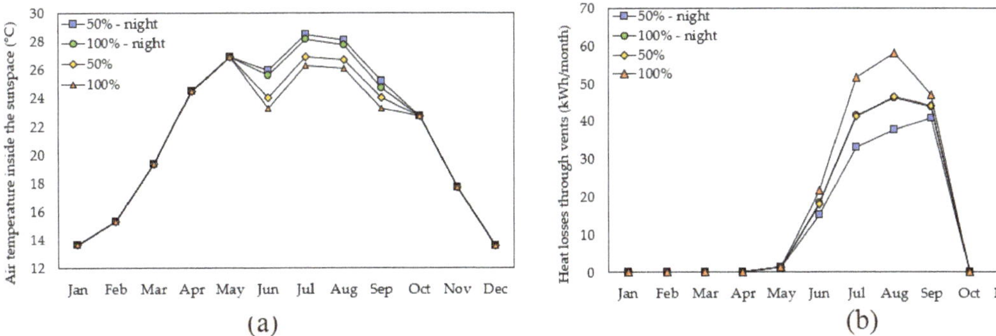

Figure 10. Ventilation analysis: (**a**) Monthly average air temperature inside the sunspace; (**b**) monthly average living room heat losses through vents.

3.1.6. Influence of Locality

The greenhouse performance analysis was evaluated by comparing the results obtained at three representative locations in Northern, Central and Southern Italy (Genoa, Rome, Capo Palinuro). The climate data for Genoa and Capo Palinuro are shown in Figure 11. The average annual outdoor temperature is 15.2 °C in Genoa and 16.3 °C in Capo Palinuro. Genoa has an average annual amount of diffuse solar radiation and direct normal radiation of 631.7 and 729.1 kWh/m^2, respectively; for Capo Palinuro these values are 644.5 and 957.8 kWh/m^2.

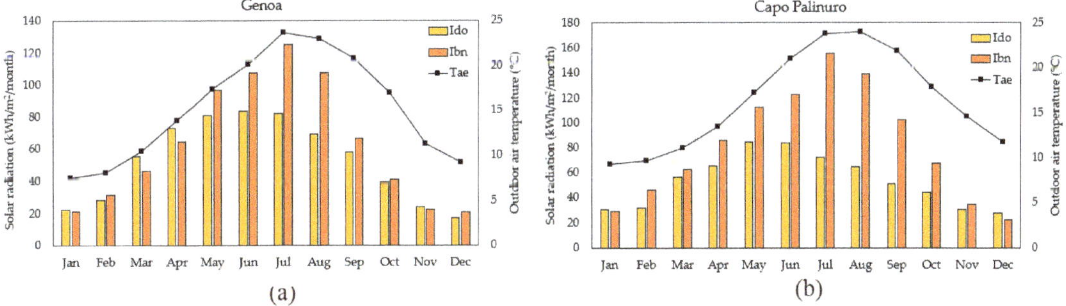

Figure 11. Monthly average climate data (diffuse irradiance on horizontal plane, direct normal irradiance and outdoor air temperature): (**a**) Genoa; (**b**) Capo Palinuro.

The results (Table 10) show that for Genoa, the benefits are lower than for Rome during both the summer and winter periods. For cities with a warmer climate, such as Capo Palinuro, the winter percentage savings are greater than in Rome, while there is a slight reduction in the summer period. In all cases, however, it should be noted that the solar greenhouse, with proper controls, also leads to a reduction in cooling requirements in all locations.

Table 10. Building's annual energy needs and percentage change relatively to the reference case varying sun space's location.

Locations	Energy Needs (Reference Building/Building with Sunspace) (kWh)		Variation from Reference Case	
	Heating	Cooling	Heating	Cooling
Genoa	1897.9/1757.5	679.8/579.7	−7.4%	−14.7%
Rome	1718.0/1570.0	954.2/805.7	−8.6%	−15.6%
Capo Palinuro	862.9/735.2	1090.1/955.9	−14.8%	−12.3%

3.2. Comparison with the Case Where the Separation Wall Is without Insulation

The results presented so far related to the performance of the greenhouse in the case of an insulated separation wall. In the following, the performance without the insulating layer in the separation wall will be analyzed. The new transmittance results show $U = 1.037$ W/m^{2}K. The absence of insulating material makes the greenhouse a hybrid system with direct gain (due to the presence of a glazed component in the separation wall) and indirect gain, in which the wall acts as a thermal storage mass.

The results in Table 11 show that, without insulation, there is an increase in energy requirements, resulting in reduced savings compared to the base case. This trend can be seen in both the winter and summer period. The saving for heating is 8.6% with the presence of insulation, and 5.9% without insulation. For cooling, the saving is 15.6% in the first case and 10.9% in the second case. The separating wall, by contributing to the accumulation of heat, helps to dampen the daily temperature fluctuations inside the greenhouse. This effect generates less heat exchange with the living room through the vents, as already discussed in the parametric analysis of the storage mass. In addition to the reduction in thermal exchange through the vents, it is necessary to take into account the counterproductive effect of the high transmittance value characterizing the non-insulated partition, which causes greater losses in the winter period and greater gains in the summer period.

Table 11. Building's annual energy needs and percentage change relative to the reference case for insulated and not insulated separation wall.

Separation Wall	Energy Needs (kWh)		Variation from Reference Case	
	Heating	Cooling	Heating	Cooling
Insulated	1570.0	805.7	−8.6%	−15.6%
Not insulated	1617.4	849.8	−5.9%	−10.9%

3.3. SPBs Electricity Production

The methodology used to calculate the electrical energy produced by photovoltaic solar shading is the one developed by Nicoletti et al. [35]. The control system adopted penalizes photovoltaic production in the winter period, as the slats of the screens are arranged parallel to the sun's rays to maximize solar gain entering the greenhouse. Figure 12 shows the monthly energy produced by the SPBs per unit of glass area. It can be seen that east- and west-facing blinds produce approximately the same amount of energy in all months. South-facing venetian blinds produce more energy throughout the year, with the exception of the months of May, June and July, when production is higher at the east and west orientations.

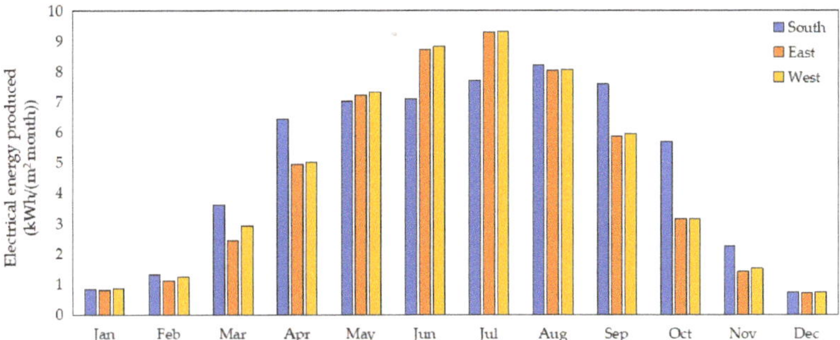

Figure 12. Monthly electrical energy produced by SPBs per unit of vertical surface.

The method [35] consideres the solar radiation incident on the photovoltaic cells to be composed of the sum of direct, diffuse and reflected radiation from the ground and from the back of the lamellas. Consequently, these contributions affect the total electricity production, as shown in Figure 13. The contribution of solar radiation reflected from the back of the louvers has zero values during the summer months, since the slats are closed or partially closed most of the time. On the other hand, during the winter period, the reflected radiation contributes to energy production, since the high inclination angle means that the view factor between the back of the slats and the outside is non-zero.

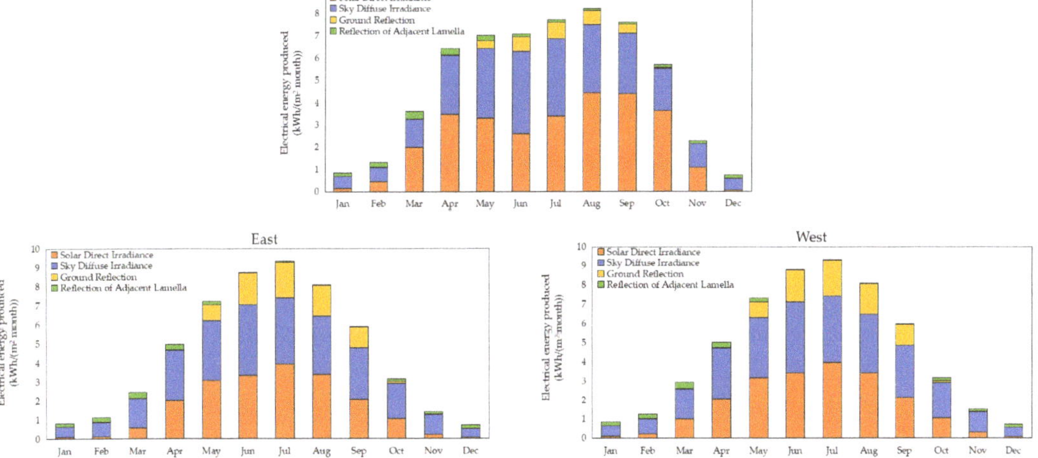

Figure 13. Contributions of electricity produced by SPBs per unit of vertical area (south, east and west orientations).

In the winter months, the contribution of reflected solar radiation from the ground is approximately zero. This occurs because the control system adjusts the inclination of the lamellas so that they are parallel to the sun's rays in the presence of direct radiation or inclined by 110° in the case of diffuse radiation only; thus, the lamellas are mainly facing upwards, with a view factor between the photovoltaic layer and the ground equal to zero. This contribution is non-zero during the summer period, when the blinds were partly or completely closed. Figure 14 shows the monthly energy produced by the three installed screens, analyzing the case where the greenhouse has a depth of 1.5 m, with the size of the south window being 4.5 m × 2.7 m and the size of the east and west windows being 1.5 m × 2.7 m. The highest energy production occurs in July (169 kWh) and the lowest in December (14.7 kWh). The electrical energy produced, taking into account the inverter efficiency, assumed to be 0.9, amounted to 1036 kWh.

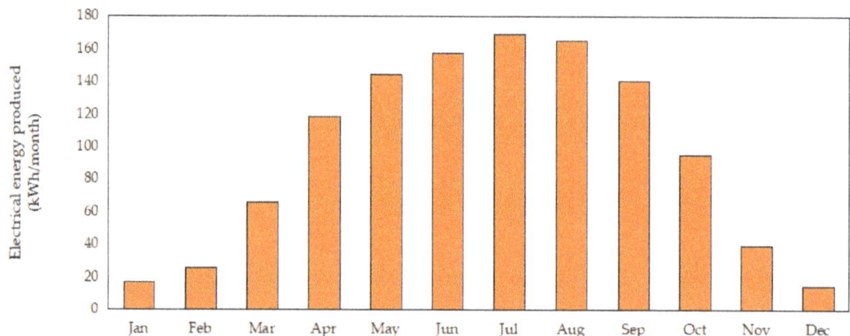

Figure 14. Overall electrical energy produced by SPBs.

3.4. Summary of Results

The analyses, which were conducted by orienting the greenhouse to the south, revealed the following main results:

- The solar greenhouse with reduced depth allows greater energy savings in winter, as the smaller amount of transparent surface area contributes to the reduction of heat loss;
- The use of low-emissivity double glazing in the construction of the greenhouse envelope results in greater gains in winter, while in summer their use is counterproductive in the absence of adequate measures to combat overheating;
- The presence of accumulation mass in the greenhouse is counterproductive in the case of an insulated separation wall, as temperatures inside the greenhouse are mitigated by reducing heat exchange through the vents;
- The use of sunscreens in summer is of paramount importance in reducing temperatures inside the greenhouse, as is an adequate level of ventilation;
- The analysis by location showed that energy savings are greater in southern Italy, as it is characterized by a higher level of solar radiation.

The use of automated control techniques, based on IoT technology, allows the isolated gain greenhouse to achieve greater energy savings than the indirect gain configuration. In particular, the automation of the vents allows the maximization of heat exchange throughout the year, leading to a reduction in energy consumption even during the summer period. The results show that cooling requirements can also be reduced. This is in agreement with studies in the literature that show that proper ventilation brings summer benefits [37,38]. In particular, many scientific articles [39,40] have emphasized how reductions in summer consumption can be achieved through shading and natural ventilation.

The analysis showed that the maximum reduction in energy needs for the city of Rome results in a thermal energy saving of 148 kWh for heating and 215 kWh for cooling. Percentage savings are 8.6% for heating and 22.6% for cooling. The optimized solar greenhouse has the following properties:

- Depth of 1.5 m;
- Floor with 5 cm thick accumulation material;
- 3/13/3 mm low-emission double glazing envelope with argon in the cavity;
- Solar shading system;
- East and west windows open 100% all day.

4. Conclusions

In this study, the impact of an attached solar greenhouse on the heating and cooling energy requirements of a residential building was evaluated. The study was conducted by means of parametric analyses, evaluating the performance of the greenhouse as the construction and operational characteristics changed.

In the winter period, the solar greenhouse contributes to greater energy savings in the case of reduced depth and the use of high-performance glass, such as low-emissivity double glazing. In the summer period, on the other hand, the greatest energy savings were found by favoring the ventilation of the greenhouse and activating solar shading to reduce the counterproductive effects of overheating. Furthermore, it appears that the presence of storage mass inside the greenhouse penalizes energy exchange in the case of insulated partition wall. Specifically, for the city of Rome, percentage savings are 8.6% for heating and 22.6% for cooling. In addition, photovoltaic solar shading contributes to the production of more than 1000 kWh/year of electrical energy.

The study, carried out on locations in the Mediterranean area, shows how a solar greenhouse constitutes an important structure for reducing energy consumption in the residential sector, contributing to the achievement of building energy self-sufficiency. The proposed system, in which the use of BIPV technologies is contemplated, can be used by planners not only in residential areas but also for offices and buildings with similar use. In addition to the reduction of energy needs for air-conditioning, this system favors the on-site production of electricity, especially if the space on the roof surfaces is insufficient for the installation of an adequate number of photovoltaic panels. Furthermore, the presence of IoT systems allows for customized management of control systems, combining the needs of maximizing energy exchange with the maintenance of comfort conditions.

The study, which was conducted by analyzing a solar greenhouse configuration that has not been extensively studied in the literature, shows that the energy savings obtained were not negligible, despite the high quality of the building envelope and the small south-facing surface area where the solar greenhouse of the reference building was assumed to be installed. Consequently, further developments could concern the analysis of the energy savings achieved by attaching an isolated gain solar greenhouse to buildings characterized by different configurations and different quality of the building envelope.

Author Contributions: Conceptualization, D.K. and F.N.; methodology, F.P. and F.N.; software, F.P.; formal analysis, F.P., P.B. and F.N.; investigation, F.P. and F.N.; resources, D.K. and N.A.; data curation, F.P. and F.N.; writing—original draft preparation, F.P.; writing—review and editing, F.P. and F.N.; visualization, F.P.; supervision, D.K., P.B. and N.A.; project administration, D.K., F.N. and N.A.; funding acquisition, F.N. and N.A. All authors have read and agreed to the published version of the manuscript.

Funding: This research received no external funding.

Institutional Review Board Statement: Not applicable.

Informed Consent Statement: Not applicable.

Data Availability Statement: Not applicable.

Conflicts of Interest: The authors declare no conflict of interest.

References

1. Arumugam, C.; Shaik, S. Air-conditioning cost saving and CO_2 emission reduction prospective of buildings designed with PCM integrated blocks and roofs. *Sustain. Energy Technol. Assess.* **2021**, *48*, 101657. [CrossRef]
2. Bevilacqua, P.; Benevento, F.; Bruno, R.; Arcuri, N. Are Trombe walls suitable passive systems for the reduction of the yearly building energy requirements? *Energy* **2019**, *185*, 554–566. [CrossRef]
3. Bruno, R.; Bevilacqua, P.; Cirone, D.; Perrella, S.; Rollo, A. A Calibration of the Solar Load Ratio Method to Determine the Heat Gain in PV-Trombe Walls. *Energies* **2022**, *15*, 328. [CrossRef]
4. Chan, H.-Y.; Riffat, S.B.; Zhu, J. Review of passive solar heating and cooling technologies. *Renew. Sustain. Energy Rev.* **2010**, *14*, 781–789. [CrossRef]
5. Szyszka, J.; Bevilaqua, P.; Bruno, R. A statistical analysis of an innovative concept of Trombe Wall by experimental tests. *J. Build. Eng.* **2022**, *62*, 105382. [CrossRef]

6. Vukadinović, A.V.; Radosavljević, J.M.; Djordjević, A.V.; Bonić, D.M. Estimation of Indoor Temperature for a Passive Solar Residential Building with an Attached Sunspace during the Heating Period. *Environ. Prog. Sustain. Energy* **2018**, *38*, 13127. [CrossRef]

7. Bruno, R.; Bevilacqua, P.; Rollo, A.; Barreca, F.; Arcuri, N. A Novel Bio-Architectural Temporary Housing Designed for the Mediterranean Area: Theoretical and Experimental Analysis. *Energies* **2022**, *15*, 3243. [CrossRef]

8. Ulpiani, G.; Giuliani, D.; Romagnoli, A.; di Perna, C. Experimental monitoring of a sunspace applied to a NZEB mock-up: Assessing and comparing the energy benefits of different configurations. *Energy Build.* **2017**, *152*, 194–215. [CrossRef]

9. Mihalakakou, G.; Ferrante, A. Energy conservation and potential of a sunspace: Sensitivity analysis. *Energy Convers. Manag.* **2000**, *41*, 1247–1264. [CrossRef]

10. Bataineh, K.M.; Fayez, N. Analysis of thermal performance of building attached sunspace. *Energy Build.* **2011**, *43*, 1863–1868. [CrossRef]

11. Grudzińska, M. Thermal and Optical Properties of the Sunspace Casing as Factors Influencing Temperature Rise in Greenhouse Systems. *Materials* **2021**, *14*, 7411. [CrossRef]

12. Sun, H.; Calautit, J.K.; Jimenez-Bescos, C. Examining the regulating impact of thermal mass on overheating, and the role of night ventilation, within different climates and future scenarios across China. *Clean. Eng. Technol.* **2022**, *9*, 100534. [CrossRef]

13. Aelenei, D.; de Azevedo Leal, H.; Aelenei, L. The Use of Attached-sunspaces in Retrofitting Design: The Case of Residential Buildings in Portugal. *Energy Procedia* **2014**, *48*, 1436–1441. [CrossRef]

14. Chiesa, G.; Simonetti, M.; Ballada, G. Potential of attached sunspaces in winter season comparing different technological choices in Central and Southern Europe. *Energy Build.* **2017**, *138*, 377–395. [CrossRef]

15. Mihalakakou, G. On the use of sunspace for space heating/cooling in Europe. *Renew. Energy* **2002**, *26*, 415–429. [CrossRef]

16. Monge-Barrio, A.; Sánchez-Ostiz, A. Energy efficiency and thermal behaviour of attached sunspaces, in the residential architecture in Spain. Summer Conditions. *Energy Build.* **2015**, *108*, 244–256. [CrossRef]

17. Givoni, B. Characteristics, design implications, and applicability of passive solar heating systems for buildings. *Sol. Energy* **1991**, *47*, 425–435. [CrossRef]

18. Mellalou, A.; Riad, W.; Mouaky, A.; Bacaoui, A.; Outzourhit, A. Optimum design and orientation of a greenhouse for seasonal winter drying in Morocco under constant volume constraint. *Sol. Energy* **2021**, *230*, 321–332. [CrossRef]

19. Ma, L.; Zhang, X.; Li, D.; Arıcı, M.; Yıldız, C.; Li, Q.; Zhang, S.; Jiang, W. Influence of sunspace on energy consumption of rural residential buildings. *Sol. Energy* **2020**, *211*, 336–344. [CrossRef]

20. Bastien, D.; Athienitis, A.K. Passive thermal energy storage, part 2: Design methodology for solaria and greenhouses. *Renew. Energy* **2017**, *103*, 537–560. [CrossRef]

21. Rempel, A.R.; Rempel, A.W.; Gates, K.R.; Shaw, B. Climate-responsive thermal mass design for Pacific Northwest sunspaces. *Renew. Energy* **2016**, *85*, 981–993. [CrossRef]

22. Daemei, A.B.; Eghbali, S.R.; Khotbehsara, E.M. Bioclimatic design strategies: A guideline to enhance human thermal comfort in Cfa climate zones. *J. Build. Eng.* **2019**, *25*, 100758. [CrossRef]

23. Kuhn, T.E.; Buhler, C.; Platzer, W.J. Evaluation of overheating protection with sun-shading systems. *Sol. Energy* **2000**, *69*, 59–74.

24. Kirimtat, A.; Kundakci Koyunbaba, B.; Chatzikonstantinou, I.; Sariyildiz, S. Rewiew of a simulation modeling for shading devices in buildings. *Renew. Sustain. Energy Rev.* **2016**, *53*, 23–49. [CrossRef]

25. Datta, G. Effect of fixed horizontal louver shading devices on thermal performance of building by TRNSYS simulation. *Renew. Energy* **2001**, *23*, 497–507.

26. Nicoletti, F.; Carpino, C.; Cucumo, M.A.; Arcuri, N. The Control of Venetian Blinds: A Solution for Reduction of Energy Consumption Preserving Visual Comfort. *Energies* **2020**, *13*, 1731. [CrossRef]

27. Carletti, C.; Sciurpi, F.; Pierangioli, L.; Asdrubali, F.; Pisello, A.L.; Bianchi, F.; Sambuco, S.; Guattari, C. Thermal and lighting effects of an external venetian blind: Experimental analysis in a full scale test room. *Build. Environ.* **2016**, *106*, 45–56. [CrossRef]

28. Kuhn, T.E. State of the art of advanced solar control devices for buildings. *Sol. Energy* **2017**, *154*, 112–133. [CrossRef]

29. Pillai, D.S.; Shabunko, V.; Krishna, A. A comprehensive review on building integrated photovoltaic systems: Emphasis to technological advancements, outdoor testing, and predictive maintenance. *Renew. Sustain. Energy Rev.* **2022**, *156*, 111946. [CrossRef]

30. Martín-Chivelet, N.; Kapsis, K.; Wilson, H.R.; Delisle, V.; Yang, R.; Olivieri, L.; Polo, J.; Eisenlohr, J.; Roy, B.; Maturi, L.; et al. Building-Integrated Photovoltaic (BIPV) products and systems: A review of energy-related behavior. *Energy Build.* **2022**, *262*, 111998. [CrossRef]

31. Yu, G.; Yang, H.; Luo, D.; Cheng, X.; Ansah, M.K. A review on developments and researches of building integrated photovoltaic (BIPV) windows and shading blinds. *Renew. Sustain. Energy Rev.* **2021**, *149*, 111355. [CrossRef]

32. Hong, S.; Choi, A.-S.; Sung, M. Development and verification of a slat control method for a bi-directional PV blind. *Appl. Energy* **2017**, *206*, 1321–1333. [CrossRef]

33. Bahr, W. A comprehensive assessment methodology of the building integrated photovoltaic blind system. *Energy Build.* **2014**, *82*, 703–708. [CrossRef]

34. Suarez López, M.J.; Gutiérrez Trashorras, A.J.; Parrondo Gayo, J.L.; Blanco Marigorta, E. Analysis of an Attached Sunspace with a Thermal Inertia Floor. *Energies* **2018**, *11*, 1136. [CrossRef]

35. Nicoletti, F.; Cucumo, M.A.; Arcuri, N. Building-integrated photovoltaics (BIPV): A mathematical approach to evaluate the electrical production of solar PV blinds. *Energy* **2023**, *263*, 126030. [CrossRef]
36. Cucumo, M.; Ferraro, V.; Kaliakatsos, D.; Marinelli, V. Simulation of the Thermal Behaviour of Buildings Equipped with low-Emissivity Glazed Components: A Performance Analysis, December 2013. *Int. J. Heat Technol.* **2013**, *31*, 111–118. [CrossRef]
37. Satola, D.; Wiberg, A.H.; Singh, M.; Babu, S.; James, B.; Dixit, M.; Sharston, R.; Grynberg, Y.; Gustavsen, A. Comparative review of international approaches to net-zero buildings: Knowledge-sharing initiative to develop design strategies for greenhouse gas emissions reduction. *Energy Sustain. Dev.* **2022**, *71*, 291–306. [CrossRef]
38. Manzano-Agugliaro, F.; Montoya, F.G.; Sabio-Ortega, A.; García-Cruz, A. Review of bioclimatic architecture strategies for achieving thermal comfort. *Renew. Sustain. Energy Rev.* **2015**, *49*, 736–755. [CrossRef]
39. Stasi, R.; Liuzzi, S.; Paterno, S.; Ruggiero, F.; Stefanizzi, P.; Stragapede, A. Combining bioclimatic strategies with efficient HVAC plants to reach nearly-zero energy building goals in Mediterranean climate. *Sustain. Cities Soc.* **2020**, *63*, 102479. [CrossRef]
40. Xuan, H.; Ford, B. Climatic applicability of downdraught cooling in China. *Arch. Sci. Rev.* **2012**, *55*, 273–286. [CrossRef]

buildings

MDPI

Article

Living Walls and Green Façades: An Implementation Code for Energy Simulation

Valeria Nesci [1], Ilaria Ballarini [1,*], Pietro Rando Mazzarino [2] and Vincenzo Corrado [1]

[1] Department of Energy "Galileo Ferraris", Politecnico di Torino, 10129 Torino, Italy; valeria.nesci@polito.it (V.N.); vincenzo.corrado@polito.it (V.C.)

[2] Department of Control and Computer Engineering, Politecnico di Torino, 10129 Torino, Italy; pietro.randomazzarino@polito.it

* Correspondence: ilaria.ballarini@polito.it

Abstract: The impacts of climate change, excessive greenhouse gas emissions, and the current energy crisis have motivated the European Union to adopt mitigation and adaptation strategies. These strategies primarily focus on the building sector due to its crucial role in addressing these issues. Among the strategies, the implementation of resilient technologies for the building envelope, such as vertical greenery systems (VGSs) is gaining ground. The literature analysis shows that existing models are not sufficiently detailed in their description of the overall thermo-physical phenomena of VGSs. The aim of this work is to overcome the research gaps by selecting and improving two mathematical models for green façades and living walls. A dedicated calculation code to estimate the effect of VGSs on a building's energy performance and indoor thermal comfort has been developed and implemented within the EnergyPlus calculation software (version 23.2). A BESTest Case from ASHRAE 140 was chosen to test the models and to assess benefits of VGSs. The results show that adopting green solutions for the building envelope can contribute to achieving the building's energy efficiency goals and that the modelling of these technologies can be easily carried out within a dynamic energy simulation of the building.

Keywords: vertical greenery systems (VGSs); green façade; living wall; numerical modelling; building energy simulation; EnergyPlus

Citation: Nesci, V.; Ballarini, I.; Rando Mazzarino, P.; Corrado, V. Living Walls and Green Façades: An Implementation Code for Energy Simulation. *Buildings* **2024**, *14*, 2040. https://doi.org/10.3390/buildings14072040

Academic Editors: Xingxing Zhang, David Coley and Apple L.S. Chan

Received: 16 April 2024
Revised: 19 June 2024
Accepted: 1 July 2024
Published: 4 July 2024

1. Introduction

The building sector faces various issues stemming from the effects of climate change, increasing urbanization, and the current energy crisis. These include rising temperatures, CO_2 emissions, increased energy consumption, and higher fuel prices for end-users. These challenges highlight the need for adopting innovative solutions to mitigate their impacts. In this regard, the European Union has adopted mitigation policies. The "Renovation Wave" [1] has two primary objectives: to increase the annual rate of energy renovation for both residential and non-residential buildings and to facilitate deep energy renovations. The deep renovation of existing buildings significantly contributes to climate change mitigation by reducing energy consumption and greenhouse gas emissions during the operational phase. However, as the impacts of climate change are expected to intensify, it becomes necessary not only to implement deep renovation interventions and mitigation measures but also to adopt adaptation solutions.

The concept of adaptation to climate change is widely studied, specifically in the building sector, where the design of resilient buildings is crucial. Studies on climate resilience are underway, particularly through initiatives such as the International Energy Agency's (IEA) Energy Building and Communities environment (EBC) research program [2], which aims to identify solutions to reduce energy and carbon emissions in the built environment. Within this program, the Annex 80 project—known as "Resilient Cooling"—focuses on energy-efficient and low-carbon cooling strategies. Among cooling strategies, green walls, known as vertical greenery systems (VGSs), which encompass green façades and living walls, are included.

To accurately assess the impact of these technologies, accurate and robust numerical models are required. The aim of this work is to enhance the existing mathematical models and develop a calculation code for energy simulation, as further detailed in the description of the objective in Section 1.3.

1.1. Vertical Greenery Systems

Vegetation of building façades has been a traditional architectural feature since ancient times. These vegetated walls were less sophisticated and basically built with self-climbing plants [3]. However, in recent years, significant technological advancements have been achieved in VGSs, making them more efficient and flexible in their application [3]. These systems have evolved considerably, and innovative techniques such as hydroponic systems, which allow plants growth without the need for a substrate, have been introduced.

With the development of these technologies, many terms have been introduced to describe them, such as vertical garden, vertical greenery, vertical green, biowalls, and vertical gardening [4]. Safikhani et al. [5] defined vertical greenery systems as plants which self-develop on vertical surfaces.

The first challenge is to identify technologies that can be classified into the VGS category. To address this issue, a VGS classification has been proposed based on a literature review, as shown in Figure 1. Vertical greenery systems are divided into two main categories: green façades and living walls.

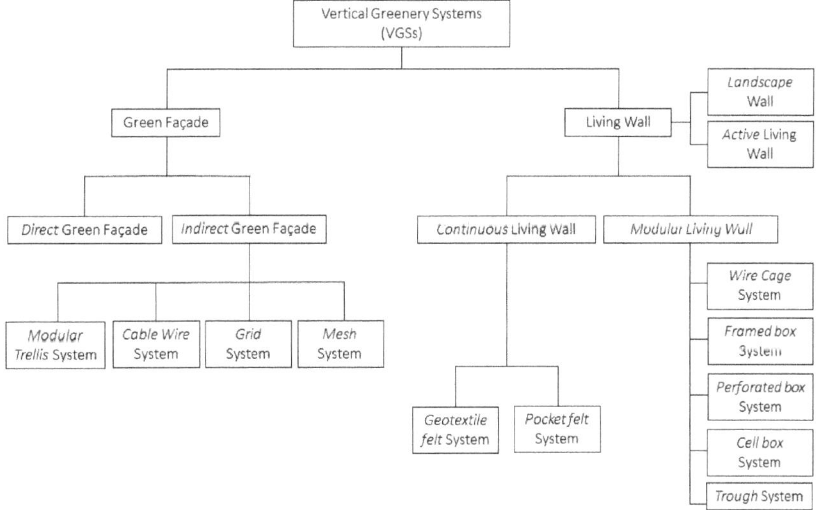

Figure 1. VGS classification.

The green façade represents the simplest typology, characterised by a vegetative layer that self-develops on the vertical surface. This can result naturally from self-climbing plants or with the help of support systems that facilitate growth across the maximum dimension of the surface [5,6]. The growing media, where plants roots obtain nourishment, is external to the wall. It typically consists of soil [5] or can be contained within planter boxes at different levels [6].

The living wall is a more complex typology compared to a green façade. Unlike a green façade, where the vegetative layer is attached to the wall, in a living wall, it is fully integrated in the building structure. This typology is characterised by the presence of a substrate that, supported by a specific system, extends across the entire façade. Typically, living walls are separated from the wall surface by a layer of waterproof membrane to protect the envelope from moisture [7].

The two typologies also differ in the plant species that can be implemented. For green façades, common plants are climbers, while for living walls almost any species, both evergreen and deciduous, can be implemented. It is important to pay attention to combining plants with similar needs.

Green façades can be further distinguished into *Direct Green Façades*, i.e., plants anchored directly to the wall, and *Indirect Green Façades*, i.e., plants with support systems to aid their development [6,8]. Based on the type of supporting structure employed, *Indirect Green Façades* can be further classified into *Modular Trellis Systems*, *Cable Wire Systems* [9], *Grid Systems*, and *Mesh Systems* [10].

Living walls can be subdivided into *Modular Living Walls*—characterised by several modules that are repeated to form the living walls—and *Continuous Living Walls*, a wall type without interruptions [6]. In addition to the two main categories, there are further types of living walls that do not fit into a specific category. On the building scale, with a different function compared to the previous types, there is the *Active Living Wall*. In the landscape context, on the other hand, there is the *Landscape Wall*.

Due to the growing interest in these solutions, several guidelines explore both their characteristics and advantages. Some significant examples include the "Growing Green Guide" [11], the "UK Guide to Green Walls" [12], and "A concise guide to safe practices for vertical greenery" [13]. These documents provide a comprehensive guide to green walls, covering various systems and offering insights on design, plant selection, irrigation, and maintenance. However, despite the availability of guidelines, there is still a lack of technical standards. For green solutions, currently, only UNI 11235:2015 [14] addresses green roofs, and technical details on VGSs are lacking.

The *Vertical greenery systems* offer a series of environmental, social, and economic benefits. They mitigate urban air and noise pollution and improve air quality and people's well-being. Plants, and, in particular, the employed species, play a crucial role in generating these benefits; for example, ivy plants (*Hedera helix*) can absorb air pollutants and fine dust as well as filter toxic chemicals from the soil, such as volatile organic compounds (VOCs) [10].

The shading effect of the vegetation layer and the process of evapotranspiration significantly lower exterior wall surface temperatures during summer, thus improving building energy performance and occupant comfort. These benefits also positively impact the economic aspect. Although installing a VGS may involve a substantial initial cost, dependent on the system chosen, the subsequent advantages, especially in terms of energy efficiency, reduce the overall energy need over time. The evapotranspirative effect of VGSs also lowers outdoor temperatures in areas surrounding these systems. In high-temperature conditions, when a building's envelope is covered with vegetation, the air temperature in the surrounding area decreases. This evapotranspirative phenomenon not only cools the surrounding environment but also mitigates urban heat island (UHI) effects [15].

VGSs act as natural noise barriers due to growing substrates and structural materials that absorb and reflect sound, significantly contributing to reducing the noise pollution. This is influenced by the depth of substrate, the materials used and the extent of the vegetation cover.

1.2. Mathematical Models: A Literature Review

Currently, VGSs are still undergoing research and development, unlike traditional solutions such as green roofs, which are widely adopted. Therefore, specific mathematical models are necessary to assess their effect on the energy performance of buildings. Mathematical models of VGSs existing in the literature are often not very detailed or incomplete due to the limited knowledge of such systems. These models mainly focus on analysing the cooling capacity of VGS and investigating the physical processes involved in heat transfer. The main physical phenomena contributing to the cooling effects include shading, resulting from the shielding effect of the vegetation layer, and evapotranspiration from both vegetation and substrate.

To assess the shielding effect of a VGS, scientific studies generally examine the transmittance of solar radiation through vegetation. A study by He et al. [16] investigated the distribution of long-wave radiation, affirming that transmitted radiation is the portion not intercepted by the leaf. The transmission capacity of long-wave radiation through vegetation is determined using Beer's law, which includes the extinction coefficient for this specific radiation. However, the study does not provide details on the method used to calculate this coefficient.

In addition to the shielding effect, the cooling effects of a VGS are also determined by evapotranspiration by the vegetation and/or substrate. For example, Stec et al. [17] state that approximately 60% of the radiation absorbed by plants is converted into latent heat, thus decreasing the leaf temperature and significantly contributing to the cooling effect. The authors proposed a simplified approach for calculating latent heat to reduce the complexity of heat and moisture transfer models associated with green walls. Nevertheless, adopting such a simplified method may lead to inaccurate results. The most widely used method for estimating the latent heat of evapotranspiration is based on the Penman–Monteith equations, with the most detailed representation described in the DHT model by Zhang et al. [18]. Malys et al. [19] also refer to the Penman–Montheith equations to describe the latent heat flow. The authors developed a heat–mass transfer model to describe green walls, focusing on the development of a hydrothermal model to study these technologies. Despite the positive results of the simulations, uncertainties remain about water balance and evapotranspiration calculation due to the lack of experimental data.

Susorova et al. [20] developed a mathematical model of an exterior wall covered with climbing vegetation to evaluate the thermal effects of plants on heat transfer through building façades. The experiment showed that a layer of plants on a façade can effectively reduce external façade surface temperatures, consequently improving indoor thermal comfort. However, the study considered the heat transfer coefficient of the vegetation-covered wall to be equal to that of a bare wall. This is a problem found in many present mathematical models due to the limited information to refer to. For the study of a living wall, some authors, including Dahanayake et al. [3], utilised the green roof model implemented in EnergyPlus as a starting point. However, the model considered by the authors does not take into account the fact that the main thermo-physical phenomena are different when analysing a vertical green wall. Only a few studies in the literature suggest changes in the main thermo-physical phenomena, such as the absorption of long-wave radiation by plants and the convective transfer flux of the vegetation and substrate. For the latter, the most detailed approaches are those proposed by Hartmann et al. [21] and García et al. [22] and Stanghellini et al. [23], respectively.

1.3. Objective of the Work

The literature analysis shows that existing models are not sufficiently detailed in the description of the overall thermo-physical phenomena of VGSs. Moving away from traditional solutions, it is necessary to accurately evaluate these technologies using detailed and robust numerical models to analyse their effects on the building energy performance. Furthermore, current detailed dynamic modelling software such as EnergyPlus only allows modelling of a few technologies, such as green roofs, while vertical greening systems are missing. Therefore, it is necessary to set up a calculation code that implements the accurate models of these technologies in simulation tools in order to make them more applicable in professional practice and to boost their use in the building design.

The present work aims to overcome the research gaps by improving existing mathematical models of VGSs as well as developing a calculation code to be integrated in EnergyPlus.

The improved models, which concern both living walls and green façades, were then applied to a BESTest Case (900FF) from ASHRAE 140/2020, adapted to the specific scope of the work, in order to test the effectiveness of the VGSs to reduce the cooling need of the building.

2. Materials and Methods

In the present study, two mathematical models were selected, one for green façades and one for living walls, due to their morphological diversity. The mathematical models presented in this paper fill gaps identified in previous models, making them among the most detailed models available for the study of VGSs. For green façades, the Dynamic Heat Transfer (DHT) model by Zhang et al. [18] was chosen. In the literature, this model describes the green façade typology and the related thermo-physical processes involved more in detail. In addition, it provides the methodological process adopted to incorporate the algorithms in EnergyPlus.

For living walls, there is no complete model available in the literature; existing models only delve into specific thermo-physical phenomena. Therefore, several scientific studies were selected by combining their specific contributions. Generally, studies simulating living walls start from the green roof model (GRM) [24], but many overlook the difference in thermo-physical phenomena between green roofs and living walls. Hence, the implemented model starts from the GRM, but it has been adapted to the specific characteristics of living walls, first of all taking into account vertical orientation. In particular, the main thermo-physical phenomena incorporated include the absorption of long-wave radiation by plants, referencing Hartmann's studies [21], convective heat transfer flux of vegetation and substrate for a vertical surface according to Stanghellini's studies [23], and the latent heat flux according to Penman–Monteith's FAO-56 model [25].

2.1. Green Façade Mathematical Model

To investigate the green façade, the Dynamic Heat Transfer (DHT) model of Zhang [18] was referenced. This model is based on the Beer–Lambert law and the Penman–Monteith equation. After developing the VGS model, a co-simulation through EnergyPlus is proposed, incorporating an additional heat source term and energy management system (EMS). The analysis starts with the heat balance of a bare external wall surface, and then the same balance when it is covered by a green façade is described. The present section provides the heat balances and the most important equations related to them; for more details, refer to the documentation [18,26].

The heat balance on the external surface wall covered by a VGS is given by:

$$q_{s,w}^{v} + q_{r,w}^{v} + q_{c,w}^{v} = q_{d,w}^{v} \tag{1}$$

where the terms, expressed in W/m^2, represent the net solar radiation flux, the net long-wave radiation flux, the convective heat flux, and the conductive heat flux on the wall covered by VGS.

The net solar radiation for a green façade differs from that of a wall without VGS due to the shading effect of the vegetation cover. The formula is:

$$q_{s,w}^{v} = q_{s,i} \cdot \tau_c \cdot \alpha_w \tag{2}$$

where $q_{s,i}$ is the total incident short-wave radiation flux (solar irradiance) $[W/m^2]$, τ_c is the total solar energy transmittance of the vegetation $[-]$ and α_w is the solar radiation absorptance of the wall surface $[-]$.

The net long-wave radiation is given by:

$$q_{r,w}^{v} = \varepsilon_w (q_{r,i} \tau_r + \varepsilon_c \sigma T_c^4 - \sigma T_{w,e}^4) \tag{3}$$

where ε_w is the emissivity of the external wall surface $[-]$, $q_{r,i}$ is the total incident long-wave radiation flux $[W/m^2]$, ε_c is the emissivity of the vegetation $[-]$, σ is the Stefan–Boltzmann constant $[W/(m^2K^4)]$, T_c is the temperature of the vegetation $[K]$, $T_{w,e}$ is the temperature of the external wall surface [K], and τ_r is the long-wave radiation transmittance through vegetation [−].

Regarding thermal convection, previous research indicates that a vertical greenery system may affect the air flow near a wall surface, which could have consequences for the convective heat transfer between that surface and the surrounding air [18]. Due to the lack of studies to refer to, it was assumed that the convective heat flow of a wall covered by VGS is equal to that of an external wall without VGS. This aspect represents a simplification of the model.

Thus, the convection heat transfer of a wall with VGS is:

$$q_{c,w}^{v} = q_{c,w}^{b} = h_c(T_{ea} - T_{w,e}) \tag{4}$$

where h_c is the convective heat transfer coefficient [W/(m²K)] and T_{ea} and $T_{w,e}$ are the external air and external wall surface temperatures, respectively [K].

Finally, the conductive heat flux of the wall covered by a VGS is the result of the sum of all the previous contributions.

In addition to the balance of the external wall surface covered by a VGS, a vegetation balance was developed. By solving the heat balance equation of the vegetation, its temperature (T_c) is obtained, and the exchange of long-wave radiation between the vegetation and the external wall surface (Equation (3)) is evaluated.

The vegetation balance is expressed by the equation:

$$q_{s,c} + q_{r,c} + q_{c,c} - q_{tr} = 0 \tag{5}$$

where the terms, expressed in W/m², represent the net solar radiation of vegetation, the net long-wave radiation of vegetation, the convective heat flux of vegetation, and finally the latent heat flux of transpiration, estimated with the Penman–Monteith equation (FAO-56).

The net solar radiation of vegetation is calculated as:

$$q_{s,c} = q_{s,i} \cdot \alpha_c \tag{6}$$

where $q_{s,i}$ is the total incident short-wave radiation flux [W/m²] and α_c is the solar radiation absorptance of vegetation [−].

The net long-wave radiation of vegetation is calculated as:

$$q_{r,c} = \varepsilon_c(q_{r,i} + \varepsilon_w \sigma T_{w,e}^4 - 2\sigma T_c^4) \tag{7}$$

where ε_c is the emissivity of the vegetation [−], $q_{r,i}$ is the total incident long-wave radiation flux [W/m²], ε_w is the emissivity of the external wall surface [−], σ is the Stefan–Boltzmann constant [W/(m²K⁴)], $T_{w,e}$ is the temperature of the external wall surface [K], and finally T_c is the temperature of the vegetation [K].

The convective heat flux of vegetation is calculated as:

$$q_{c,c} = 2LAI\frac{\rho_{ea}c_{ea}}{r_a}(T_{ea} - T_c) \tag{8}$$

where LAI is the leaf area index [m²/m²], ρ_{ea} is the external air density [kg/m³], c_{ea} is the specific heat of external air at constant pressure [J/(kg·K)], and r_a is the aerodynamic resistance [s/m].

Finally, the latent heat flux of transpiration is calculated using the Penman–Monteith equation:

$$q_{tr} = \frac{\Delta(q_{s,c} + q_{r,c}) + \rho_{ea}c_{ea}(e_{vs} - e)/r_a}{\Delta + \gamma + r_s\gamma/r_a} \tag{9}$$

where Δ is the slope of the vapour saturation pressure curve as a function of temperature [kPa/K], e_{vs} is the saturated vapour pressure [kPa], e is the real vapour pressure [kPa], γ is the psychrometric constant [kPa/K], and r_s is the stomatal resistance [s/m].

The additional heat source term for the co-simulation approach of the external wall surface covered by VGS was calculated using the equation:

$$q_{add} = q_{s,w}^{v} + q_{r,w}^{v} - q_{s,w}^{b} - q_{r,w}^{b} \tag{10}$$

where $q_{s,w}^{v}$ and $q_{r,w}^{v}$ are the fluxes of net short-wave radiation and net long-wave radiation on the VGS covered wall, $[W/m^2]$, and $q_{s,w}^{b}$ and $q_{r,w}^{b}$ are the fluxes of absorbed solar radiation and net long-wave radiation on the surface of the bare external wall, $[W/m^2]$, respectively.

2.2. Living Wall Mathematical Model

Starting from the green roof model (GRM), the main equations that modify the model's balances based on previous studies are given to be adapted to a living wall. To study the phenomenon of the absorption of long-wave radiation by leaves and the substrate, reference is made to Hartmann's study [21]. The GRM of EnergyPlus considers the absorption of long-wave radiation from the sky, the emission of radiation by the leaves, and multiple reflections between the plant layer and the substrate. In the case of a vertical green wall, however, the radiant exchange is not only towards the substrate and the sky but also towards the ground. Therefore, the view factors for the sky (F_{sky}) and ground (F_{gr}) were added to the long-wave radiation balance [21,22]; it is calculated for vegetation and substrate, respectively, as:

$$q_{r,c} = \sigma_c \left[F_{c,gr} \varepsilon_c \varepsilon_{gr} \sigma (T_{gr}^4 - T_c^4) + F_{c,sky} \varepsilon_c (I_r^{\downarrow} - \sigma T_c^4) \right] + \frac{\sigma_c \varepsilon_s \varepsilon_c \sigma}{\varepsilon_s + \varepsilon_c - \varepsilon_s \varepsilon_c} (T_s^4 - T_c^4) \tag{11}$$

$$q_{r,s} = (1 - \sigma_c) \cdot \left[F_{s,gr} \varepsilon_s \varepsilon_{gr} \sigma (T_{gr}^4 - T_s^4) + F_{s,sky} \varepsilon_s (I_r^{\downarrow} - \sigma T_s^4) \right] - \frac{\sigma_c \varepsilon_s \varepsilon_c \sigma}{\varepsilon_s + \varepsilon_c - \varepsilon_s \varepsilon_c} (T_s^4 - T_c^4) \tag{12}$$

Convective heat transfer of both vegetation and substrate is the phenomenon that significantly distinguishes green walls from green roofs. The convective sensible heat flow can be calculated using Newton's law:

$$q_c = h_c (T_{ea} - T_c) \tag{13}$$

where q_c is the sensible convective heat flux $[W/m^2]$, and h_c is the convective heat transfer coefficient $[W/(m^2 K)]$.

To determine the sensible convective heat flux, it is necessary to find the convective heat transfer coefficient h_c, which is often determined using the dimensionless Nusselt number:

$$h_c = Nu \, \lambda_{ea} \, d^{-1} \tag{14}$$

where λ_{ea} is the thermal conductivity of external air $[W/(m\ K)]$ and d is the characteristic dimension of the component in exam. For the identification of the Nusselt number, there are several possibilities of determination; among them, however, the Stanghellini resolution (Equation (15)) [23] was selected, as it is one of the few that addresses the convective exchange between the air and the leaf layer and can be applied in the case of a vertical green wall.

$$Nu = 0.405 \left(Pr \cdot Gr + 6.29 \, Pr \cdot Re^2 \right)^{0.25} \tag{15}$$

Pr, *Gr*, and *Re* are the dimensionless Prandtl, Grashof, and Reynolds numbers, respectively.

Lastly, the latent heat flux for foliage and substrate can be modelled based on the FAO-56 Penman–Monteith equation [25], as seen for the green façade (Equation (9)). For the substrate, Equation (9) is modified by replacing the short- and long-wave radiation of the substrate. The modelling of living wall behaviour does not take into account the contribution of latent heat due to freeze–thawing of the substrate.

3. Software Module Development: General Methodology

After the models were selected and examined in detail, they were implemented and simulated in EnergyPlus to evaluate their impact on the building's energy performance. Since EnergyPlus lacks objects to describe the vertical green components, the aforementioned models were externalised and embedded into Python code.

To facilitate this integration, two Python plugins for EnergyPlus were developed. Python plugins are a novel feature in EnergyPlus that enable the seamless integration of Python code into the traditional EnergyPlus execution workflow. This is achieved by overriding a base class that contains a set of callbacks, which are invoked at various stages of the EnergyPlus simulation process.

The two developed plugins differ in their internal algorithmic approach but interact with the EnergyPlus calculations following the same approach. To account for the presence of vertical greenery components, the approach involves modifying the thermal heat balance at the specific surface to which the vertical greenery component is attached. This modification reflects the unique thermal properties and interactions associated with vertical greenery systems. Thus, looking at Figure 2, it is possible to see the generic calling points that allow both Python plugins and EnergyPlus to be interfaced. Indeed, after the initialization, at each simulation step, the EnergyPlus workflow is paused until the Python plugin obtains the needed values, performs its calculation, and sets new boundary conditions for the surfaces of interest. In this work, the calculation for the vertical green components is translated into an additional heat term that is provided to EnergyPlus to properly modify the heat balance on the external surfaces.

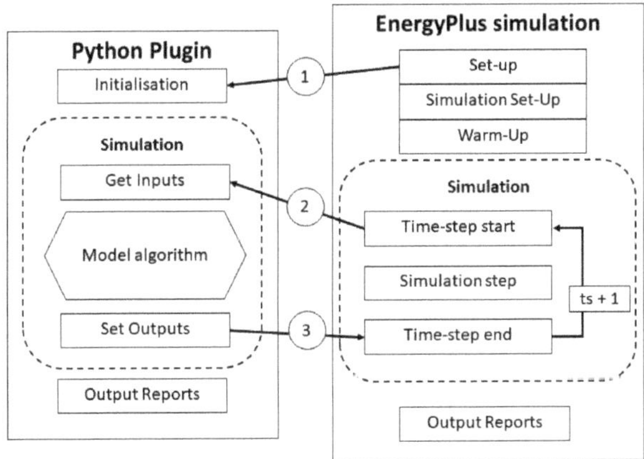

Figure 2. Interaction between Python (version 3.8) and EnergyPlus environments.

To run the simulation with the proposed Python plugins, additional information is needed with respect to the EnergyPlus input file. Parameter configuration for the simulation is then performed partly in EnergyPlus through the Material:Roofvegetation object and partly externally through a configuration file to integrate information not already included in the EnergyPlus object.

Figure 3 shows the details of the two implemented models. Figure 3a reports the steps for the green façade algorithm, while Figure 3b shows the steps for the living wall. Looking at Figure 3a, it is possible to see that the workflow is divided into solar and thermal radiation calculations; the first allows direct estimation of incident solar radiation on the external surface behind the vertical green component, while the latter needs to perform an heat balance on the green façade in order to calculate the temperature at the leaves and then the incident thermal radiation on the external surface. Once both incident solar and thermal radiation are estimated, it is possible to calculate the additional heat term to be set as a boundary condition to the external surface balance inside EnergyPlus.

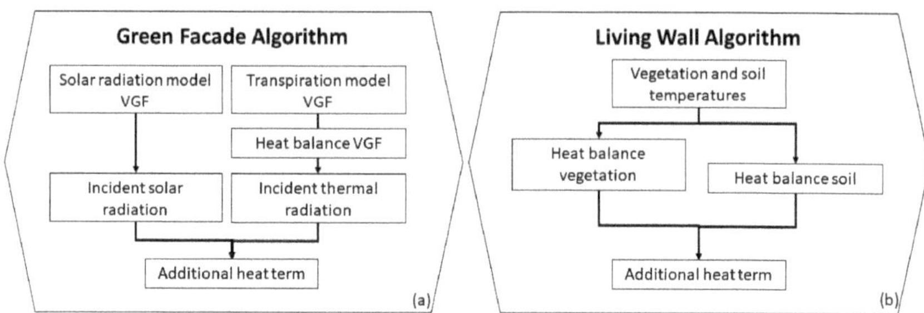

Figure 3. Model algorithm schema: (**a**) green façade, (**b**) living wall.

Figure 3b shows the living wall algorithm steps; the primary objective is the estimation of temperatures at the vegetation and soil layers. This is performed by recursively solve the heat balances at both layers in order to obtain temperatures and heat components. From this, it is possible to determine the heat components involving the external surface behind the living wall and thus set the additional heat term for the EnergyPlus calculations.

In the balances of the green façade and living wall, the vegetation and substrate temperatures represent the final outputs. To solve the balances, balance equations are linearised by expressing the fourth powers of the temperatures involved at timestep i, according to the following formula:

$$T_i^4 = T_{i-1}^4 + 4\,(T_{i-1})^3(T_i - T_{i-1}) \tag{16}$$

where T_{i-1} represents the temperature [K] in the previous timestep, while T_i represents the temperature [K] in the current timestep, and the unknown term in the equation.

A user manual was produced, in which the procedures for first use and the installations required to operate the plugins are explained.

4. Case Study

The reference case study used for dynamic simulations of mathematical models is a BESTest Case of the ANSI/ASHRAE 140-2020 standard: "Method of Test for Evaluating Building Performance Simulation Software" [27]. The selected case study is the 900FF: free floating temperature test for base case of a high mass building. The basic case study (Figure 4) consists of a single rectangular room with no internal partitions and two openings on the south side. It has a high mass construction without a heating or cooling system.

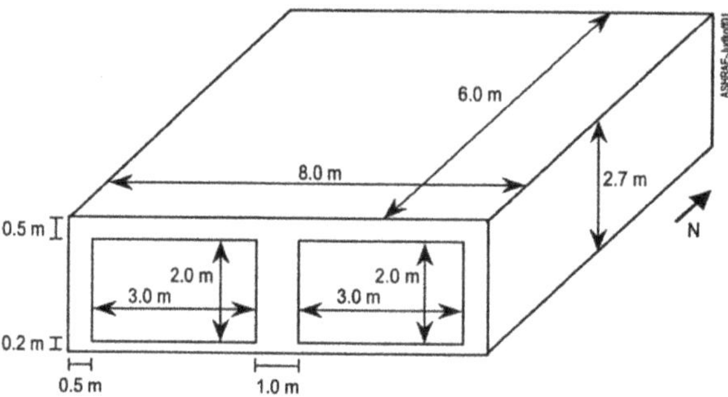

Figure 4. Base case study (900FF BESTest Case of ANSI/ASHRAE 140-2020 standard).

Table 1 shows the main geometric data of the case study, taken from the ANSI/ASHRAE 140-2020 standard.

Table 1. Main geometric data of the case study.

Parameter	Value
Gross volume, V [m^3]	130
Floor area, A_{fl} [m^2]	48

To neutralise the effect caused by the presence of transparent components and to apply the surface finishing under analysis, the initial case study was modified as follows. The south-facing openings were eliminated, and only the concrete block for the vertical walls and the concrete slab for the floor were maintained in the building elements' stratigraphy constituting the opaque envelope. For the upper floor, the standard configuration was replaced with a concrete slab to ensure uniformity in the structure. The dimensions of the geometric model remained unchanged. The decision to simplify the initial model was driven by the need to assess the effects of the applied technologies without being influenced by the presence of a more complex configuration, as proposed in the initial case, or the presence of a heating or cooling system.

Figure 5 shows the final configuration of the case study used for the simulation and validation of the mathematical models for VGSs. Finally, the green technology was applied to the south wall of the case study. For dynamic simulations, climatic data of Turin (Italy) were employed, neglecting both ventilation and internal gains.

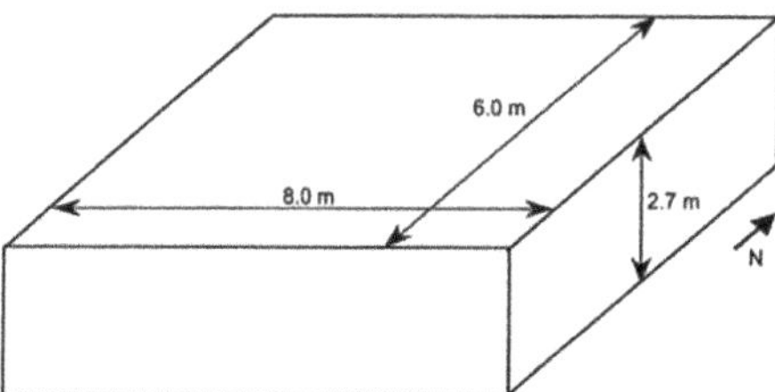

Figure 5. Final geometrical configuration of the case study.

The outputs of the simulations were the external surface temperature of the wall and the operating temperature of the internal environment in the free-floating condition. Finally, the heating and cooling loads of the space were calculated, simulating an ideal system with infinite heat capacity.

5. Results and Discussion

In the analysis, two specific days were considered: 10 December and 17 August. The selection aimed to examine the behaviour of green walls at crucial times of the year. These days were chosen as representative of winter and summer seasons, based on the daily average of global solar radiation and outdoor temperature. For each day, the performance of the south façade, with and without the VGS, was evaluated by assessing the wall's external surface temperature profiles in free-floating conditions. The annual cooling and heating needs were also assessed with and without the green technology. Additionally, a sensitivity analysis was conducted by varying the most significant parameters for the applied technologies. In the dynamic simulation performed, the leaf area index (*LAI*) was not kept constant throughout the year but adjusted to the specific values for each of the selected days, i.e., *LAI* equal to 1 in winter and 5 in summer. This approach aimed to enhance the accuracy of the simulations and bring them closer to reality. The most significant results are presented below.

On the winter day (Figure 6), a notable decrease in the external wall surface temperature is observed compared to that of the bare wall, especially in the central hours of the day due to the shielding effect produced by the applied solution.

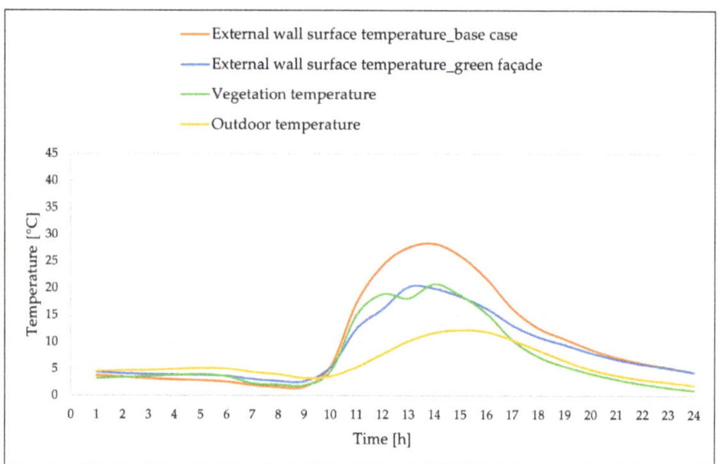

Figure 6. Simulation of the green façade. Comparison of external surface temperature profiles—10 December.

During the summer day (Figure 7), the shielding from solar radiation and the evapotranspiration provided by the presence of the green solution cause a significant decrease in the external surface temperature of the wall, which is higher in summer than in winter as a result of the growth of foliage ($LAI = 5$).

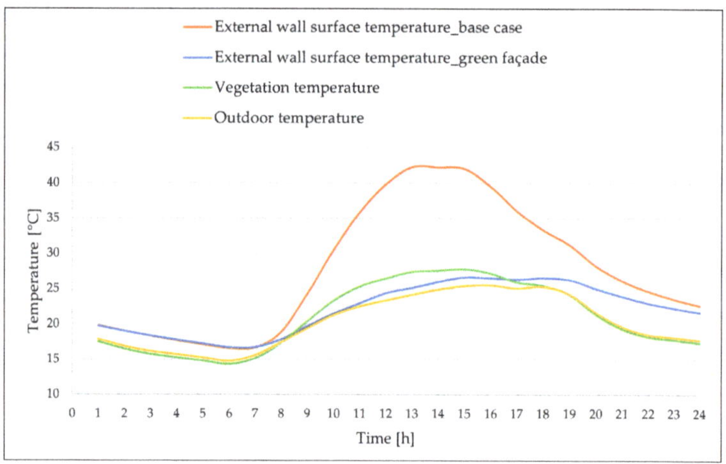

Figure 7. Simulation of the green façade. Comparison of external surface temperature profiles—17 August.

Concerning the ideal thermal energy need of the building, in winter (Figure 8), the cooling of the external surface when the technology is present causes an increase, however minimal (1.0%), in the need for space heating compared to the base case without the green façade. Similarly, in summer, there is a decrease in the need for cooling (15.2%). Limited to the analysed climate and defined boundary conditions, on an annual basis, the green façade causes a greater decrease in cooling need than the increase in heating need.

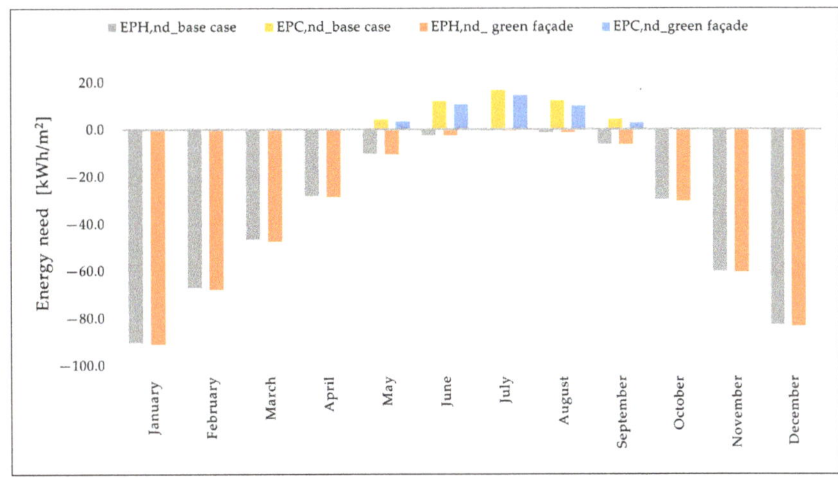

Figure 8. Simulation of the green façade. Comparison of the ideal thermal energy need for heating ($EP_{H,nd}$) and for cooling ($EP_{C,nd}$) on a monthly basis.

The results obtained are in line with the literature. Referring to the work of Zhang et al. [18], the reduction in the cooling load in the summer season is between 11.7% and 18.4%, applying the solution on all the external walls. In the present work, a reduction of 8.2% in the cooling load has been achieved by applying the solution only on the south wall of the case study.

As shown in the work of Zhang et al. [18] and in this work, the temperature of the vegetation is higher at certain times of the day when compared to that of the external wall in the presence of the green solution.

Finally, for the green façade, the result of the sensitivity analysis is presented (Figure 9).

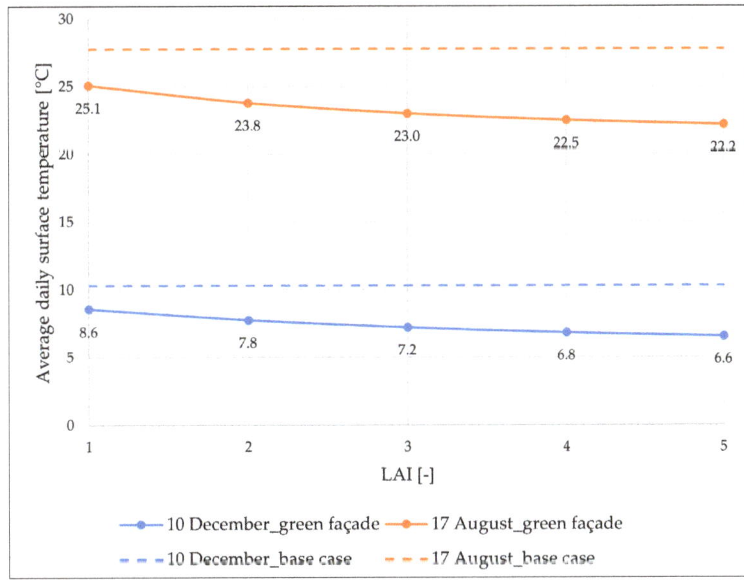

Figure 9. Simulation of the green façade. Sensitivity analysis: LAI variation.

The parameter subject to variation is the leaf area index (*LAI*). The *LAI* influences two important factors: the amount of solar radiation incident on the external wall behind the leaves and evapotranspiration. Therefore, from a thermo-physical point of view, the *LAI* value affects the external surface temperature of the wall on which the vegetation is applied. Figure 9 shows the variation in the average daily temperature on the surface to which the green façade solution is applied by varying *LAI*. The decrease in the surface temperature resulting from the increase in *LAI* is noticeable, and it is more significant in summer than in winter.

Also, for the living wall, in the winter day and summer day, there is a decrease in the external wall surface temperature in the presence of the green solution (Figures 10 and 11).

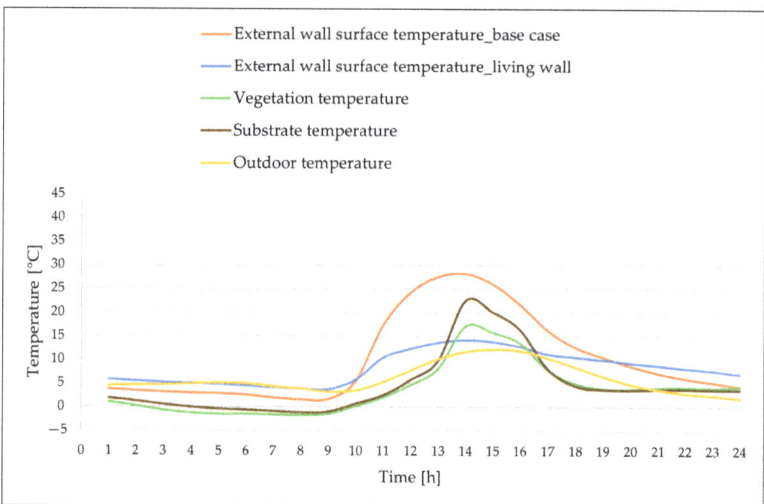

Figure 10. Simulation of the living wall. Comparison of external surface temperature profiles—10 December.

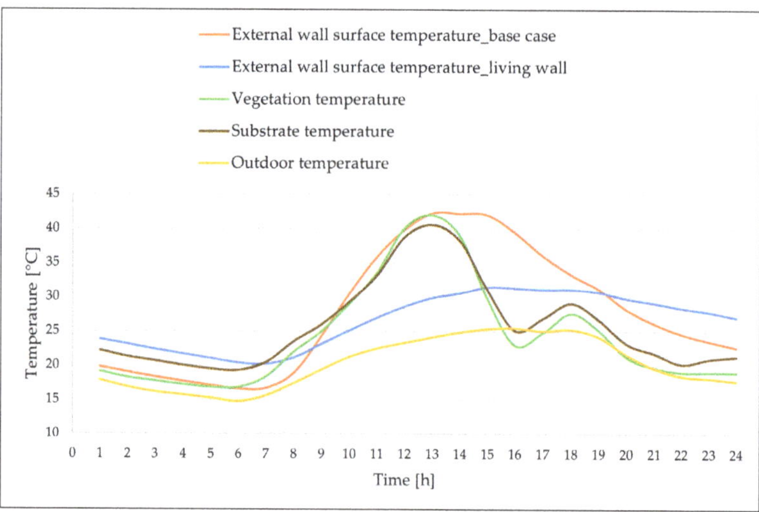

Figure 11. Simulation of the living wall. Comparison of external surface temperature profiles—17 August.

In terms of the ideal heating need, the living wall solution (Figure 12), compared to a wall without substrate and vegetation, results in a reduction—however minimal—in the thermal energy need for heating (3.2%) and a more significant reduction in that for cooling (8.5%). The reduction in heating need would be attributed to a higher thermal resistance of the wall due to the presence of the substrate. For the living wall, the parameters that varied were the *LAI* and the thickness of the substrate.

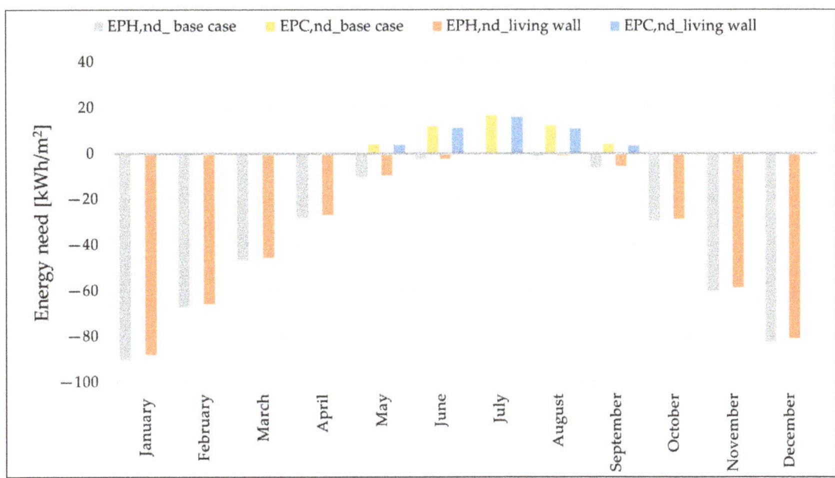

Figure 12. Simulation of the living wall. Comparison of the ideal thermal energy need for heating ($EP_{\mathrm{H,nd}}$) and for cooling ($EP_{\mathrm{C,nd}}$) on a monthly basis.

Figure 13 shows the average daily temperature on the surface to which the living wall is applied, as the *LAI* varies, while maintaining the thickness of the substrate constant. For both winter and summer days, the average daily temperature on the surface remains almost constant as the *LAI* varies. The presence of the adopted solution determines a reduction in the average daily temperature on the surface compared to the base case, although it is not influenced by the variation in the amount of foliage.

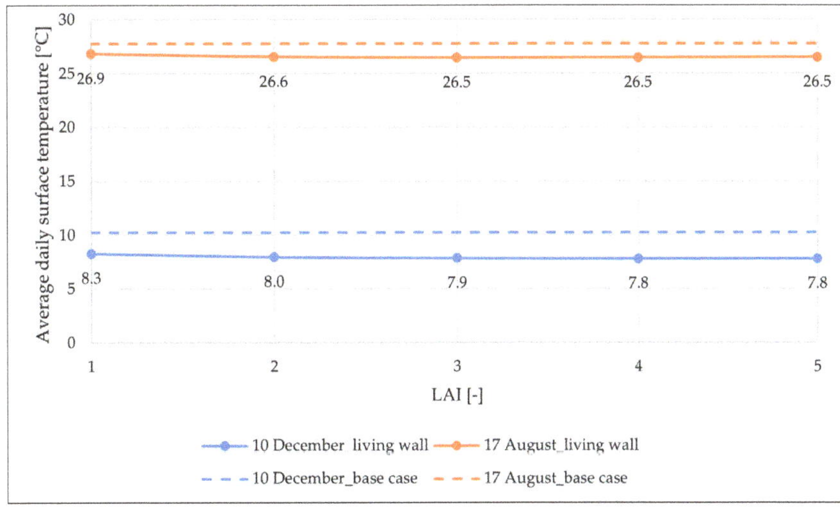

Figure 13. Simulation of the living wall. Sensitivity analysis: LAI variation.

Figure 14 shows the variation in the daily average temperature on the surface to which the substrate is applied, maintaining the *LAI* fixed and varying the thickness of the substrate. In this case, it is evident that the substrate and the variation in its thickness influence the daily mean temperature of the outer surface of the wall.

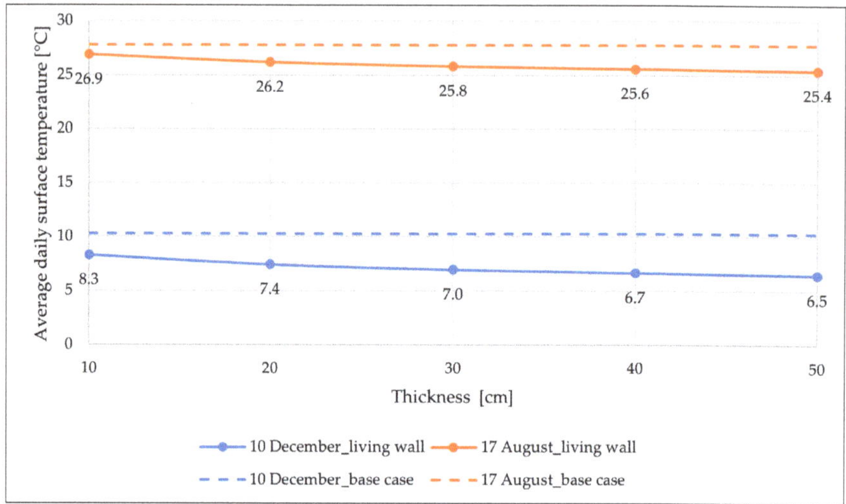

Figure 14. Simulation of the living wall. Sensitivity analysis: thickness substrate variation.

Table 2 summarises the annual thermal energy needs for heating and cooling for the base case, the case with the green façade, and the case with the living wall, highlighting the variation with respect to the base case.

Table 2. Annual thermal energy need and percentage change—comparison between the base case, green façade, and living wall.

	$\lvert EP_{H,nd} \rvert$ [kWh/m^2]	$\Delta \lvert EP_{H,nd} \rvert$ [%]	$\lvert EP_{C,nd} \rvert$ [kWh/m^2]	$\Delta \lvert EP_{C,nd} \rvert$ [%]
Base case	426	—	50	—
Green façade	430	+1.0	42	−15.2
Living wall	412	−3.2	46	−8.5

6. Conclusions

The impacts of climate change, excessive greenhouse gas emissions, and the current energy crisis have motivated the European Union to adopt mitigation and adaptation strategies, focusing mainly on the building sector due to the crucial role it represents. To achieve this goal, the implementation of resilient technologies for the building envelope, such as vertical greenery systems (VGSs), particularly living wall and green façades, is gaining ground. These systems are considered cooling strategies because, acting as a shading, they partially mitigate solar radiation and, through evapotranspiration, induce a reduction in the surface temperature of the component on which they are installed and of the surrounding environment.

Dedicated mathematical models from the literature were improved to simulate their actual performance, and special calculation codes were created in the present work. Their implementation and simulation in EnergyPlus produced some interesting results.

Firstly, it was found that both technologies, the green façade and the living wall, applied to a test case study contribute positively to the summer season decreasing the surface temperature of the external wall to which they are applied. The shielding effect reduces the incident solar radiation on the back wall, and the evapotranspirative effect, both of vegetation and substrate, based on the technology applied, involves cooling in terms of external surface temperature and immediate surroundings.

In fact, considering the green façade solution effect, the external wall surface temperature in summer decreases by 41%, while for the living wall solution it decreases by 30%. At the same time, in the winter season, always for the reasons expressed earlier, there is a decrease in the external surface temperature of 26% with green façade solution and 18% with living wall, which in that case implies a disadvantage.

Analysis of the thermal energy need for both solutions shows that although the VGSs are not particularly advantageous in winter, the thermal energy need for heating in such months does not change significantly: for green façades, it increases by 1.0% and for living walls by 3.2%. Instead, in summer, the thermal energy need for cooling decreases for green façades and for living walls by 15.2% and 8.5%, respectively.

The sensitivity analysis showed how crucial it is to manage the main data characterising green walls, such as the leaf area index (*LAI*), especially for the green façade, and the substrate thickness in the living wall. In accordance with the existing literature, these are the two most influential parameters. In particular, given the influence of *LAI*, in future studies, measured data could be used.

By setting up improved models of VGSs and related calculation codes, the outcomes of this research could effectively contribute to the enhancement of the standardisation activity by introducing new numerical methods in the overall energy performance assessment framework of buildings. The need to model such technologies emerges from the new European directives, in particular, EU Directive 2024/1275 [28], which promotes interventions to make buildings resilient to climate change.

Finally, the implemented tool demonstrated an easy application with no increase in the computational time compared to a standard simulation. This is an advantageous element in incentivising professionals to use the tool.

The scalability of VGS and the study of their long-term performance under different climatic conditions will be a prerogative for future developments of this work. Case studies resulting from different geographical locations and building types will be considered in order to increase the generalizability of the findings. In addition, future research will focus on the assessment of the accuracy of the developed models by comparison with empirical data.

Author Contributions: Conceptualization, I.B. and V.C.; methodology, V.N., I.B., P.R.M. and V.C.; software, P.R.M.; validation, V.N., I.B., P.R.M. and V.C.; formal analysis, V.N. and P.R.M.; investigation, V.N. and P.R.M.; writing—original draft preparation, V.N. and P.R.M.; writing—review and editing, I.B. and V.C.; visualization, V.N.; supervision, I.B. and V.C.; project administration, V.C.; funding acquisition, V.C. All authors have read and agreed to the published version of the manuscript.

Funding: This research was funded by the Italian National Agency for New Technologies, Energy and Sustainable Economic Development (ENEA), Electrical System Research, Implementation Plan 2022–2024, Line of activity: LA2.7, "Implementation of a calculation code for energy modelling of Vertical Greenery Systems".

Data Availability Statement: The raw data supporting the conclusions of this article will be made available by the authors on request.

Conflicts of Interest: The authors declare no conflicts of interest.

Nomenclature

Symbol	Quantity	Unit
A	Area	m^2
c	Specific heat	J/(kg K)
d	Characteristic dimension	m
e	Vapour pressure	kPa

EP_{nd}	Areic thermal energy need	kWh/m^2
Gr	Grashof number	—
h	Heat transfer coefficient	$\text{W/(m}^2\text{K)}$
LAI	Leaf area index	—
Nu	Nusselt number	—
Pr	Prandtl number	—
q	(Areic) energy flux	W/m^2
r_{a}	Aerodynamic resistance	s/m
r_{s}	Stomatal resistance	s/m
Re	Reynolds number	—
T	Temperature	K
V	Volume	m^3
α	Absorptance	—
γ	Psychometric constant	kPa/K
Δ	Slope of the vapour saturation pressure curve	kPa/K
ε	Emissivity	—
λ	Thermal conductivity	W/(m K)
ρ	Density	kg/m^3
σ	Stefan–Boltzmann constant	$\text{W/(m}^2\text{K}^4\text{)}$
σ_{c}	Fractional vegetation coverage	—
τ	Transmittance	—

Superscripts

b	Bare
v	Vegetated

Subscripts

add	Additional
C	Cooling
c	Convection
c	Canopy
d	Conduction
e	External
ea	External air
fl	Floor
gr	Ground
H	Heating
i	Incident
i	Index
r	Long-wave radiation (infrared)
s	Short-wave radiation (solar)
s	Substrate
sky	Sky vault
vs	Saturated vapour
w	Wall

References

1. European Commission. *A Renovation Wave for Europe—Greening Our Buildings, Creating Jobs, Improving Lives*; European Commission: Brussels, Belgium, 2020.
2. IEA EBC. International Energy agency's Energy in Buildings and Communities Programme. Available online: https://www.iea-ebc.org/ (accessed on 10 April 2024).
3. Dahanayake, K.W.D.; Kalani, C.; Chow, C.L. Studying the potential of energy saving through vertical greenerysystems: Using EnergyPlus simulation program. *Energy Build.* **2017**, *138*, 47–59. [CrossRef]
4. Susca, T. Nature-Based Solutions Applied to the Built Environment to Alleviate Climate Change: Benefits, Co-benefits, and Trade-offs in a Geographical Multi-scale Perspective. In *Handbook of Climate Change Mitigation and Adaptation*; Lackner, M., Sajjadi, B., Eds.; Springer: Cham, Switzerland, 2022; pp. 2118–2153.
5. Safikhani, T.; Abdullah, A.M.; Ossen, D.R.; Baharvand, M. A review of energy characteristic of vertical greenery systems. *Renew. Sustain. Energy Rev.* **2014**, *40*, 450–462. [CrossRef]
6. Manso, M.; Castro-Gomes, J. Green wall systems: A review of their characteristics. *Renew. Sustain. Energy Rev.* **2015**, *41*, 863–871. [CrossRef]

7. Susorova, I.; Bahrami, P. Facade-integrated vegetation as an environmental sustainable solution for energy-efficient buildings. *MADE Res. J. Cardiff Univ. Cardiff UK* **2013**, 7–14. Available online: https://www.researchgate.net/publication/265593368_Facade-integrated_vegetation_as_an_environmental_sustainable_solution_for_energy-efficient_buildings (accessed on 10 April 2024).

8. Perini, K.; Ottelé, M.; Fraaij, A.L.A.; Haas, E.M.; Raiteri, R. Vertical greening systems and the effect on air flow and temperature on the building envelope. *Build. Environ.* **2011**, *46*, 2287–2294. [CrossRef]

9. The American Institute of Architects Continuing Education System (AIA/CES). Introduction to Green Walls Technology, Benefits and Design. Green Roofs for Healthy Cities. 2008. Available online: https://www.calameo.com/read/00135160798c31a1b7998 (accessed on 10 April 2024).

10. Radic, M.; Dodig, M.B.; Auer, T. Green Facades and Living Walls—A Review Establishing the Classification of Construction Types and Mapping the Benefits. *Sustainability* **2019**, *11*, 4579. [CrossRef]

11. Francis, J.; Hall, G.; Murphy, S.; Rayner, J. Growing Green Guide: A Guide to Green Roofs, Walls and Facades in Melbourne and Victoria, Australia. February 2014. Available online: https://www.melbourne.vic.gov.au/community/greening-the-city/green-infrastructure/Pages/growing-green-guide.aspx (accessed on 10 April 2024).

12. Urban greening. UK Guide to Green Walls: An Introductory Guide to Designing and Constructing Green Walls in the UK. 2013. Available online: https://media.wix.com/ugd/c0a820_877fc4a9e779472c53296c58fdfb8d20.pdf (accessed on 10 April 2024).

13. National Parl Board. A Concise Guide to Safe Practices for Vertical Greenery. 2013. Available online: https://www.nparks.gov.sg/~/media/srg/publications/A_Concise_Guide_to_Safe_Practices_for_Vertical_Greenery.pdf (accessed on 10 April 2024).

14. *UNI 11235*; Instructions for the design, execution, control and maintenance of green roofs. Italian Organisation for Standardisation (UNI): Milan, Italy, 2015.

15. Loh, S. *Living Walls—A Way to Green the Built Environment*; Environment Design Guide 2008, TEC 26; Royal Australian Institute of Architects: Melbourne, Australia, 2008; pp. 1–7.

16. He, Y.; Yu, H.; Ozaki, A.; Dong, N.; Zheng, S. An investigation on the thermal and energy performance of living wall system in Shangai area. *Energy Build.* **2017**, *140*, 324–335. [CrossRef]

17. Stec, W.J.; van Paassen, A.H.C.; Maziarz, A. Modelling the double skin façade with plants. *Energy Build.* **2005**, *37*, 419–427. [CrossRef]

18. Zhang, Y.; Zhang, L.; Meng, Q. Dynamic heat transfer model of vertical green façades and its co-simulation with a building energy modelling program in hot-summer/warm-winter zones. *J. Build. Eng.* **2022**, *58*, 105008. [CrossRef]

19. Malys, L.; Musy, M.; Inard, C. A hydrothermal model to assess the impact of green walls on urban microclimate and building energy consumption. *Build. Environ.* **2014**, *73*, 187–197. [CrossRef]

20. Susorova, I.; Angulo, M.; Bahrami, P.; Stephens, B. A model of vegetated exterior facades for evaluation of wall thermal performance. *Build. Environ.* **2013**, *67*, 1–13. [CrossRef]

21. Hartmann, M.; Alsaad, H.; Mosayebzadeh, M.; Voelker, C. The reduction of building energy demand by a living wall system in a Central European climate. In Proceedings of the BS2023, the 18th Conference of IBPSA, Shanghai, China, 4–6 September 2023. [CrossRef]

22. Garcia, M.; Vera, S.; Bustamante, W.; Rouault, F. Modelling and Validation of two Heat and Mass Transfer Model of Living Walls and Evaluation of Their Impact on the Energy Performance of a Supermarket in a Semiarid Climate. In Proceedings of BS2019, the 16th Conference of IBPSA, Rome, Italy, 2–4 September 2019. [CrossRef]

23. Stanghellini, C. *Transpiration of Greenhouse Crops an Aid to Climate Management*; Instituut voor Mechanisatie, Arbeid en Gebouwen (IMAG): Wageningen, The Netherlands, 1987.

24. Sailor, D.J. A green roof model for buildings energy simulation programs. *Energy Build.* **2008**, *40*, 1466–1478. [CrossRef]

25. Zotarelli, L.; Dukes, M.D.; Romero, C.C.; Migliaccio, K.W.; Morgan, K.T. *Step by Step Calculation of the Penman-Monteith Evapotranspiration (FAO-56 Method)*; University of Florida: Gainesville, FL, USA, 2010.

26. Zhao, C.; Zhang, L.; Yang, Y.; Liu, M.; Yan, J.; Zhao, L. Long-wave infrared radiation properties of vertical green façades in subtropical regions. *Build. Environ.* **2022**, *223*, 109518. [CrossRef]

27. *ASHRAE Standard 140-2020*; Method of Test for Evaluating Building Performance Simulation Software. American Society of Heating Refrigerating and Air-Conditioning Engineers (ASHRAE): Peachtree Corners, GA, USA, 2020.

28. European Union. Directive (EU) 2024/1275 of the European Parliament and of the Council of 24 April 2024 on the energy performance of buildings (recast). Official Journal of the European Union, L. 8.5.2024. Available online: https://www.google.com.hk/url?sa=t&source=web&rct=j&opi=89978449&url=https://eur-lex.europa.eu/legal-content/EN/TXT/?uri=CELEX%253A32024L1275&ved=2ahUKEwiYwd6vm4eHAxX8lK8BHXXTABgQFnoECBIQAQ&usg=AOvVaw13VMZWlcEen2EF3z4EUF6r (accessed on 10 April 2024).

 **buildings**

Article

CFD Analysis of Different Ventilation Strategies for a Room with a Heated Wall

Y Quoc Nguyen [1,2,*], Viet T. Nguyen [1], Long T. Tran [3,4] and John C. Wells [5]

[1] Computational Engineering and Design Research Group, School of Engineering and Technology, Van Lang University, Ho Chi Minh City 70000, Vietnam

[2] Faculty of Engineering and Technology, Binh Duong Economics and Technology University, Binh Duong 75200, Vietnam

[3] Faculty of Civil Engineering, Ho Chi Minh City University of Technology, 268 Ly Thuong Kiet Street, District 10, Ho Chi Minh City 70000, Vietnam

[4] Vietnam National University Ho Chi Minh City, Linh Trung Ward, Thu Duc District, Ho Chi Minh City 70000, Vietnam

[5] Faculty of Science & Engineering, Ritsumeikan University, Shiga 525-8577, Japan

* Correspondence: y.nq@vlu.edu.vn or y.nq@ktkt.edu.vn

Abstract: Solar chimneys can help to reduce solar heat gain on a building envelope and to enhance natural ventilation. In this work, we proposed three configurations of two solar chimneys combined with a heated wall for the natural ventilation of a room: (I) the chimneys are connected serially, (II) the chimneys are parallel and exhaust air at two separate outlets, and (III) the chimneys are parallel, but the outlets are combined. The airflow rate achieved with each configuration was predicted with a Computational Fluid Dynamics model. The results show the effects of the heat flux in each channel and the geometries of the channels. Configuration (II) shows the highest flow rate. Particularly, the proposed configurations enhance the flow rate significantly and up to 40% when compared to the typical setup with a single channel solar chimney. The findings offer a novel design option for building façades for reducing solar heat gain and enhancing natural ventilation.

Keywords: solar chimney; heated wall; CFD; natural convection; airflow rate

Citation: Nguyen, Y.Q.; Nguyen, V.T.; Tran, L.T.; Wells, J.C. CFD Analysis of Different Ventilation Strategies for a Room with a Heated Wall. *Buildings* **2022**, *12*, 1300. https://doi.org/10.3390/buildings12091300

Academic Editors: Gianpiero Evola and Elena Lucchi

Received: 19 July 2022
Accepted: 18 August 2022
Published: 25 August 2022

Publisher's Note: MDPI stays neutral with regard to jurisdictional claims in published maps and institutional affiliations.

1. Introduction

Thermal insulation of building envelopes, particularly walls, is important in green and energy-efficient buildings to reduce energy demand for heating and cooling. In Viet Nam, the current regulation for energy-efficient buildings, QCVN 09:2017/BXD requires the minimum thermal resistance of 0.56 $m^2 \cdot K/W$ for opaque walls. To satisfy this value, concrete walls with hollow bricks must have a minimum total thickness of 220 mm, which is equivalent to two brick layers. Meanwhile, most residential buildings in Viet Nam have concrete walls with one brick layer whose total thickness of about 110 mm and the equivalent thermal resistance of 0.383 $m^2 \cdot K/W$. Therefore, to comply with this standard, the thermal insulation of such single layer concrete walls must be improved by adding additional insulation layers or external shading devices [1].

Among the most common methods for reducing the solar gain of building envelopes is the solar chimney. This device is based on the buoyancy effects of the air warmed by solar radiation. The buoyancy-driven airflow transports the absorbed heat in the walls and discharges to the ambient atmosphere. As a result, the thermal insulation of the building is enhanced. Miyazaki et al. [2], Al Touma and Ouahrani [3], Hong et al. [4], and Ma et al. [5] reported that the reduction in the annual energy consumption for houses and buildings with solar chimneys was between 2.3% and 77.8%. A combined solar chimney-window of a room can eliminate 11.4% of the daily heat gain [3].

As the induced flow rate is an important performance parameter of a solar chimney [6], it has been the focus of several studies. Previous works have shown that the flow rate is strongly influenced by the heat flux and the dimensions of the cavity. As the buoyancy effects are enhanced with the heat flux and the height of the air channel, the induced flow rate also increases with those two parameters [7–18]. Increasing the gap also boosts the flow rate [7,8,10–12,16,18]. However, when the gap is large compared to the height, there is a reverse flow at the top the air channel. The reverse flow obstructs the main flow and reduces the flow rate [8,11,19–22].

The designs of solar chimneys differ based on the types of buildings they are integrated into. In Singapore, the solar chimneys of the BCA Zero Energy Building are composed of roof cavities connected to circular tubes [23]. In Syria, the Lycée Charles de Gaulle school utilizes vertical solar chimneys on the top of two-story buildings for natural ventilation. In Viet Nam, the Deutsches Haus Ho Chi Minh City has walls with open double glass layers which function as solar chimneys. Tall buildings, such as the GSW Headquarters in Germany and Manitoba Hydro Place in Canada, were designed with large-scale solar chimneys [24]. Consequently, the performance of each solar chimney needs to be examined with its specific configuration [6].

On the building envelope, the chimney effect also takes place on a ventilated façade [25]. Air cavities behind the external claddings function as solar chimneys and increase the thermal resistance of the wall. Rahiminejad and Khovalyg [26] reported that the total effective resistance of a cladding and a cavity was up to nine times the thermal resistance of the cladding alone. Particularly, the total resistance increased with the induced flow rate in the cavity. This again confirms the important role of the flow rate.

The problem of the natural ventilation of a room with a wall heated by solar radiation has also been considered in many previous studies. With a simple opening at the top of the room with one vertical wall heated, Marcias-Melo et al. [27] showed that the most efficient case for removing heat was when the opening was on the top of the heated wall. To enhance the ventilation rate, Vazquez-Ruiz et al. [28] added a solar chimney to the top of the room. Their results showed the highest ventilation rate was achieved with the chimney positioned nearest to the air inlet, which was on the wall perpendicular to the heated one. In these studies, the heat from the wall was transferred directly into the room air. In addition, there was also no chimney to exploit the heat transfer on the outer surface of the wall.

Hernandez–Lopez et al. [29] utilized a chimney covering the outer surface of a heated wall to induce airflow for ventilation. Although this system successfully reduced the room air temperature by 9.1 °C, the air in the room was heated significantly because of the lack of a heat barrier on the inner surface of the heated wall. By adding a roof vertical chimney to a system identical to that of Hernandez–Lopez et al. [29], Wang et al. [30] reported that the flow rate through the wall chimney was enhanced in most cases. However, the performance of the optimized designs of the wall chimney did not change with the addition of the roof one. In addition, their system still allowed the heat transfer from the heated wall into the room.

The main research objective of this study was to exploit the heat transfer on both sides of the heated wall with two solar chimneys, covering the outer and inner surfaces of the wall, respectively. The outer chimney collects the heat loss from the outer surface of the wall while the inner chimney functions as a heat barrier to prevent the heat transfer from the wall into the room. Therefore, the proposed system can overcome the limitations of the above-mentioned works.

The performance of the proposed system was examined by a Computational Fluid Dynamics (CFD) model. CFD can provide details of the flow and thermal dynamics which are not obtained with typical building energy simulation models. Zhai and Chen [31] achieved more accurate solutions of an energy model coupled with a CFD model which could offer better predictions of heat transfer on the walls. In this study, as the performance of the proposed system depends strongly on the natural convection in the air cavity, the CFD technique was selected.

2. Description of the System

The proposed system, as seen in Figure 1, consists of two solar chimneys for a dual purpose: (i) releasing solar heat gain on a wall and preventing its transfer into the room, and (ii) natural ventilation of the room. The room has a window and a wall exposed to solar radiation (the left wall in Figure 1). The wall can be on the west or the south side of the room. Solar radiation absorbed on the wall may be conducted into the room. The conductive heat flux should depend on the thermal properties of the wall materials.

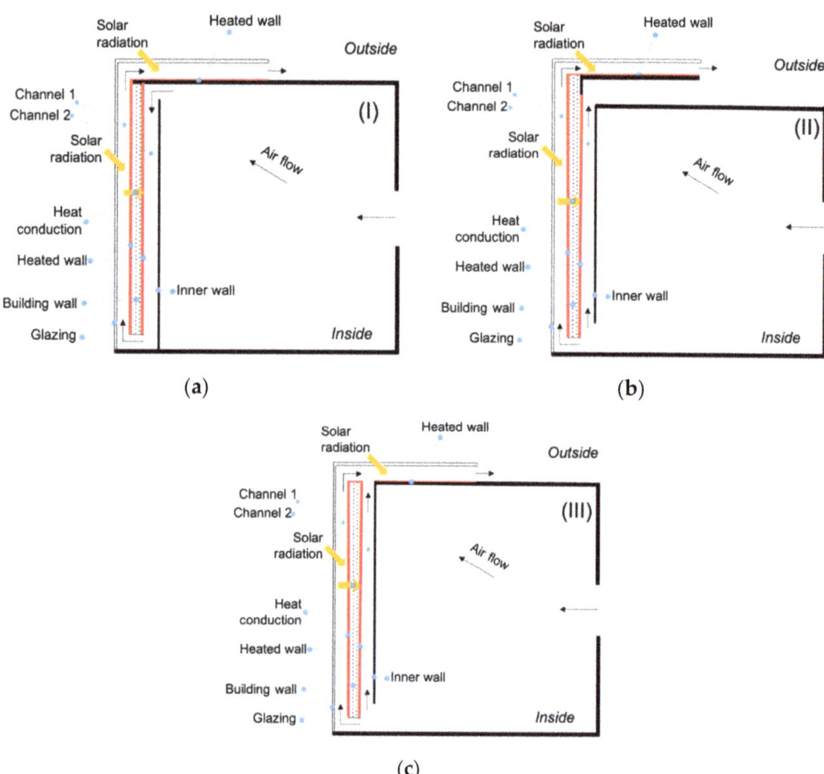

Figure 1. The proposed system: (**a**) Configuration (I), (**b**) Configuration (II), and (**c**) Configuration (III).

To satisfy the above two functions, the proposed system must be able to capture the heat from both sides of the wall. Accordingly, there are two air channels inside and outside of the wall. The outer channel (Channel 1) consists of a vertical section connected to a horizontal part. This channel collects warmed air along the outer surface of the wall and on the roof of the room. The horizontal part of the channel is assumed to cover half of the roof as the rest of the roof is saved for other structures. The inner channel (Channel 2) is formed by the inner surface of the wall and a partition, which is denoted as the "inner wall" in Figure 1. The two channels are arranged in three configurations, as seen in Figure 1.

Configuration (I) (Figure 1a): Channel 2 is connected to Channel 1 on its lower end. As the warm air in Channel 1 rises, air from the room enters Channel 2 from its upper opening, flows into Channel 1, and escapes to the outside environment.

Configuration (II) (Figure 1b): Channel 2 is separated from Channel 1. Air enters both channels at their lower openings and exits at their upper ends. At the outlets, the two airflows do not merge.

Configuration (III) (Figure 1c): It is similar to Configuration (II), but the upper end of Channel 2 is connected into the horizontal part of Channel 1. The airflow in Channel 2 merges with that in Channel 1 at the junction. The merged airflow escapes at the outlet of Channel 1.

Among three systems, Configuration (I) can extract stalled air and contaminant, such as smoke, in the upper zone of the room. However, this design is unfavorable for the flow in Channel 2, as it must move oppositely to the thermal effect. Configuration (II) can only suck air from the lower zone of the room, but the flows in both channels are in the same direction as the thermal effects. However, there must be an exhaust tube to discharge air in Channel 2. Configuration (III) does not require a discharge tube such as in Configuration (II).

It is assumed that there is ignorable heat conduction through the roof and the inner wall into the room. This is practically possible by using roof materials with good thermal resistance. The heat transfer between two walls of Channel 2 is mainly under radiation mode, which can be reduced significantly by using polished materials.

The performance of the configurations in Figure 1 was examined with a Computational Fluid Dynamic (CFD) model. The induced flow rate through the room was computed at different dimensions of the chimneys. To model the effects of the thermal resistance of a building wall, different heat flux ratios in two channels were also considered.

The remainder of this paper is divided into three parts. The computational model is described in the section, "CFD Model". The main findings are reported in the section, "Results and Discussion". Lastly, the section "Conclusions" summarizes the main results.

3. CFD Model

For assessing the ventilation performance of the configurations in Figure 1, the airflow rate in each case was computed with a model based on Computational Fluid Dynamics (CFD). CFD has been employed extensively for simulations of indoor air and ventilation [9,14,15,28,32–36]. Following the CFD models for solar chimneys in the literature, a two-dimensional model was built with the following assumptions:

The flow and heat transfer were steady;
The flow was incompressible and turbulent;
The ambient air was at atmospheric pressure;
The flow in a solar chimney becomes turbulent when the Rayleigh number, $Ra = g\beta q_t L^4/\nu\alpha\lambda$, is above 10^{10} [8,22,33]. In this study, the Rayleigh number was about $10^{12} - 10^{13}$. Accordingly, the flow in the cavity was assumed to be turbulent.

The governing equations in the Reynolds-Averaged Navier–Stokes (RANS) forms are as follows:

$$\frac{\partial(\rho u_i)}{\partial x_i} = 0 \tag{1}$$

$$\frac{\partial(\rho u_i u_j)}{\partial x_j} = -\frac{\partial p}{\partial x_i} + \frac{\partial}{\partial x_j}\left(\mu\left(\frac{\partial u_j}{\partial x_i} + \frac{\partial u_i}{\partial x_j}\right) - \rho\overline{u_i'u_j'}\right) - \rho g_i\beta(T - T_r) \tag{2}$$

$$\frac{\partial(\rho u_i T)}{\partial x_i} = \frac{\partial}{\partial x_i}\left(\frac{\lambda}{c_p}\frac{\partial T}{\partial x_i} - \rho\overline{u_i'T'}\right) \tag{3}$$

The turbulence terms ($-\rho\overline{u_i'u_j'}$ and $\rho\overline{u_i'T_j'}$) in Equations (3) and (4) were solved with the RNG $k - \epsilon$ model. Common RANS turbulence models (standard $k - \omega$, standard $k - \varepsilon$, RNG $k - \varepsilon$, Low Reynolds number $k - \omega$) were successfully used for simulations of solar chimneys [9,17,37,38]. Particularly, Gan [9] and Hinojosa et al. [37] reported that the RNG $k - \varepsilon$ model offered the best performance at $Ra \approx 10^{10} - 10^{12}$, which is close to the Rayleigh numbers in this study. Therefore, the RNG $k - \epsilon$ model was selected. Details of the formulations can be seen in [9,14,15,28,32,33,37].

Based on the above formulations, a numerical model was built for the configurations in Figure 1. Its main components included the computational domain, mesh, discretization methods for the governing equations, and suitable boundary conditions. All these settings were conducted with the CFD software ANSYS Fluent, the version for Academic 2021R2.

Figure 2 shows the computational domain of configuration (II). It was composed of the room, two channels, and the extensions to the ambient space beyond the window of the room and the outlet of Channel 1. The extensions of the domain for simulations of the natural convection flow in a heated cavity was first proposed by Gan [9], and later applied in the works by Gagliano et al. [39], Pasut and De Carli [40], Deblois et al. [41], Tong and Li [42], and Nguyen and Wells [14,15]. These extensions are important as they ease the adaption of the flow to the local dynamic conditions at the openings of the cavity. Accordingly, more accurate solutions can be obtained [9,40].

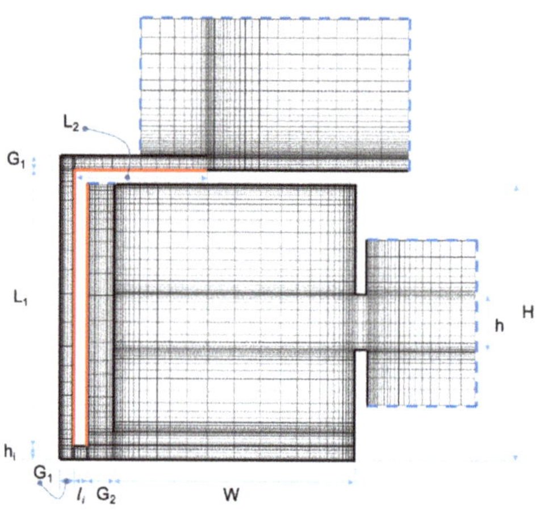

——— Wall with heat source ——— Wall or solid boundary — — — Open boundary

Figure 2. Configuration (II): Computational domain and mesh.

Details of the dimensions in Figure 2 are presented in Table 1. Similar settings of the computational domains and dimensions were also applied to the other configurations.

Table 1. The main dimensions of the room and the solar chimneys.

Dimension	Value (m)	Dimension	Value (m)
H	2.0	G_2	0.05–0.2
W	1.9	h_i	0.1
L_1	2.1	l_i	0.1
L_2	1.0	h	0.4
G_1	0.05–0.2		

Figure 2 also displays the mesh structure. A non-uniform mesh of rectangular cells was used. The mesh size decreases toward the solid walls. It was smallest next to the surfaces of the channels. To find an appropriate mesh pattern in each case, different mesh sizes were tested and the airflow rate through the room was compared. Table 2 shows a test for Configuration I. The number of cells increased from 16,460 to 184,620, and the maximum non-dimensional distance of the first nodes on the heated walls, or $y^+ = \Delta_1 u_\tau / \nu$, decreased from 3.9 to 0.95. The flow rate obtained at each grid resolution was compared to that of the finest one (Test 5 in Table 2). It was found that with $y^+ < 1.5$, the flow rate changed less than 1.0% when the mesh was refined. This observation agrees with the results by Zamora and Kaiser [22]. Therefore, a mesh yielding $y^+ < 1.5$ was selected. The corresponding maximum and minimum cell sizes were below 42 mm and 0.29 mm, respectively.

Table 2. Mesh independence test.

Test	No. of Cells	Max. Mesh Size (mm)	Min. Mesh Size (mm)	y^+	Q (kg/s)	Difference (%)
1	16,460	90	1.1	3.9	0.05121	1.26
2	42,820	57	0.55	2.25	0.05118	1.32
3	120,665	42	0.29	1.3	0.05153	0.64
4	152,250	38	0.23	1.1	0.05169	0.32
5	184,620	35	0.2	0.95	0.05186	0.00

The boundary conditions are also presented in Figure 2. On the open boundaries of the extended regions, the pressure and temperature were applied with those of the ambient atmosphere which are zero-gauge pressure and 20 °C, respectively. The turbulence level at the open boundaries was assumed to be low, with an intensity of 2.0%. The heat source was distributed on the right and the left walls of Channels 1 and 2, respectively. Two walls of each channel also exchanged radiative heat transfer which was calculated with the S2S model bundled in ANSYS Fluent. It was assumed that there was no heat transfer on other solid surfaces.

The Finite Volume Method was employed for the discretization of the governing equations with the following settings:

- SIMPLEC method for the coupling of pressure and velocity;
- PRESTO! method for the pressure interpolation on the mesh faces;
- Second order scheme for all equations.

These settings were also employed for simulations of solar chimneys in previous studies [9,14,15,42,43].

The experiment by Burek and Habeb [7] was used to validate the CFD model. The chimney was a vertical rectangular cavity which was open to the ambient air at the upper and lower ends. Its side walls were enclosed. The height of the chimney was 1.025 m. The width was 0.925 m. The gap changed from 0.02 m to 0.11 m. The heat source was distributed on one wall of the air channel at the flux of 600 W/m^2, which was around the median value in Singapore [23] and Vietnam [44]. The back side of the air channel was well insulated. The air velocity was measured at the center of the air channel near the inlet. The mass flow rate was then computed from the air velocity, the cross-sectional area of the cavity, and the air density.

In the experiment, as the chimney was stand-alone, air freely entered the air cavity, rose due to the thermal effect, and escaped to the ambient atmosphere from the top of the air cavity. To reproduce the experimental conditions, the computational domain also covered both the air cavity and ambient air. The extensions from the cavity walls to the external domains were 10 times of the cavity gap, as proposed by Gan [9]. The boundary conditions of atmospheric temperature and pressure were applied on the external boundary. The heat flux, which was the same as in the experiment, was applied on one side of the air cavity. The rest of the setup was the same as described above.

The computed flow rate was compared to the measured data. The comparison is displayed in Figure 3. It shows that the predicted flow rate agrees well with the measured one. The discrepancies were from 0% at $G = 40$ mm to 9.7% at $G = 20$ mm. Considering possible measurement errors in the experiment, this maximum difference is acceptable. Therefore, the CFD model is considered reliable.

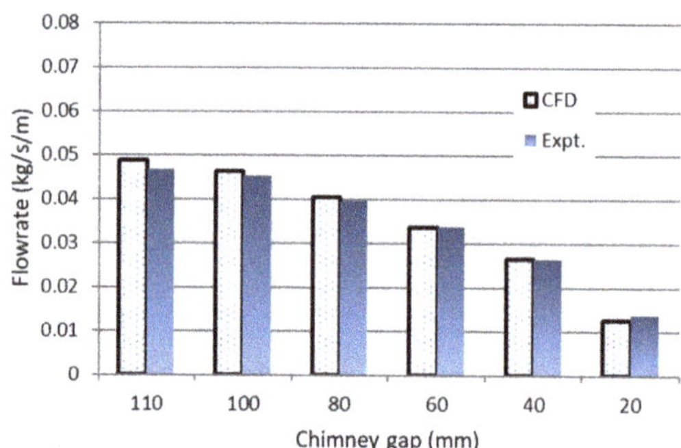

Figure 3. The CFD results were compared to the experimental ones by Burek and Habeb [7].

4. Results and Discussion

The ventilation performance of the proposed configurations was evaluated in two cases:

Firstly, G_2 were fixed to 0.1 m while G_1 changed. Other dimensions are presented in Table 1. The total heat flux in the two channels was $q_t = 600$ W/m^2. The heat flux in Channel 1, q_1, and Channel 2, q_2, changed, but $q_1 + q_2 = q_t$. The flow rate was computed for different ratios of q_2/q_1 to model different heat fluxes conducted through the building wall depending on its thermal conductivity.

Secondly, the effects of changing the heat flux and the gap G_2 of Channel 2 were evaluated.

4.1. Changing G_1

The flow velocity and temperature for $q_1 = 500$ W/m^2 and $q_2 = 100$ W/m^2 are presented in Figure 4 ($G_1 = G_2 = 0.1$ m). It is assumed that a total solar radiation flux of $q_t = 600$ W/m^2 is transmitted through the outer glass and absorbed in the building wall; then a heat flux of $q_1 = 500$ W/m^2 is transferred into Channel 1 and $q_2 = 100$ W/m^2 is conducted through the wall and transferred into Channel 2. Figure 4a shows that for Configuration (I), the airflow moves down in Channel 2, then along Channel 1, and discharges to the ambient air from the outlet of Channel 1. The flow temperature increases along the flow path as it receives more heat. In Figure 4b for Configuration (II), the air from the room enters both channels. As more heat is supplied to Channel 1 ($q_1 > q_2$), the flow speed in Channel 1 is higher. In addition, the temperature at the outlet of Channel 1 is also higher. In Figure 4c for Configuration (III), the airflows in both channels merge at the junction. As a result, the flow speed from the junction to the outlet of Channel 1 is higher than those in other parts of the channels. The lower-temperature flow in Channel 2 penetrating Channel 1 is seen to obstruct the flow in Channel 1.

As the induced flow rate was calculated at the window of the room, it is the total flow rate, Q, through both channels. In Configuration (I), it is also the flow rate through Channels 1 and 2, i.e., $Q = Q_1 = Q_2$. In Configurations (II) and (III), it is the sum of Q_1 and Q_2.

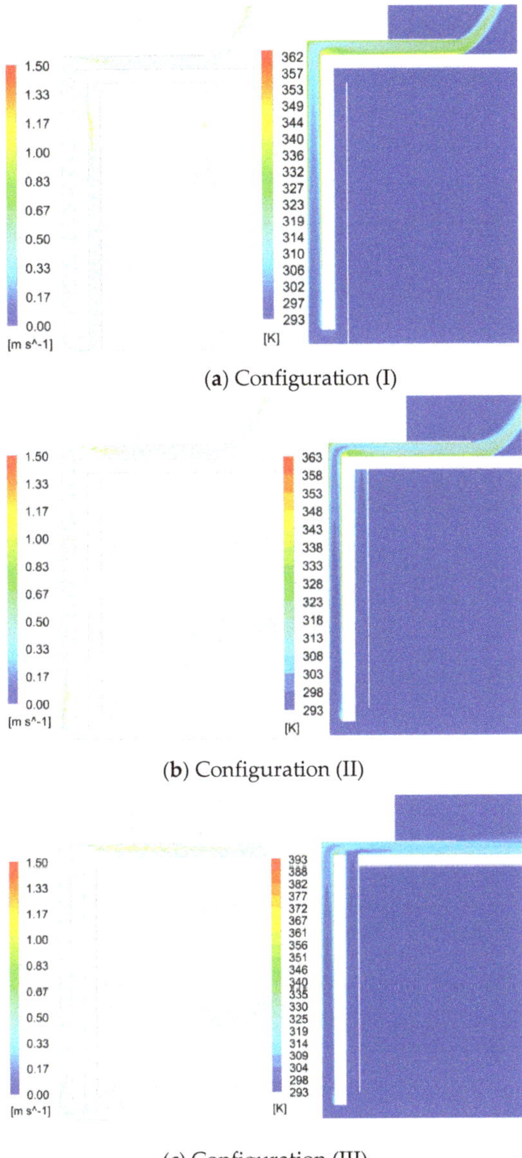

(**a**) Configuration (I)

(**b**) Configuration (II)

(**c**) Configuration (III)

Figure 4. Distributions of the flow velocity and temperature ($q_1 = 500$ W/m^2, $q_2 = 100$ W/m^2, $G_1 = G_2 = 0.1$ m).

The induced flow rates, Q, of three configurations are plotted in Figures 5–7. In Figure 5 for Configuration (I), the flow rate is higher as the gap increases. Significant improvement of the flow rate of up to 51% is obtained when the gap changes from 0.05 m to 0.1 m. However, increasing the gap from 0.1 m to 0.2 m only increases the flow rate by 8%. As q_2 increases, the flow rate does not change significantly, and the change is only within 6%. This is because, although q_2 increases, the total heat flux, q_t, that the flow receives in Configuration (I) is constant. Accordingly, a minor change in the flow rate with q_2 can be expected.

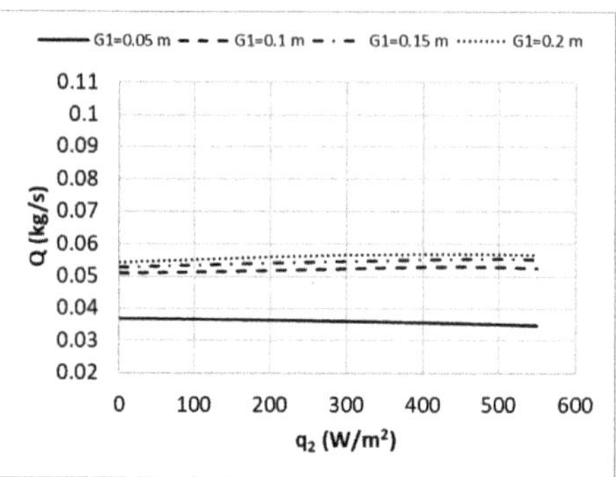

Figure 5. Mass flow rate of Configuration (I) ($G_2 = 0.1$ m) at different values of q_2 and G_1.

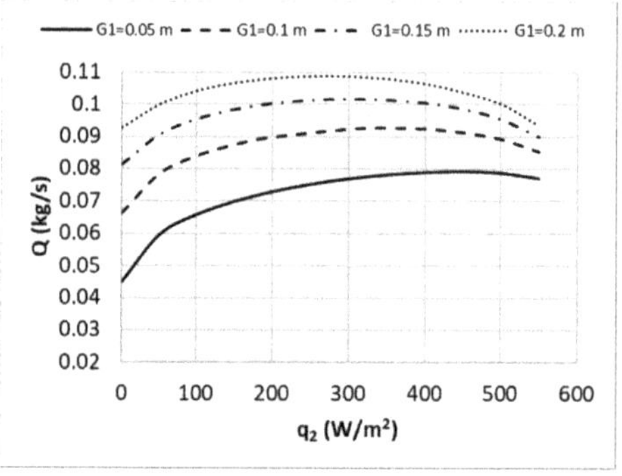

(**a**) Mass flow rate

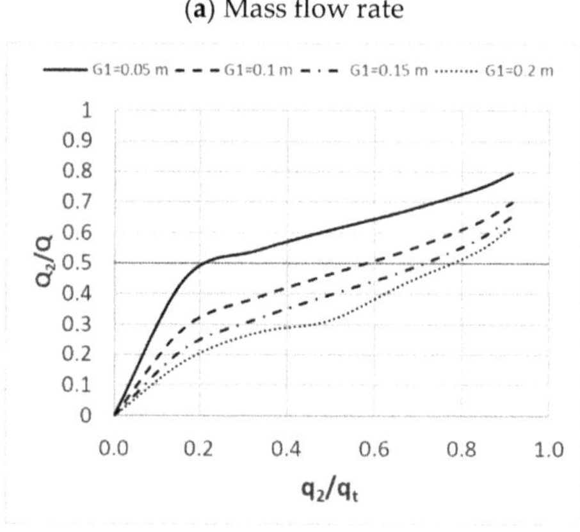

(**b**) Normalized mass flow rate

Figure 6. Mass flow rate of Configuration (II) ($G_2 = 0.1$ m) at different values of q_2 and G_1.

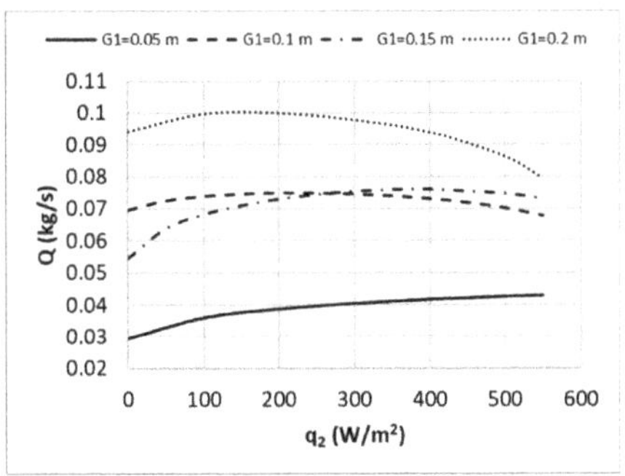

(**a**) Mass flow rate

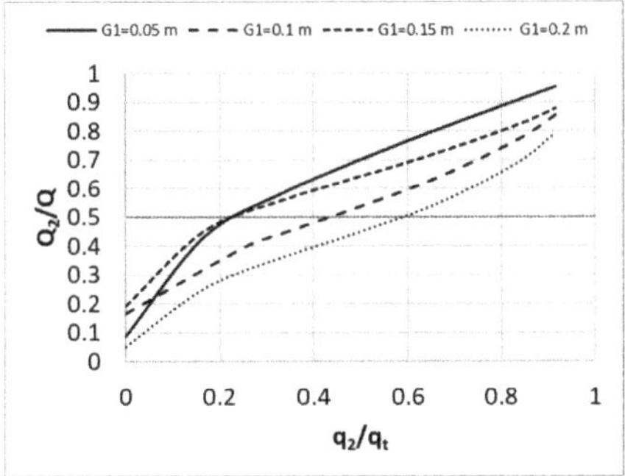

(**b**) Normalized mass flow rate

Figure 7. Mass flow rate of Configuration (III) ($G_2 = 0.1$ m) at different values of q_2 and G_1.

Figure 6a shows the airflow rate of Configuration (II). The flow rate is higher for the larger gap G_1. As q_2 increases, the flow rates first increase then decrease. The maximum flow rate happens when q_2 is from 300 to 500 W/m². In Figure 6b, the airflow rate in Channel 2 is compared to the total one and plotted as the functions of the ratio of q_2/q_t. At a given q_2/q_t, the ratio of Q_2/Q is lower for the higher gaps. It may be that because as G_1 is small, the flow in Channel 1 experiences more flow resistance, resulting in a lower flow rate in Channel 1. Q_2/Q also increases with q_2, as more heat is supplied to Channel 2.

The flow rate of Configuration (III) in Figure 7a is enhanced significantly, up to 3.2 times, when G_1 increases from 0.05 m to 0.2 m. However, for the gaps of $G_1 = 0.1$ m and $G_2 = 0.15$ m, the flow rate of $G_1 = 0.1$ m is higher for $q_2 < 250$ W/m²; otherwise, the flow rate of $G_1 = 0.15$ m is higher. Figure 7b shows that at all values of q_2/q_t, increasing the gap G_1 from 0.05 m to 0.2 m demonstrates a clear decrease of Q_2/Q. However, increasing G_1 from 0.1 m to 0.15 m results in an increase of Q_2/Q.

Previous studies have shown a consistent increase in the airflow rate versus the air gap [7,8,10–12,16,18]. Therefore, it is expected that when the gap G_1 increases, the flow rate in Channel 1, Q_1, gradually dominates over Q_2. This point is obvious for Configuration (II) in Figure 6b, and Configuration (III) with $G_1 = 0.05$ m and 0.2 m in Figure 7b. To explore why the flow rate in Configuration (III) does not always increase as G_1 increases from 0.1 m to 0.15 m in Figure 7b, the streamlines for $q_2 = 100$ W/m² of Configuration (III) are displayed in Figure 8. It shows that when the flow in Channel 2 merges with the one in Channel 1 in the marked region in Figure 8, the effective flow areas of both flows decrease significantly. As a result, a higher flow resistance is expected at the junction and downstream of it. In addition, the presence of the recirculation area in the marked region in Figure 8 may also cause the complex behavior of the airflow rate in the cases of $G_1 = 0.1$ m and 0.15 m in Figure 7. Similar effects were also reported by Zamora and Kaiser [22] and Kim et al. [20].

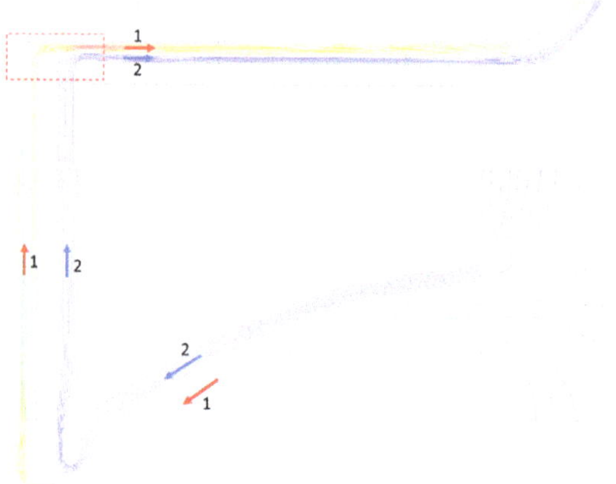

Figure 8. Streamlines of Configuration (III) in Figure 6 for $q_2 = 100 \ \text{W/m}^2$.

Comparing the mass flow rates in Figures 5, 6a and 7a shows that the total flow rate is always the highest with Configuration (II) and the lowet with Configuration (I). With $G_1 = 0.2$ m, the maximum flow rate of Configurations (I), (II), and (III) are 0.057, 0.109, and 0.1 kg/s, respectively.

4.2. Changing the Total Heat Flux

Figure 9 shows the induced flow rate as the heat flux changes for $G_1 = G_2 = 0.1$ m. The flow rate, Q, is normalized by that with $q_2 = 0$, Q_0. The ratios of Q/Q_0 for different heat fluxes of each configuration match well. The maximum scatter is for Configuration (II), and only about 4%. Consequently, the aero-thermal behaviors of the flows are seen to be identical for different values of q_t.

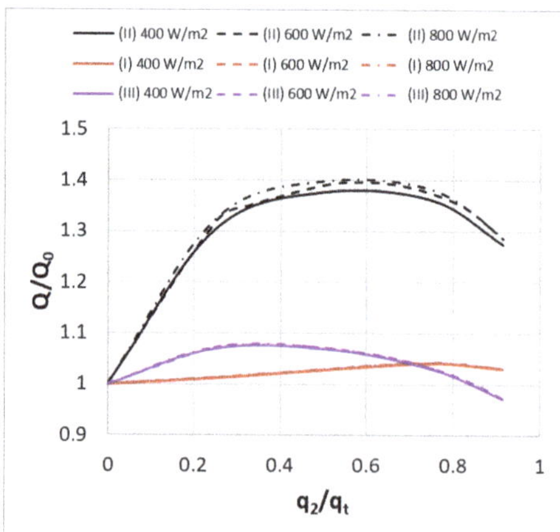

Figure 9. Changing the heat flux for three configurations with $G_1 = G_2 = 0.1$ m.

Figure 9 also shows that, for all configurations, increasing q_2 results in higher, or identical, flow rates compared with those for $q_2 = 0$. The least enhancement of 4% is with Configuration (I) and a slight decrease of 2.0% is seen for Configuration (III) at $q_2/q_t = 0.9$.

The maximum increase of 40% is seen with Configuration (II). The ratios of q_2/q_t where the Q/Q_0 peaks are 0.75, 0.55, and 0.35 for Configurations (I), (II), and (III), respectively.

4.3. Changing G_2

As seen in Figures 5–7 and 9, Configuration (II) offers the highest flow rate among the three configurations. Therefore, it is selected to examine the effects of changing Channel 2's gap, G_2. Figure 10 presents the flow rate of Configuration (II) as G_2 and q_2 change. As expected, Q increases with G_2 (Figure 10a). The flow rate also peaks at a specific $q_{2,c}$ for each G_2. On the other hand, $q_{2,c}$ also increases with G_2. They are 200, 300, 400, and 500 W/m^2 for $G_2 = 0.05, 0.1, 0.15$, and 0.2 m, respectively. Figure 10b reveals that the ratio of Q_2/Q becomes higher as G_2 increases. However, the increasing rate of Q_2/Q versus G_2 decreases with G_2. For example, at $q_2/q_t = 0.5$, increasing G_2 from 0.05 m to 0.1 m boosts Q_2/Q up to 27%, but increasing G_2 from 0.15 m to 0.2 m enhances Q_2/Q by only 6%. In conclusion, raising G_2 increases not only the total flow rate but also $q_{2,c}$ and the flow rate, Q_2, in Channel 2.

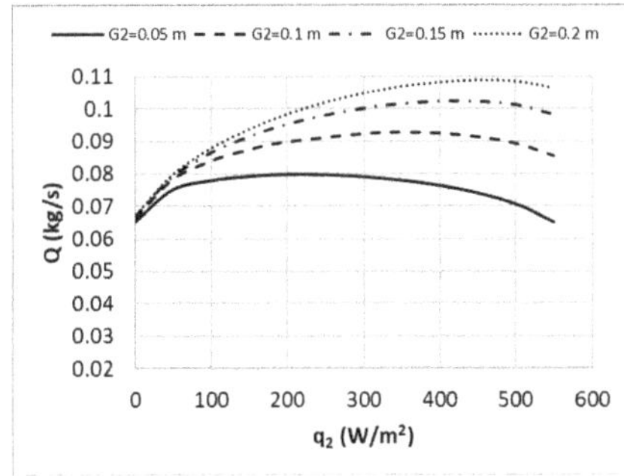

(a) Mass flow rate.

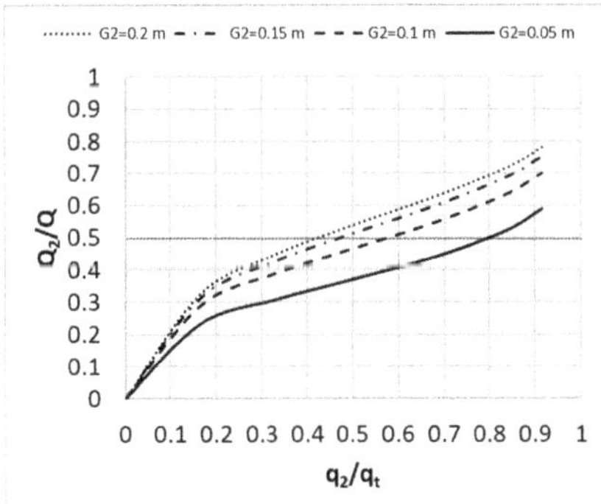

(b) Normalized mass flow rate.

Figure 10. The flow rate at different gaps G_2 for Configuration (II).

4.4. Discussion

The fact that Configuration (II) offers the highest flow rate among three proposed configurations, as seen in Figures 5–7 and 9, agrees well with the experiments by Macias-Melo et al. [27] and Vaquez-Ruiz et al. [28]. They reported that for a room with a heated wall, the highest airflow rate is obtained when the opening or the roof vertical solar chimney is right on top of the heated wall. However, as there was not a barrier wall inside the room in their models, heat is transferred into the whole room in their experiments.

Hernadez-Lopez et al. [29] investigated a wall solar chimney for a room 2.55 m high. Their configuration is also without an internal barrier wall. They showed an airflow rate of 0.0144 kg/s/m for a gap of 0.1 m, and a heat flux of 363 W/m². Configuration (II) in this study at 400 W/m² and a gap of 0.1 m for both channels offered the maximum flow rate of 0.0805 kg/s/m when $q_1 = q_2 = 200$ W/m². As the flow rate in this study is about 5.6 times higher than that in Hernadez-Lopez et al. [29], it clearly shows the advantage of using a heat barrier wall inside a room, i.e., the "inner wall" in Figure 1.

A comparison of Figures 6 and 9 shows that changing either G_1 or G_2 while the other is fixed, results in identical maximum flow rates. The difference between the two cases is the value of $q_{2,c}$ where the flow rate is peak. The ratio of $q_{2,c}/q_t$ decreases with G_1 while it increases with G_2. Therefore, in practical designs, it is optional to increase G_1 or G_2 to achieve the desired ventilation rate. Channel 2 takes space inside the room but should have less construction cost as it is structurally simpler than Channel 1.

5. Conclusions

Three configurations consisting of two solar chimneys for a room's natural ventilation with a heated wall were proposed and examined numerically. The flow rate in each channel, and the total flow rate through both channels, were evaluated. In most cases, the flow rate increases with the channel gaps. When the heat flux, q_2, in Channel 2 increases, the airflow rate of Configuration (I) changes insignificantly, but those of the two other configurations increase, peak, then decrease. The ratio of Q_2/Q increases with q_2/q_t but decreases with G_1 when G_2 is fixed. When G_1 is fixed and G_2 changes, Q_2/Q increases with both q_2/q_t and G_2.

Among the three proposed configurations, Configuration (II) offers the highest flow rate, which is 91% and 9% higher than those of Configurations (I) and (III), respectively. As q_2 increases, the maximum increase in the flow rate of Configuration (II) (40%) is also higher than those of Configurations (I) (4%) and (III) (8%). Therefore, the best ventilation performance is seen with Configuration (II). In addition, with Configuration (II), changing either G_1 or G_2 results in similar maximum flow rates.

Based on these findings, design engineers may consider various solutions for natural ventilation and solar load-reducing methods of rooms or houses with solar-heated walls, as follows:

For exhausting stalled air in the upper zone of the room, Configuration (I) is preferred, but the ventilation rate decreases to about 52% of that of Configuration (II). Moreover, as the performance of Configuration (I) does not depend on q_2, it eases the selection of the material of the building wall.

For maximizing the ventilation rate, Configuration (II) is the best choice. However, as Configuration (III) performs closely to Configuration (II) and the handling of the outlet is simpler, Configuration (III) is preferred.

In future works, experiments with the proposed configurations can be conducted. Such experiments can offer real heat flux ratios, q_2/q_1, in each channel obtained with real wall materials.

Author Contributions: Data curation, V.T.N.; Methodology, L.T.T.; Writing—original draft, Y.Q.N.; Writing—review & editing, J.C.W. All authors have read and agreed to the published version of the manuscript.

Funding: Van Lang University, Vietnam.

Institutional Review Board Statement: Not applicable.

Informed Consent Statement: Not applicable.

Data Availability Statement: The data are not publicly available due to institutional property rights.

Acknowledgments: This study was financially supported by Van Lang University, Vietnam.

Conflicts of Interest: The authors declare no conflict of interest.

Nomenclature

c_p	specific heat capacity of air (J/kg·K)
G_1	gap of Channel 1 (m)
G_2	gap of Channel 2 (m)
H	room height (m)
h	window height (m).
h_i	height of inlet of Channel 1 (m)
L	cavity height (m)
L_1	height of vertical section of Channel 1 (m)
L_2	length of horizontal section of Channel 1 (m)
l_i	length of inlet of Channel 1 (m)
P	pressure (Pa)
Q	total mass flow rate per meter width (kg/s)
Q_1	mass flow rate per meter width in Channel 1 (kg/s)
Q_2	mass flow rate per meter width in Channel 2 (kg/s)
q_1	heat flux in Channel 1 (W/m^2)
q_2	heat flux in Channel 2 (W/m^2)
q_t	total heat flux (W/m^2)
T, T', T_r	temperature (mean, fluctuation, and ambient) (K)
u, u'	velocity (mean and fluctuation) (m/s)
u_τ	friction velocity (m/s)
W	room width (m)
y^+	Non-dimensional distance from a wall
β	thermal expansion of air (1/K)
Δ_1	distance of the first node from the wall (m)
λ	thermal conductivity of air (W/mK)
μ:	dynamic viscosity of air (Pa.s)
ν	kinematic viscosity of air (m^2/s)
ρ	density of air (kg/m^3)

References

1. Thollander, P.; Karlsson, M.; Rohdin, P.; Wollin, J.; Rosenqvist, J. *Introduction to Industrial Energy Efficiency*; Academic Press: New York, NY, USA, 2020.
2. Miyazaki, T.; Akisawa, A.; Kashiwagi, T. The effects of solar chimneys on thermal load mitigation of office buildings under the Japanese climate. *Renew. Energy* **2006**, *31*, 987–1010. [CrossRef]
3. Al Touma, A.; Ouahrani, D. Performance assessment of evaporatively-cooled window driven by solar chimney in hot and humid climates. *Sol. Energy* **2018**, *169*, 187–195. [CrossRef]
4. Hong, S.; He, G.; Ge, W.; Wu, Q.; Lv, D.; Li, Z. Annual energy performance simulation of solar chimney in a cold winter and hot summer climate. *Build. Simul.* **2019**, *12*, 847–856. [CrossRef]
5. Ma, Q.; Fukuda, H.; Wei, X.; Hariyadi, A. Optimizing energy performance of a ventilated composite Trombe wall in an office building. *Renew. Energy* **2019**, *134*, 1285–1294. [CrossRef]
6. Shi, L.; Zhang, G.; Yang, W.; Huang, D.; Cheng, X.; Setunge, S. Determining the influencing factors on the performance of solar chimney in buildings. *Renew. Sustain. Energy* **2018**, *88*, 223–238. [CrossRef]
7. Burek, S.A.M.; Habeb, A. Air flow and thermal efficiency characteristics in solar chimneys and Trombe Walls. *Energy Build.* **2007**, *39*, 128–135. [CrossRef]
8. Chen, Z.D.; Bandopadhayay, P.; Halldorsson, J.; Byrjalsen, C.; Heiselberg, P.; Li, Y. An experimental investigation of a solar chimney model with uniform wall heat flux. *Build. Environ.* **2003**, *38*, 893–906. [CrossRef]
9. Gan, G. Impact of computational domain on the prediction of buoyancy-driven ventilation cooling. *Build. Environ.* **2010**, *45*, 1173–1183. [CrossRef]
10. Hou, Y.; Li, H.; Li, A. Experimental and theoretical study of solar chimneys in buildings with uniform wall heat flux. *Sol. Energy* **2019**, *193*, 244–252. [CrossRef]
11. Jing, H.; Chen, Z.; Li, A. Experimental study of the prediction of the ventilation flow rate through solar chimney with large gap-to-height ratios. *Build. Environ.* **2015**, *89*, 150–159. [CrossRef]
12. Liu, B.; Ma, X.; Wang, X.; Dang, C.; Wang, Q.; Bennacer, R. Experimental study of the chimney effect in a solar hybrid double wall. *Sol. Energy* **2015**, *115*, 1–9. [CrossRef]
13. Mathur, J.; Mathur, S. Summer-performance of inclined roof solar chimney for natural ventilation. *Energy Build.* **2006**, *38*, 1156–1163. [CrossRef]

14. Nguyen, Y.Q.; Wells, J.C. Effects of wall proximity on the airflow in a vertical solar chimney for natural ventilation of dwellings. *J. Build. Phys.* **2020**, *44*, 225–250. [CrossRef]

15. Nguyen, Y.Q.; Wells, J.C. A numerical study on induced flowrate and thermal efficiency of a solar chimney with horizontal absorber surface for ventilation of buildings. *J. Build. Eng.* **2020**, *28*, 101050. [CrossRef]

16. Villar-Ramos, M.M.; Macias-Melo, E.V.; Aguilar-Castro, K.M.; Hernández-Pérez, I.; Arce, J.; Serrano-Arellano, J.; Díaz-Hernández, H.P.; López-Manrique, L.M. Parametric analysis of the thermal behavior of a single-channel solar chimney. *Sol. Energy* **2020**, *209*, 602–617. [CrossRef]

17. Zavala-Guillén, I.; Xamán, J.; Hernández-Pérez, I.; Hernández-Lopéz, I.; Gijón-Rivera, M.; Chávez, Y. Numerical study of the optimum width of 2a diurnal double air-channel solar chimney. *Energy* **2018**, *147*, 403–417. [CrossRef]

18. Zhang, T.; Yang, H. Flow and heat transfer characteristics of natural convection in vertical air channels of double-skin solar façades. *Appl. Energy* **2019**, *242*, 107–120. [CrossRef]

19. Khanal, R.; Lei, C. Flow reversal effects on buoyancy induced air flow in a solar chimney. *Sol. Energy* **2012**, *86*, 2783–2794. [CrossRef]

20. Kim, K.M.; Nguyen, D.H.; Shim, G.H.; Jerng, D.-W.; Ahn, H.S. Experimental study of turbulent air natural convection in open-ended vertical parallel plates under asymmetric heating conditions. *Int. J. Heat Mass Transf.* **2020**, *159*, 120135. [CrossRef]

21. Ren, X.-H.; Liu, R.-Z.; Wang, Y.-H.; Wang, L.; Zhao, F.-Y. Thermal driven natural convective flows inside the solar chimney flush-mounted with discrete heating sources: Reversal and cooperative flow dynamics. *Renew. Energy* **2019**, *138*, 354–367. [CrossRef]

22. Zamora, B.; Kaiser, A.S. Thermal and dynamic optimization of the convective flow in Trombe Wall shaped channels by numerical investigation. *Heat Mass Transf.* **2009**, *45*, 1393–1407. [CrossRef]

23. Tan, A.Y.K.; Wong, N.H. Natural ventilation performance of classroom with solar chimney system. *Energy Build.* **2012**, *53*, 19–27. [CrossRef]

24. Al-Kodmany, K. Sustainability and the 21st Century Vertical City: A Review of Design Approaches of Tall Buildings. *Buildings* **2018**, *8*, 102. [CrossRef]

25. Petresevics, F.; Nagy, B. FEM-Based Evaluation of the Point Thermal Transmittance of Various Types of Ventilated Façade Cladding Fastening Systems. *Buildings* **2022**, *12*, 1153. [CrossRef]

26. Rahiminejad, M.; Khovalyg, D. Thermal resistance of ventilated air-spaces behind external claddings; definitions and challenges (ASHRAE 1759-RP). *Sci. Technol. Built Environ.* **2021**, *27*, 788–805. [CrossRef]

27. Macias-Melo, E.; Aguilar-Castro, K.; Xamán, J.; Hernández-Pérez, I. Experimental study of convective heat transfer in a ventilated rectangular cavity. *J. Build. Phys.* **2018**, *42*, 388–415. [CrossRef]

28. Vazquez-Ruiz, A.; Navarro, J.M.A.; Hinojosa, J.F.; Xamán, J.P. Effect of the solar roof chimney position on heat transfer in a room. *Int. J. Mech. Sci.* **2021**, *209*, 106700. [CrossRef]

29. Hernández-López, I.; Xamán, J.; Zavala-Guillén, I.; Hernández-Pérez, I.; Moreno-Bernal, P.; Chávez, Y. Thermal performance of a solar façade system for building ventilation in the southeast of Mexico. *Renew. Energy* **2020**, *145*, 294–307. [CrossRef]

30. Wang, Q.; Zhang, G.; Wu, Q.; Li, W.; Shi, L. A combined wall and roof solar chimney in one building. *Energy* **2022**, *240*, 122480. [CrossRef]

31. Zhai, Z.J.; Chen, Q.Y. Performance of coupled building energy and CFD simulations. *Energy Build.* **2005**, *37*, 333–344. [CrossRef]

32. Abdeen, A.; Serageldin, A.A.; Ibrahim, M.G.E.; El-Zafarany, A.; Ookawara, S.; Murata, R. Experimental, analytical, and numerical investigation into the feasibility of integrating a passive Trombe wall into a single room. *Appl. Therm. Eng.* **2019**, *154*, 751–768. [CrossRef]

33. Altaç, Z.; Uğurlubilek, N. Assessment of turbulence models in natural convection from two- and three-dimensional rectangular enclosures. *Int. J. Therm. Sci.* **2016**, *107*, 237–246. [CrossRef]

34. Lechowska, A.; Szczepanik-Ścisło, N.; Schnotale, J.; Stelmach, M.; Pyszczek, T. CFD modelling of transient thermal performance of solar chimney used for passive ventilation in a building. *IOP Conf. Ser. Mater. Sci. Eng.* **2018**, *415*, 012049. [CrossRef]

35. Nielsen, P.V. Fifty years of CFD for room air distribution. *Build. Environ.* **2015**, *91*, 78–90. [CrossRef]

36. Szczepanik, N.; Schnotale, J. CFD simulations and measurements of carbon dioxide transport in a passive house. In Proceedings of the 24th IIR International Congress of Refrigeration, Yokohama, Japan, 16–22 August 2015.

37. Hinojosa, J.F.; Orozco, D.A.; Xaman, J. Experimental and Numerical Study of a Ventilated Room with Located Heat Sources. *J. Energy Eng.* **2020**, *146*, 04020024. [CrossRef]

38. Zamora, B.; Kaiser, A.S. Optimum wall-to-wall spacing in solar chimney shaped channels in natural convection by numerical investigation. *Appl. Therm. Eng.* **2009**, *29*, 762–769. [CrossRef]

39. Gagliano, A.; Patania, F.; Nocera, F.; Ferlito, A.; Galesi, A. Thermal performance of ventilated roofs during summer period. *Energy Build.* **2012**, *49*, 611–618. [CrossRef]

40. Pasut, W.; De Carli, M. Evaluation of various CFD modelling strategies in predicting airflow and temperature in a naturally ventilated double skin façade. *Appl. Therm. Eng.* **2012**, *37*, 267–274. [CrossRef]

41. DeBlois, J.; Bilec, M.; Schaefer, L. Simulating home cooling load reductions for a novel opaque roof solar chimney configuration. *Appl. Energy* **2013**, *112*, 142–151. [CrossRef]

42. Tong, S.; Li, H. An efficient model development and experimental study for the heat transfer in naturally ventilated inclined roofs. *Build. Environ.* **2014**, *81*, 296–308. [CrossRef]

43. Al-Kayiem, H.H.; Sreejaya, K.V.; Chikere, A.O. Experimental and numerical analysis of the influence of inlet configuration on the performance of a roof top solar chimney. *Energy Build.* **2018**, *159*, 89–98. [CrossRef]
44. Nguyen, Y.Q. An experiment on an energy-saving air cooling system combining a solar chimney and an indirect evaporative cooling method in real weather conditions of Hochiminh city. *J. Sci. Technol. Univ. Danang-Vietnam* **2015**, *9*, 69–73.

 *buildings*

MDPI

Article

FEM-Based Evaluation of the Point Thermal Transmittance of Various Types of Ventilated Façade Cladding Fastening Systems

Fanni Petresevics and Balázs Nagy *

Department of Construction Materials and Technologies, Faculty of Civil Engineering, Budapest University of Technology and Economics, Műegyetem rkp. 3, 1111 Budapest, Hungary; petresevics.fanni@edu.bme.hu
* Correspondence: nagy.balazs@emk.bme.hu; Tel.: +36-1-463-1175

Abstract: The prevalence of ventilated façade systems is not only due to their aesthetic properties but also due to the fact they provide mechanical and acoustic protection for the façade and reduce the energy demand of the building. However, it is essential to mention that the point thermal bridges of the fastening system with brackets and anchors are often neglected during simplified energy performance calculations and practical design tasks. The reason practitioners do not consider the brackets in the calculation is the lack of standards for the simplified calculation of point thermal transmittances, or there being no comprehensive, manufacturer-independent thermal bridge catalogue available. This study aims to evaluate the point thermal transmittances created by the brackets and anchors of the ventilated façade claddings by using 3D numerical thermal modelling. A broad point thermal bridge catalogue was created, considering multiple factors of the ventilated facades. The FEM-based results show that thermal breaks/isolators could reduce the point thermal transmittances by only 2 to 28%, depending on the material of the brackets and the isolators. The brackets' material and geometrical properties/parameters could cause up to 70% of difference between corrected and uncorrected thermal transmittance values, as well as significant differences between the results if the brackets were applied to different kinds of masonry walls or reinforced concrete walls.

Keywords: building physics; point thermal bridges; ventilated facade claddings; brackets and anchors; thermal break; three-dimensional numerical thermal modelling

Citation: Petresevics, F.; Nagy, B. FEM-Based Evaluation of the Point Thermal Transmittance of Various Types of Ventilated Façade Cladding Fastening Systems. *Buildings* **2022**, *12*, 1153. https://doi.org/10.3390/buildings12081153

Academic Editors: Gianpiero Evola and Elena Lucchi

Received: 10 July 2022
Accepted: 29 July 2022
Published: 2 August 2022

Publisher's Note: MDPI stays neutral with regard to jurisdictional claims in published maps and institutional affiliations.

1. Introduction

The ventilated façade system is a popular cladding system to decrease energy consumption in most recently built office buildings or buildings under renovation. Through its construction, it protects and keeps the walls and the thermal insulation of the building dry, as well as reducing the heat transfer of the walls, thus ensuring a longer life for the installation. Since the cladding is anchored to the wall with brackets, cracks in the cladding caused by building movement can be avoided. Due to the "mass-spring-mass" principle, it has advantageous sound protection properties and is made with dry technology, so it can be constructed all year round and requires little maintenance [1,2]. The most crucial measure of the building envelope in energy performance calculations is its thermal transmittance, which must be corrected due to various inhomogeneities and thermal bridges. In the case of ventilated façade claddings, the point thermal bridges which need to be considered within the thermal transmittance of the building envelope are caused by the elements of the fastening system, namely the brackets that punctuate the thermal insulation and the anchors and dowels holding the brackets.

As mentioned above, one of the first efforts to consider the heat losses caused by the fixings was published in 1984 [3]. Later, in 2006, BREE published a guideline [4] for calculating U-values, including the effect of ventilated façade claddings as a non-derivative recommendation, in case it is not possible to determine the thermal transmittance by numerical modelling. The standards nowadays do not provide any current guidance to account for the thermal effects of fasteners except for finite element simulations or empirical modelling, which is why several professionals have tried to develop thermal bridge catalogues [5] and simplified calculation methods [6].

Analysing the available scientific literature on the topic, most of the studied literature agrees that the neglect of point thermal bridges caused by fasteners can result in a significant difference of up to 5 to 30% in the calculation of heat losses in the studied building. A good example is a study by García et al. [7], which examined the results calculated based on empirical, numerical and experimental methods in a real environment, in different façade configurations. The research confirmed that the results obtained with the simplified method differ significantly from the experimental hotbox measurement results, especially when using high thermal conductivity brackets. A study by Levinskytė et al. [8] concluded that the use of the simplified method could be misleading when looking at the results since while in the case of brackets with low thermal conductivity, we get the results calculated from the numerical simulation with a difference of only 3.6%, in the case of brackets with higher thermal conductivity this difference is of 70 to 130.4%. They also examined the difference between the results of empirical and numerical calculation methods. They said that in all cases, the empirical calculation method according to ISO 6946 [9] showed much higher thermal transmittance values than the method according to ISO 10211 [10] using 3D simulation software.

Most of the previously published research has examined the effect of stainless steel, steel and aluminium brackets [8,11] on point thermal transmittances, but in [12], they also dealt with perforated brackets. Glass fibre-reinforced brackets were also examined [8]. A catalogue for carbon steel and "thermo" brackets made of polymer composite material was also created [13]. The studies mentioned above showed that choosing a bracket with the appropriate thermal conductivity, such as stainless steel or steel composite, can significantly reduce the U-values calculated for the structure by up to 40%. In the case of aluminium brackets [14], the tests have shown that an increase in thermal conductivity of the material of the supporting layer and the thickness of the thermal insulation layer may increase the U-value of the entire wall up to 35% as a result of the effect of point thermal bridge. Hilti [15], one of the largest European manufacturers of brackets, also used numerical models to investigate how thermal insulation properties and brackets affected the value of point thermal transmittance and created a thermal bridge catalogue for their own fastening systems. Another main aspect of the investigation in most of the research is the effect of thermal insulation properties and the wall on point thermal transmittance. They mainly examined local materials typical of the country of the study. Thus, the primary materials of the walls are concrete, reinforced concrete and masonry wall, but also cellular concrete [12], red clay brick [11] and silicate block [16] were examined. Theodosiou et al. [17–19] examined the effects of different materials and geometric properties on the point thermal transmittance of brackets. They concluded that neither the thermal insulation nor the thermal breaks could effectively reduce the effects of the point thermal bridges generated by the brackets. In most scientific literature, only the brackets were examined, and their anchoring and dowels were neglected. However, in [18], steel, plastic and chemical anchoring were examined, respectively, but only steel and chemical anchors were analysed in comparison. In [16], 3D numerical simulations were used to create temperature distribution diagrams for walls and thermal insulation with different materials and thermal conductivities. They pointed out the problems of thermal bridges caused by brackets and dowels. To reduce the point thermal bridges caused by the fastenings, Ingeli et al. [20] patented the use of plastic-coated anchors, which can effectively reduce the effect of point thermal bridges, and thus the heat loss of the buildings. However, most studies only deal with the extent to which parameters influence the value of the thermal transmittance but only came to the evaluation of the data, and no simplified calculation method was developed. Šadauskienė et al. [21] have tried to develop a simplified calculation method. However, the system of equations determining the point thermal transmittance they create applies only to the modelled ranges and contains non-independent parameters.

The current research aims to create a comprehensive and manufacturer-independent thermal bridge catalogue that can be used internationally to help practitioners and academics calculate ventilated façades. Firstly, the possible values of the different parameters needed for modelling were collected. Then, a total of 60 different parametric geometric models were created to be able to handle all significant cases. The point thermal transmittances of the ventilated façades on different walls using different brackets, thermal breaks, anchors and dowels were modelled and evaluated. The values were summarised in a thermal bridge catalogue, and a simplified method was developed based on the current ISO standards.

2. Materials and Methods

2.1. Construction of the Ventilated Façade System

The general structure of ventilated façade cladding systems is shown in Figure 1. The material of the cladding systems changes on an extensive scale; there are glazed ceramics, metal sheets, fibre cement, stone slabs, composites and plastic boards. The outer crust/layer operates on a "parasol-umbrella" principle (see Figure 1). It protects the wall and the thermal insulation against solar radiation like a parasol. It protects against precipitation like an umbrella, hence the common term "rainscreen cladding" [22]. The outer layer also includes protection against external mechanical effects and meteorological loads, such as wind. The next layer is the open ventilation air gap, in which the air flows from the bottom upwards due to the "chimney effect", so that in winter, the flowing air transports the diffused moisture from the interiors [23]. At the same time, in summer, it acts as a heat shield, thus improving the thermal insulation of the façade. To ensure this effect and the ventilation, the recommended thickness is between 3 and 5 cm considering the moisture removal capacity of the ventilated air channel [24], but this also depends on, among other things, ambient conditions, type of materials used, the height of the building and the width of the façade wall [25].

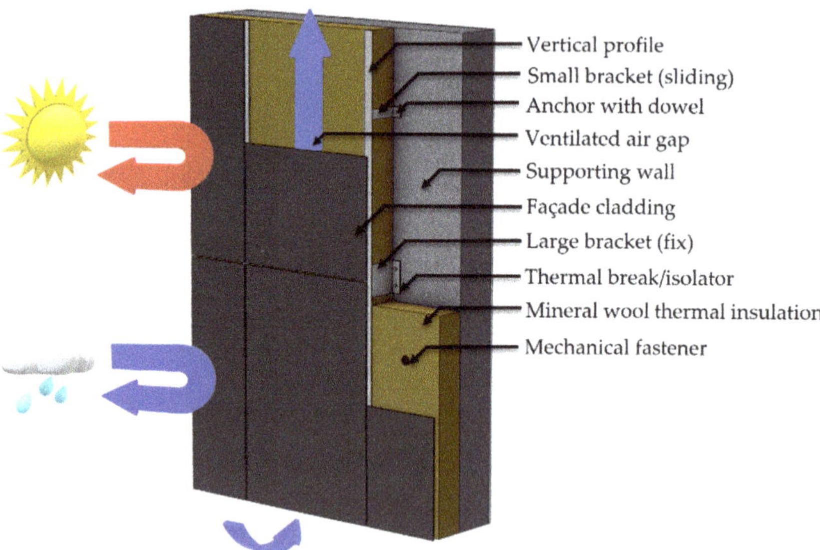

Figure 1. Structure of a ventilated façade cladding acting as a parasol-umbrella.

Choosing the appropriate air layer thickness is also important and cannot be neglected, as here, we can compensate for the differences in dimensional tolerances between the retaining wall and cladding boards. It is important to note that due to the chimney effect, great attention must be paid to fire protection, as fire can spread much more quickly in these ventilated air gaps due to the air flowing upwards [26]. Hence, it is crucial to choose non-combustible thermal insulation. Behind the intensively ventilated air layer is the thermal insulation; in most cases, mineral wool. It is recommended to select a higher density version so that the air in the thermal insulation remains calm and does not start flowing, thus impairing the thermal insulation effect [27].

2.2. Parameters and Geometry

The first step to being able to model the ventilated façade claddings' fastening systems was to collect their elements and determine their geometrical and thermal properties. It is important to note that during the modelling, the effect of the exterior cladding and the ventilated air layer was neglected. This research investigates the fastening system's thermal impact penetrating the thermal insulation. Researchers have dealt with air gaps and the experimental and numerical modelling of the air flowing in the air gap before [28–31]; however, the thermal effects of the support brackets are typically neglected in these studies. In this current study, we focus on the effects of the fastening systems. ISO 6946 [9] states that if an air layer is intensively ventilated, it can be excluded from the thermal calculations with all the other external layers. Many previously mentioned studies excluded the cladding and the ventilated air layer from their study. The above-mentioned standard also states that the heat transfer coefficient should be applied on the surface of the thermal insulation, and it shall not be corrected due to the brackets sticking out since all the brackets have higher than 2.5 W/(m $\times$ K) thermal conductivity.

Therefore, during the modelling, it is sufficient to consider elements in direct contact with fasteners such as the wall, plaster/mortar, thermal insulation, brackets and dowels. Since more than 20 independent parameters were required to build the geometric model, to reduce the number of combinations, the lower and upper limits and, in some cases, an intermediate value, were specified as input parameters. Even after collecting and reducing the parameters, nearly 100,000 possible combinations would have been impossible to handle; thus, the geometry had to be split into parts. A distinction was made between the material of the wall and the other separations along the presence of the thermal breaks/isolators, the wall and the thermal insulation thicknesses and the number of dowels. This process resulted in creating 60 parameterised geometric models using Comsol Multiphysics finite element numerical modelling software [32], in which the rest of the parameters were controlled by presets. A selected numerical model, including a reinforced concrete (RC) wall, a bracket fixed with two anchors and a thermal break with its simplifications, is shown in Figure 2 compared to a commercially available ventilated façade system containing similar elements.

An example of a parameterised geometric model is shown in Figure 2b. We selected the components and materials for the study to correspond to the most common setup of fastening systems based on [15], and we used top and bottom thermal conductivities to represent the variability of the materials, where it was expected to matter. All models were created to handle 1 m^2 of construction to simplify the calculations later and ensure that the multidimensional heat flows can develop during the modelling to their full extent; therefore, the model size does not affect the results. The geometry of the used elements was simplified to be able to be handled and meshed during finite element modelling.

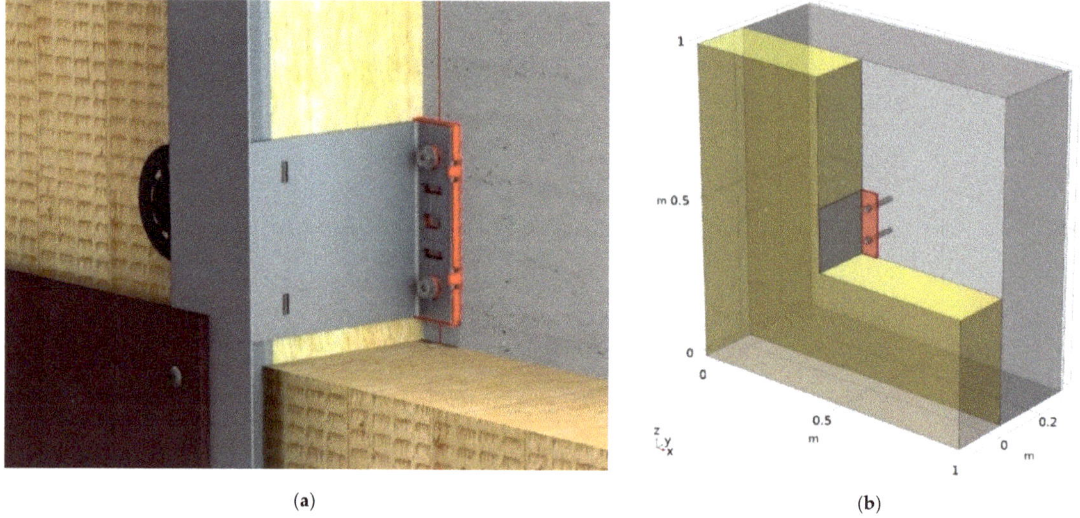

Figure 2. (**a**) Ventilated façade construction with large aluminium brackets using thermal break [15], (**b**) parametric geometry model of concrete wall with a large aluminium bracket and thermal break created in Comsol used for the numerical modelling.

In the modelled scenarios, the following parameters were considered and listed in Table 1. The wall is constructed using masonry or reinforced concrete. Masonry walls were considered with thicknesses between 25 cm and 38 cm, using different thermal conductivities available in Hungary, e.g., 0.07 W/(m × K) to represent a modern thermal insulation-filled masonry block or 0.72 W/(m × K) to include small solid masonry bricks. The RC walls were 15, 20, 25 and 30 cm thick using 2.0 and 2.5 W/(m × K) thermal conductivity. Masonry walls included both 1.5 cm thick internal and 1 cm thick external plaster, while RC walls only had 1.5 cm thick internal plaster since it is considered airtight without external plastering. The thermal insulation was mineral wool with a thickness of 10, 20 or 30 cm, using 0.03 W/(m × K) or 0.04 W/(m × K), the available range of nowadays' mineral wool thermal conductivity.

The L-shaped supporting brackets were modelled using stainless steel, steel or aluminium. The heights of the brackets were 6 cm for small (fix) and 20 cm for large (sliding) brackets. The base of the brackets was selected as 6 cm for the small bracket and either 6 cm in width or 10 cm for the large bracket. The thickness of the brackets was also parameterised and changed between 4 mm and 8 mm. The length of the brackets was changed accordingly to the thermal insulation's widths; therefore, 10, 20 and 30 cm were also tested.

Under the brackets, thermal breaks were included in some cases. When thermal breaks were applied, the material was polyamide (PA) or high-density polyethylene (HDPE) with a 5 mm or 20 mm thickness. The brackets were fixed to the wall using stainless steel or steel anchors and PA dowels. To fix the brackets, either 1 anchor for the small or 2 anchors for the large brackets were used in the models, and the fixing depth was also changed between 5 cm and 15 cm, where applicable.

Table 1. Parameters used in the study.

Component	Thickness (cm)/ Width × Height (cm)	Material	Thermal Conductivity (W/(m×K))
Internal plaster	1.5	Lime–cement	0.8
Wall	25		0.25
	30	Masonry	0.07, 0.19, 0.64
	38		0.07, 0.19, 0.72
	15, 20, 25, 30	Reinforced concrete	2, 2.5
External plaster	1.5	Lime–cement	0.8
Insulation	10, 20, 30	Mineral wool	0.03, 0.04
Dowels	0.2	PA	0.25
Anchors	5, 15	Stainless steel	17
	5, 15	Steel	50
Thermal break	0.5, 2	PA	0.25
	0.5, 2	HDPE	0.5
Bracket	0.2, 0.4/6 × 6, 6 × 20, 10 × 20	Stainless steel	17
	0.2, 0.4/6 × 6, 6 × 20, 10 × 20	Steel	50
	0.2, 0.4/6 × 6, 6 × 20, 10 × 20	Aluminium	160

2.3. Numerical Modelling Methodology

Using numerical modelling, we can perform detailed calculations considering the effect of multidimensional heat fluxes [33]. According to [34], in the case of window installations, the effect of point fixings, such as brackets, was compared with 2D and 3D simulations, and the conclusion that in the case of point fixings, 3D numerical simulations should be used. Although there is a method for estimating point thermal transmittance with 2D simulations [35], in our research, we used 3D numerical simulation to handle point thermal transmittances of the fastening systems.

In the frame of this study, Comsol Multiphysics 5.6 software was used to solve the 3D steady-state heat conduction equations to determine the point thermal transmittance considering the effect of point thermal bridges. The calculation methodology, boundary conditions and required accuracy for solids are specified in ISO 10211 [10]. The partial differential equation of steady-state heat conduction is the following:

$$\nabla \mathbf{q} = \nabla \left(\lambda_{eff} \times \nabla T \right) = 0 \tag{1}$$

The boundary conditions are set using Equations (2) and (3):

$$- \mathbf{n} \times \mathbf{q} = h_{ci} + \varepsilon \times 4 \times \sigma \times T_{m,i}{}^3 \tag{2}$$

$$- \mathbf{n} \times \mathbf{q} = h_{ce} + \varepsilon \times 4 \times \sigma \times T_{m,e}{}^3 \tag{3}$$

where in Equation (2), h_{ci} is the internal convective surface heat transfer coefficient (2.5 W/(m^2 × K)), ε is the longwave emissivity of the surface (0.9), σ is the Stefan–Boltzmann constant (5.67 × 10^{-8} W/(m^2 × K^4)) and $T_{m,i}$ is the mean thermodynamic temperature of the internal surface and its surroundings set to 293.15 K according to MSZ 24140 [36]. In Equation (3), $h_{ce} = 4 + 4 \cdot v$, where v is the wind speed in [m/s] according to [9]. Wind speed was neglected since the cladding protects the surface of the thermal insulation from the wind loads. $T_{m,e}$ is the mean thermodynamic temperature of the external surface and its surroundings in Kelvin set to 268.15 K according to [36].

A mesh independence test on one of the most complex geometrically constructed models (see Figure 2b) was created to select the most suitable finite element mesh density in terms of the accuracy of the results and the run times. This model contained a large bracket with thermal break and was fixed with two anchors. The basis of the error calculation was the value of the point thermal transmittance obtained by applying the existing highest density mesh. As it is possible to set the automatic mesh manually within the Comsol Multiphysics software, a mesh with increased density has been selected for brackets, dowels and thermal breaks, then homogeneously modelled elements such as the wall, thermal insulation and plaster. The meshing and simulations were performed by a workstation including AMD Ryzen Threadripper 2950X CPU (Advanced Micro Devices, Inc., Santa Clara, USA), 128 GB DDR4 RAM (ADATA Technology Co., Ltd., Taiwan), Nvidia Quadro RTX 4000 GPU (NVIDIA Corporate, USA) and 2 TB m.2 SSD (ADATA Technology Co., Ltd., Taiwan). The mesh statistics are summarised in Table 2 for automatic meshing cases.

Table 2. Mesh statistics of automatic meshing.

Mesh Type	Elements	DoF	Meshing Time	Calculation Error
Extremely fine	2,455,556	3,313,544	690 s	-
Extra fine	849,732	1,154,228	99 s	0.48%
Finer	375,266	513,094	33 s	1.33%
Fine	207,507	285,978	17 s	2.37%
Normal	125,816	174,791	12 s	3.24%
Coarse	58,688	82,881	6 s	5.41%
Coarser	27,244	39,233	5 s	7.98%
Extra coarse	13,403	19,605	4 s	11.45%
Extremely coarse	4503	6699	4 s	20.00%

We also created user-controlled meshing based on automated ones. However, we increased the density of the mesh only for the components of the fastening system, e.g., brackets, thermal breaks, anchors and dowels. Table 3 shows that a user-controlled mesh was performed with a similar error as the automated mesh using finer settings. The software even created the same number of elements and the degree of freedom (DoF). However, this mesh setting resulted in 26 s meshing time instead of 33 s. Since we performed thousands of runs, we preferred the user-controlled meshing to save computational time on the meshing.

Table 3. Mesh statistics of user-controlled meshing.

Mesh Type	Elements	DoF	Meshing Time	Calculation Error
Fine + extra fine	207,507	285,978	13 s	2.37%
Finer + extra fine	375,266	513,094	26 s	1.33%

2.4. Calculation of the Point Thermal Transmittances

According to Hungarian TNM decree 7/2006 [33], when examining a wall's U-value, the effect of the mechanical fastenings must be taken into account based on the equation below during detailed calculations, according to ISO 6946 [9]:

$$U \left[\frac{W}{m^2 K} \right] = \frac{1}{R_{se} + \sum_i \frac{d_i}{\lambda_i} + R_{si}} + \Delta U_f + \Delta U_g + \Delta U_r + \sum_k n_k \times \chi_k \qquad (4)$$

where R_{se} is the surface resistance of the external surface $((m^2 \times K)/W)$, R_{si} is the resistance of the internal surface $((m^2 \times K)/W)$, d_i is the thickness of a composing layer (m), λ_i is the thermal conductivity of a composing layer $(W/(m \times K))$, ΔU_f is the correction factor of the mechanical fasteners ($W/(m^2 \times K)$), ΔU_g is the correction factor of the joint gaps ($W/(m^2 \times K)$), ΔU_r is the correction factor of the reverse layered flat roofs ($W/(m^2 \times K)$), n_k is the number of point thermal bridges ($1/m^2$) and χ_k is the point thermal transmittance value (W/K) considered for other kinds of point thermal bridges (e.g., tie-rods, brackets, etc.).

In this research, we investigated the thermal effect of the fastening elements of the façade cladding. Therefore, we dealt with the correction factor for the mechanical fastenings of the brackets. This is important since ISO 6946 Annex F [9] only contains a simplified calculation for ΔU_f in the case of cylindrical-shaped mechanical fasteners when fixing the thermal insulation, and it does not deal with point thermal bridges caused by the anchors of the brackets or the L-shaped brackets, which have very different geometries [8].

If we want to take point thermal bridges into account for transmission losses during our calculations other than cylindrical fasteners, then in the absence of a simplified calculation method, we must use numerical modelling, as presented earlier. The behaviour of point thermal bridges can be described by the point thermal transmittance, χ (W/K), which in this case shows how much additional heat flow (W) is caused by a piece of fastening element as a result of a unit temperature difference ($1/K$).

Its calculation is based on the difference between the heat fluxes calculated for the entire 3D element during the numerical simulation and the heat fluxes obtained by neglecting point thermal bridges:

$$\chi \left[\frac{W}{K}\right] = L_{3D} - \sum_{i=1}^{N_i} U_i \times A_i \tag{5}$$

where χ is the point thermal transmittance (W/K), L_{3D} is the thermal coupling coefficient from the three-dimensional calculation ($W/(m^2 \times K)$), U_i is the thermal transmittance value calculated by neglecting 3D point thermal bridges ($W/(m^2 \times K)$) and A_i is the area of the examined element (m^2).

Heat losses caused by fasteners can be considered with a correction to the thermal transmittance value. Since the fixing elements repeatedly occur on the façade, their thermal effect can be calculated with a correction factor ΔU ($W/(m^2 \times K)$), if we specify how many fixing elements (n_f) pierce the thermal insulation layer per 1 m^2 of the tested surface:

$$\Delta U_f = n_f \times \chi \tag{6}$$

For simplification, we examined 1 m^2 of wall surface during the modelling on which we placed a single bracket. In this case, according to Equation (5), the value of the point thermal transmittance (χ) at the point is the same as the difference between the uncorrected thermal transmittance (U) and point thermal transmittance calculated by the finite element program, which also takes into account the corrections (L_{3D}):

$$\Delta U_f = L_{3D} - U = \chi \times 1/m^2 \tag{7}$$

3. Results and Discussion

3.1. Visualisation of Results

After the numerical thermal modelling, the temperature distribution and heat flux density can be visualised in 3D. However, the best visibility is provided using 2D sections (see Figure 3).

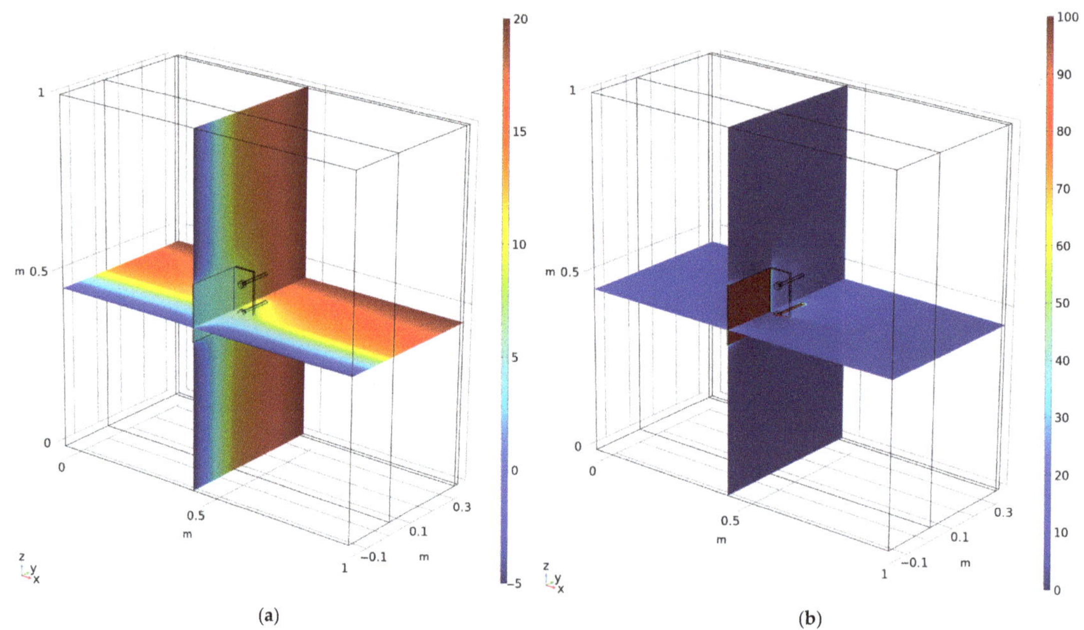

Figure 3. 3D models showing the results using vertical and horizontal sections for (**a**) temperature distribution (°C) and (**b**) heat flux density (W/m^2).

In Figures 4 and 5, included for illustrational purposes of the evaluation, we selected masonry wall-based geometry to show the possibilities of visualisation. The temperature distribution shows that a bracket causes vast disturbance in the temperature field, especially in the vertical direction, as visible in Figure 4. The brackets are much warmer than the thermal insulation and create thermal bridges in the structure. It is also observable that the brackets, due to their high thermal conductivity, have almost the same temperature along their structure, while in the thermal insulation, the temperature distribution is visible between significantly more extensive temperature ranges. We can also conclude that a single bracket, despite causing temperature disturbance in the thermal insulation, near the base of the bracket and near the anchors and dowels, does not cause any significant changes on the internal surface. Therefore, no condensation risk can occur on the internal surface due to the applying of brackets to fix ventilated façades. This conclusion correlates with Arregi et al. [37]. Their study used different types of brackets made from stainless steel. They said that even 40 mm of thermal insulation is enough to raise the internal surface temperature above the dew point temperature to avoid mould growth. We extend this conclusion by claiming that one can even use aluminium brackets without thermal breaks. The internal temperature will not be significantly affected when using walls with 10–30 cm thermal insulation. Besides examining the temperature distribution, the heat flux density is also visualised in Figure 5. The heat flux density along the bracket is the highest, from which it can be concluded that the heat flows at a significantly higher rate from the structure along the brackets than through the thermal insulation. It is also visible that the surrounding of the base of the bracket also has a higher heat flux density in the masonry wall.

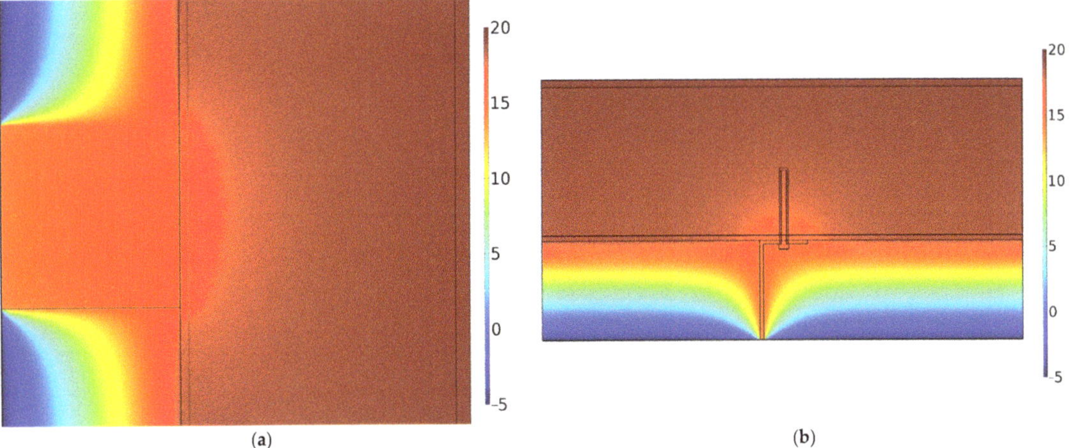

Figure 4. Temperature distribution (°C) in 2D vertical (**a**) and horizontal (**b**) sections.

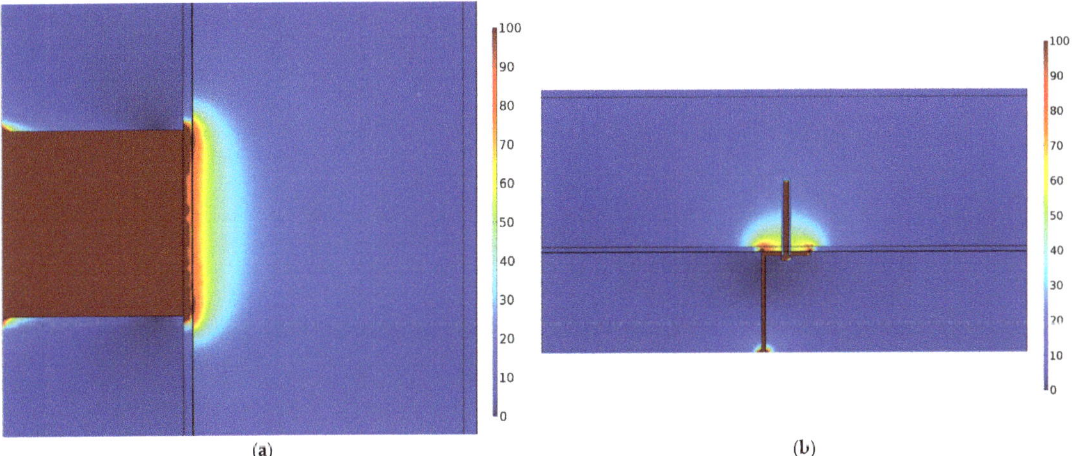

Figure 5. Heat flux density (W/m^2) in 2D vertical (**a**) and horizontal (**b**) sections.

It is also possible to visualise the direction and magnitude of the heat flow vectors shown in Figure 6. Examining several models with different bracket materials, it can be said that the direction of the arrows representing the heat flow vectors is similar and does not change significantly with the bracket material, only the magnitude of the heat flow changes. It can be seen from the figures that in the present case, the heat flows in the horizontal direction from the bracket which can be said to be one-dimensional, and then moving closer, we can speak of 3D heat flows. In the vertical direction, within a distance of about 25–30 cm, the heat flows change from one-dimensional to three-dimensional. The figure in both directions shows that the change in heat flow also affects the thermal insulation layer. Based on these, it can be said that the modelled one square meter area examined during the numerical modelling is sufficient for the study, as there are already one-dimensional heat flows on the boundary surfaces. However, the three-dimensional heat flows around the fixings support the need for 3D numerical simulation. We can also conclude that the ventilated façade claddings' fastening brackets can be modelled using point thermal bridges since the brackets cause thermal bridges around them and cannot be represented simply by a 2D model.

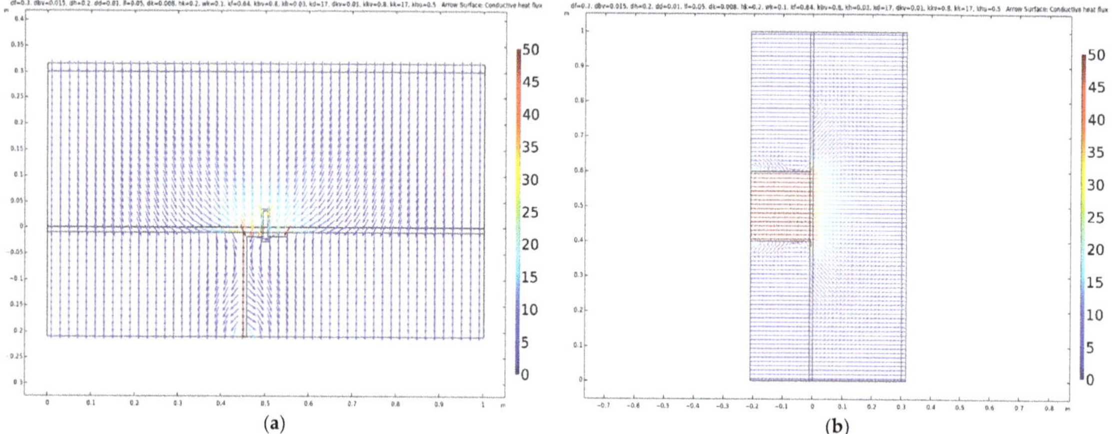

Figure 6. Heat flux vectors around the brackets in the vertical (**a**) and horizontal (**b**) section.

3.2. Effect of the Properties of the Wall

To represent the effect of different wall thicknesses and wall types, we created a figure that shows cases with similar brackets and thermal insulations. In Figure 7, we show cases with 20 cm thermal insulation applied on the walls (0.04 W/(m × K)), and small aluminium brackets with thermal brakes were also considered. Figure 7 shows that as the thermal resistance of the wall (thickness of the wall divided by the thermal conductivity of the wall) increases, the point thermal transmittance decreases significantly, similar to the results of [14], where one of the main findings said the point thermal bridge may decrease the U-value of the entire wall up to 28% regarding the increase of supporting layer thickness and using insulation materials with higher thermal conductivity. We also reached a similar result with [11] in the case of a reinforced concrete wall because due to its high thermal conductivity, we obtain a much higher point thermal transmittance than in masonry walls. This means that the material, the thermal conductivity of the supporting wall, affects the point heat transfer coefficient.

However, Figure 7 also shows that the properties of the anchoring, such as the material of the anchors, the number of anchors and the length of the drill hole, influence the value of the point thermal transmittance, especially when masonry walls were examined. However, anchors have a much smaller influence on the value of the point thermal transmittance than the wall material and may be neglected. This can be explained by the material of the dowels of the anchors because they are made of PA, which essentially acts as a thermal isolator for the anchors. This effect is also examined by [18] including chemical anchoring, and they concluded that this type of anchoring should be chosen whenever possible. Chemical anchoring also creates a thermal isolator for the anchors, similarly to PA. However, in their manuals, fastening system producers usually advise using PA dowels for the anchors when possible since it is much less costly, more known and easier to construct.

Evaluating the reinforced concrete walls with different thicknesses and thermal conductivities, it can be seen that in cases with the same thickness and thermal conductivity but with different anchors, the largest difference between the point thermal transmittance is 3.2%. As the thermal conductivity of the supporting wall increases, the difference between cases with different anchors decreases. In the case of the tested masonry wall with the highest thermal conductivity (0.72 W/(m × K), which belongs to solid ceramic masonry bricks), the difference is also only 3% for the various anchors. However, in the case of masonry walls with low thermal conductivity (0.07 W/(m × K), which belongs to modern, thermal insulation-filled masonry blocks [38]), this difference is of almost 20%.

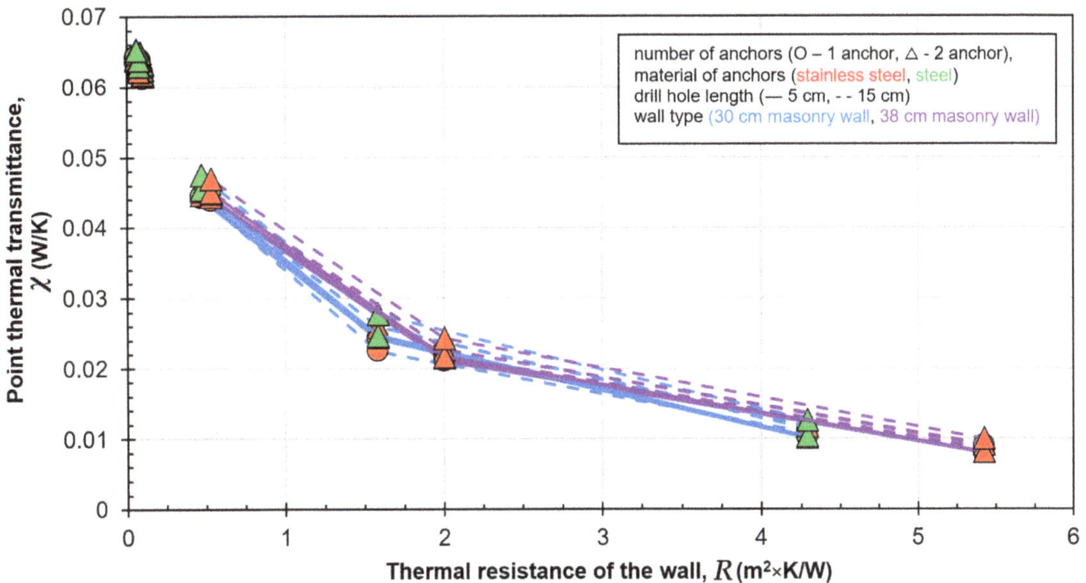

Figure 7. The effect of wall thickness and type and anchors. 20 cm, λ = 0.04 W/(m × K) thermal insulation penetrated with small (0.06 m × 0.06 m × 0.004 m) aluminium brackets equipped with 5 mm PA thermal breaks.

3.3. Effect of the Properties of the Brackets and Thermal Insulation

The effect of different thicknesses and thermal conductivity of thermal insulations on the point thermal transmittances were also examined. In Figure 8, we show two cases, Figure 8a shows a reinforced concrete wall and Figure 8b a masonry wall. In both cases, different brackets were fixed in a sliding position using a single anchor. Brackets with a different geometry were represented using different markers, while thicknesses were represented with different colours (Figure 8). There are considerable differences in the point thermal transmittances due to the thickness of the brackets. The 8 mm thick brackets tend to have much larger values than the 4 mm thick brackets. Therefore, choosing thinner brackets can save a lot of energy, primarily if RC walls or masonry walls with lower thermal resistance are used, and if mechanical calculations allow them to be used. The thickness of the thermal insulation does not cause large differences in the point thermal transmittance considering the thickness of the brackets. However, with more thermal insulation on the supporting wall construction, differences in the point thermal transmittance caused by the thickness of the brackets are reduced slightly in the case of RC wall (Figure 8a) and increased slightly in the case of masonry wall (Figure 8b).

Comparing the effect of adding thermal insulation, there is a significant difference between the different wall materials. In the case of RC walls, the decrease can be up to 30% if we choose 30 cm thermal insulation instead of 10 cm. Using large brackets, the relative difference is smaller than using small brackets. While adding more thermal insulation to the RC wall, it can decrease the point thermal transmittances, and more thermal insulation will increase the point thermal transmittances on modern masonry walls. This effect can be explained by the difference in the wall materials' thermal resistance.

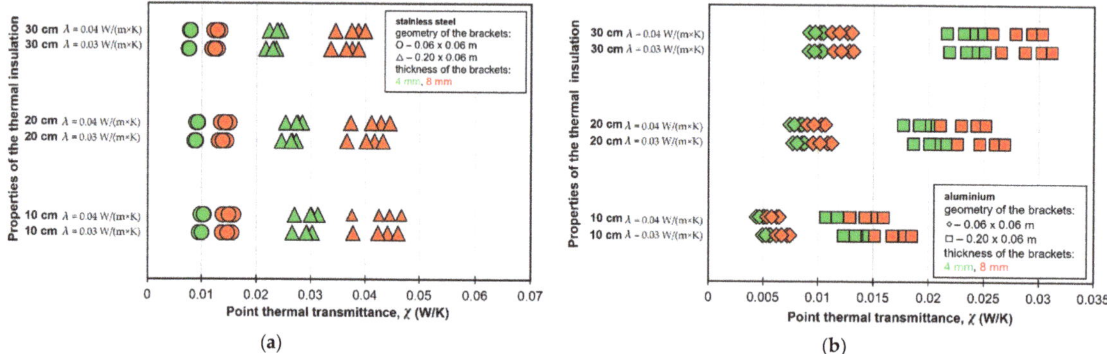

Figure 8. (**a**) Stainless steel brackets with thermal breaks fixed with a single 5 cm long anchor steel anchor in 20 cm thick RC wall. (**b**) Aluminium brackets with thermal breaks fixed with a single 5 cm long steel anchor in 30 cm thick masonry wall.

This effect was also seen in [5,6,8,12,39]. We can extend their statements by concluding that choosing thicker thermal insulation in the case of ceramic masonry walls can increase the point thermal transmittance by more than 45% for both small and large brackets. Therefore, it is crucial to deal with the heat loss of brackets in detail since simply choosing thicker thermal insulation does not solve the case of point thermal transmittances.

3.4. Effect of the Properties of the Brackets and Thermal Breaks

The effect of the bracket and thermal break properties, such as its material, thickness and geometry, was also evaluated. A case with an RC wall and 20 cm thermal insulation were displayed in Figure 9. Previous studies [8,11] also investigated the effect of different metal brackets on the point thermal transmittance. However, they only examined brackets made of galvanised steel with several lengths, whereas we examined brackets of 3 different materials (stainless steel, steel and aluminium) with various geometries.

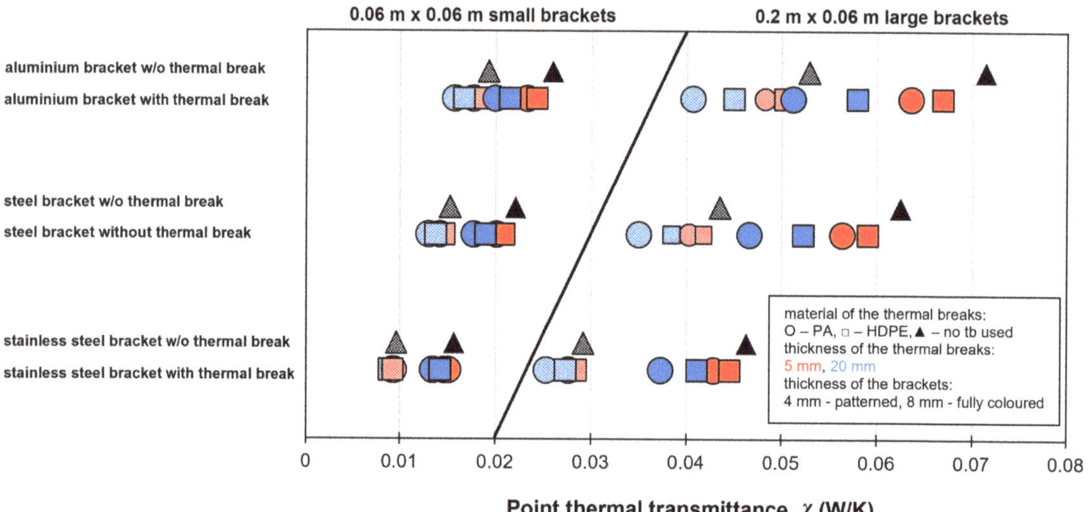

Figure 9. Brackets fixed using a single 5 cm long steel anchor in 20 cm thick reinforced concrete wall with 20 cm thick $\lambda = 0.04$ W/(m $\times$ K) thermal insulation.

After evaluating the results of the numerical modelling, it can be said that the point thermal transmittance of the brackets is significantly affected by the geometry, material and thickness of the brackets. Thicker brackets can increase χ by at least 26% (small stainless-steel brackets) and up to 60% (large aluminium brackets). Brackets with higher thermal conductivity can increase the point thermal transmittance by at least 27% (large brackets) and up to 48% (small brackets). Choosing large brackets over small ones can increase the point thermal transmittance value by 61 to 67%. Although we did not examine the mechanical performance of the brackets within the framework of the research, based on the results, it can be said that during the design process, it is worth choosing the smallest and thinnest statically appropriate brackets since this way the heat losses on the façade surfaces can be significantly reduced. It is also more favourable based on economic and sustainability aspects since less material is used.

Based on the evaluation of the thermal breaks, also illustrated in Figure 9, it can be said that the existence, material and thickness of the thermal breaks do not terminate the heat loss of the brackets. The latter statement is, unfortunately, a widespread belief of practitioners, but it is rebuttable by performing the numerical modelling as some previous studies also concluded [8,12], especially for stainless steel brackets. Besides our study, researchers also examined glass fibre-reinforced brackets [8]. They concluded that the difference in results of 3D simulations with thermal breaks and without them is almost equal to zero (0.0–0.3%). Using stainless steel brackets, the difference is up to 0.9%, steel brackets up to 5.5% and aluminium brackets vary from 4.8% to 5.7%. Therefore, we only tested PA and HDPE thermal breaks with 5 mm to 20 mm thickness.

Based on our models, we can say that 20 mm thick thermal breaks can reduce the χ by at least 2% for small stainless-steel brackets and up to 24% for large aluminium brackets. If the thermal breaks were constructed from 5 mm thick PA, they could reduce the large aluminium brackets by up to 12% point thermal transmittance. Comparing the different thermal breaks available, even a 20 mm thick PA isolator can save only 28% on an RC wall. Therefore, it is visible that if we only consider thermal aspects, the use of thermal isolators may only be worth the effort with aluminium brackets. However, we recommend calculating the economic aspects of using thermal breaks before applying them.

3.5. Thermal Bridge Catalogues

We created a comprehensive thermal bridge catalogue based on the total 41,118 FEM-based numerically modelled point thermal transmittance values (see Figure 10). In order to compile the results into a manageable catalogue, two result summary tables were created; one containing the point thermal transmittance obtained from the simulation of models without thermal breaks and the other with thermal breaks. The materials were sorted along the wall constructions of different materials and widths, insulations of different thicknesses, thermal conductivity and the anchor length. A drop-down list-based search list also facilitates the finding of the appropriate values for the given parameters. The thermal bridge catalogue can be found as a supplementary materials to the article in Excel format to help readers and practitioners use the study results.

Point thermal transmittance, χ [W/K]								Properties of fasteners			
								λ of brackets [W/(m×K)]			
								17			
								Geometry of brackets [W x H x T], [m]			
								0.06 x 0.06 x 0.004			
Design of layers								λ of thermal breaks [W/(m×K)]			
								0.25		0.5	
Properties of anchors		Anchor length [m]	Properties of supporting wall			Properties of insulation		Thickness of thermal breaks [m]			
number of anchors	λ [W/(m×K)]		type of material	width [m]	λ [W/(m×K)]	thickness [m]	λ [W/(m×K)]	0.005	0.02	0.005	0.02
1	17	0.05	*masonry wall*	0.25	0.25	0.1	0.03	0.005414	0.005010	0.005563	0.005377
							0.04	0.004959	0.004539	0.005109	0.004906

Figure 10. Thermal bridge catalogue (extract).

3.6. Simplified Method

In addition to the thermal bridge catalogue, a new simplified method was also created based on the results of the numerical models and the calculation of the effect of the mechanical fasteners according to ISO 6946 Annex F [9]. The new simplified method is created to handle the brackets' effect with only slight modifications to the original equation in [9]. The basic structure of the equation has remained, but a new "α" multiplication correction factor is created for the equation that considers the effect of the brackets, as well as some of the components (A_b, λ_b) now represent the brackets instead of the cylindrical mechanical fasteners as follows in Equation (8):

$$\Delta U_{f,b} \left[\frac{W}{m^2 \times K} \right] = \alpha \times \frac{n \times A_b \times \lambda_b}{d_0} \times \left(\frac{R_1}{R_{th}} \right)^2 \tag{8}$$

where $U_{f,b}$ is the point thermal transmittance of the brackets (W/(m^2 × K)), n is the number of the brackets (1/m^2), A_b is the surface area of the bracket penetrating the thermal insulation layer (m^2), λ_b is the thermal conductivity of the bracket's material (W/(m × K)), d_0 is the thickness of the penetrated thermal insulation (m), R_1 is the thermal resistance of the penetrated thermal insulation ((m^2 × K)/W) and R_{th} is the total thermal resistance of the cross-section without correction factors ((m^2 × K)/W). The "α" coefficient was created to get the numerical modelling results back with less than ±10% deviation compared to the numerical models in case of no thermal breaks applied. As visible in Equation (8), this simplified method neglects the existence and properties of thermal breaks. However, if this method calculates the correction of brackets with thermal breaks, the results will deviate in favour of safety.

In tabular form, we compared point thermal transmittance for certain parameter combinations. We checked the difference between the results calculated in the original form of the correction formula and the χ-values obtained from the simulation. During the investigation, we concluded that changing the geometry and material of the bracket causes significant differences in the values of the point thermal transmittance at the point, changing the thermal conductivity of the thermal insulation and the properties of the dowels, and the existence of the thermal breaks only results in differences within 10%. Based on these, we chose the strategy of examining the cases belonging to the specific thermal insulation, wall and bracket properties in separate groups and finding what constant multiplier we can use to achieve for the correction formula to return the results of the simulation within an error margin under ±10% in case of no thermal breaks applied. We tested and created multiplication correction factors for both evaluated bracket types; therefore, α factors can handle aluminium, steel and stainless-steel brackets. Since we concluded that thermal breaks have relatively small effects on the point thermal transmittance, we neglected them. We considered this simplification to halve the number of the multiplication correction factors. The multiplication correction factors were also created to handle a range between insulation thicknesses (0.1, 0.2 and 0.3 m) and thermal conductivity of the insulation (0.3 and 0.4 W/(m $\times$ K), as well as small (0.06 m) to large (0.2 m) brackets with different thicknesses (0.004 to 0.008 m). Any values between these values can be calculated using linear interpolation.

In the case of masonry walls, seven tables were created containing "α" multiplication correction factors, considering all the walls' tested wall thicknesses and thermal conductivity factors. For example, Table 4 shows the table for a 30 cm thick ceramic masonry wall with a thermal conductivity of $\lambda = 0.19$ W/(m $\times$ K). In the top row of the table, we can find the values for the different thermal insulation properties. In contrast, in the leftmost column, we can find the values for the different thermal conductivity of the brackets, depending on the geometry of the brackets.

Table 4. "α" multiplication correction factors for 30 cm thick masonry wall ($\lambda = 0.19$ W/(m $\times$ K)).

Thermal Conductivity of the Brackets [W/(m$\times$K)]	The Geometry of the Brackets [h $\times$ t]	Thickness of Insulation [m]					
		0.1		0.2		0.3	
		Thermal Conductivity of Insulation [W/(m$\times$K)]					
		0.03	0.04	0.03	0.04	0.03	0.04
17 (stainless steel)	0.06 $\times$ 0.004	0.320	0.208	2.000	1.200	5.200	3.200
	0.06 $\times$ 0.008	0.224	0.144	1.440	0.920	4.400	2.560
	0.2 $\times$ 0.004	0.224	0.144	1.440	0.920	4.400	2.560
	0.2 $\times$ 0.008	0.176	0.104	1.120	0.720	3.440	2.000
50 (steel)	0.06 $\times$ 0.004	0.128	0.080	0.960	0.560	2.800	1.760
	0.06 $\times$ 0.008	0.088	0.088	0.640	0.400	2.000	1.280
	0.2 $\times$ 0.004	0.104	0.056	0.720	0.440	2.240	1.440
	0.2 $\times$ 0.008	0.064	0.064	0.480	0.296	1.520	0.960
160 (aluminium)	0.06 $\times$ 0.004	0.044	0.029	0.344	0.224	1.120	0.720
	0.06 $\times$ 0.008	0.029	0.019	0.216	0.144	0.720	0.480
	0.2 $\times$ 0.004	0.034	0.022	0.256	0.160	0.880	0.560
	0.2 $\times$ 0.008	0.022	0.014	0.160	0.104	0.560	0.320

In the case of a reinforced concrete wall, it was sufficient to create only one table (see Table 5) because the tests showed that due to the relatively high thermal conductivity of the RC wall (2 to 2.5 W/(m $\times$ K)), the wall thickness and the thermal conductivity do not have a significant effect on the point thermal transmittance.

Tables 4 and 5, and the other six tables (Tables S1–S6) for masonry walls with other thicknesses, can be found as Supplementary Materials to the article in Excel format.

To validate the simplified method, we performed a comparison represented in Figure 11. We examined the differences between the results obtained from the numerical simulation and the calculated ones with the chosen multiplication factors in the case of reinforced concrete walls and masonry walls. Based on the graphs, it can be said that for both wall types, in the cases without thermal breaks, the deviation was kept within the 10% margin of error; therefore, the simplified method met its expectations. However, when thermal breaks were added, some deviations, especially at higher point thermal transmittance values, were slightly greater than 20%. However, the deviation of the simplified method was in the positive direction in favour of safety.

Table 5. "α" multiplication correction factors for reinforced concrete wall.

Thermal Conductivity of the Brackets [W/(m×K)]	The Geometry of the Brackets [h × t]	Thickness of Insulation [m]					
		0.1		0.2		0.3	
		Thermal Conductivity of Insulation [W/(m×K)]					
		0.03	0.04	0.03	0.04	0.03	0.04
17 (stainless steel)	0.06 × 0.004	0.320	0.208	2.000	1.256	5.600	3.360
	0.06 × 0.008	0.240	0.160	1.680	1.040	4.800	2.880
	0.2 × 0.004	0.296	0.184	1.840	1.080	5.200	3.200
	0.2 × 0.008	0.224	0.136	1.480	0.880	4.160	2.560
50 (steel)	0.06 × 0.004	0.136	0.088	1.040	0.640	3.200	2.080
	0.06 × 0.008	0.096	0.064	0.760	0.480	2.400	1.520
	0.2 × 0.004	0.120	0.080	0.920	0.560	2.800	1.760
	0.2 × 0.008	0.088	0.056	0.680	0.400	2.080	1.280
160 (aluminium)	0.06 × 0.004	0.048	0.032	0.400	0.264	1.360	0.880
	0.06 × 0.008	0.032	0.021	0.272	0.176	0.920	0.560
	0.2 × 0.004	0.040	0.028	0.336	0.224	1.120	0.720
	0.2 × 0.008	0.030	0.019	0.232	0.144	0.800	0.480

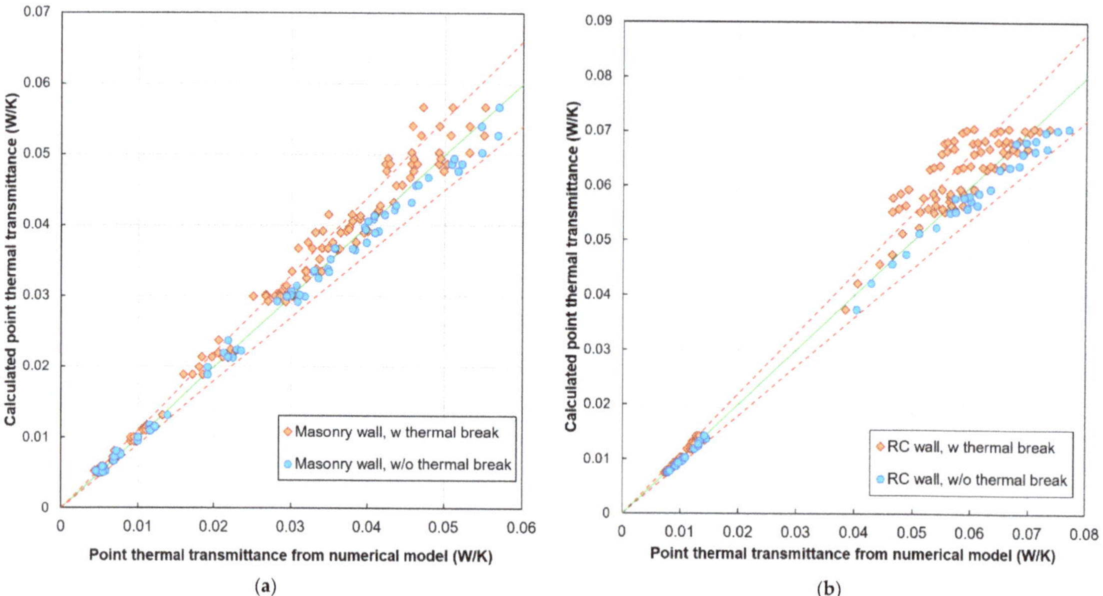

Figure 11. Comparison of calculated and numerical modelled point thermal transmittance in the case of masonry (**a**) or RC (**b**) wall.

4. Conclusions

During the research, the effect of the parameters was investigated on a broader scale than most of the previous research revealed during the literature review. Numerical simulations have been carried out to predict the effects of metal fasteners on the thermal performance of the building envelope. The main findings can be summarised as follows within the range of the tested parameters:

1. The thermal resistance (thickness of the wall/thermal conductivity of the wall) of the wall significantly affects the point thermal transmittance (χ) of the brackets.

2. The point thermal transmittances (χ) of the brackets are not significantly affected by the number, the material of anchors, dowels, and drill hole length.
3. The point thermal transmittances (χ) of the brackets are not significantly affected by the thermal conductivity of the thermal insulation.
4. The point thermal transmittances (χ) of the brackets are significantly affected by the thickness of the thermal insulation.
5. The point thermal transmittances (χ) of the brackets are significantly affected by their geometry (size), material and thickness of the brackets.
6. The thermal breaks can reduce the point thermal conductivity much less than expected, especially when stainless-steel brackets are used.

Overall, it can be said that the results of the numerical simulations clearly show that, for mechanical fixings, only considering the anchors and the doweling is not sufficient. The effect of the brackets on the point thermal transmittance is also significant. Using the results of tens of thousands of numerical models, we created a point thermal bridge catalogue. In the research framework, besides creating the catalogue, we created a simplified method using a multiplication correction factor in tabular form. The resulting catalogue and simplified method can be used in cost-effectiveness studies and promotes the further development of load-bearing structures and the development of their design. There are also further research opportunities such as multiphysical tests (e.g., combined heat and moisture transport) or structural analysis (e.g., effect of construction inaccuracies, deformation of the bracket).

Supplementary Materials: The following supporting information can be downloaded at: https://www.mdpi.com/article/10.3390/buildings12081153/s1. FP-BN_Thermal_bridge_catalogue_v1.0.zip.

Author Contributions: Conceptualisation, B.N.; methodology, B.N.; investigation, F.P. and B.N.; data curation, B.N.; writing—original draft preparation, F.P. and B.N.; writing—review and editing, B.N.; validation, B.N.; visualisation, F.P. and B.N.; supervision, B.N. All authors have read and agreed to the published version of the manuscript.

Funding: The first author was supported by the ÚNKP-21-1 New National Excellence Program of the Ministry for Innovation and Technology from the source of the National Research, Development and Innovation Fund. The corresponding author was supported by the ÚNKP-21-4 New National Excellence Program of the Ministry for Innovation and Technology from the source of the National Research, Development and Innovation Fund. Project FK_128663 has been implemented with the support provided from the National Research, Development and Innovation Fund of Hungary, financed under the FK_18 funding scheme.

Institutional Review Board Statement: Not applicable.

Informed Consent Statement: Not applicable.

Data Availability Statement: Not applicable.

Conflicts of Interest: The authors declare no conflict of interest.

References

1. Urbán, D.; Roozen, N.B.; Zatko, P.; Rychtáriková, M.; Tomašovič, P.; Glorieux, C. Assessment of sound insulation of naturally ventilated double skin facades. *Build. Environ.* **2016**, *110*, 148–160. [CrossRef]
2. Zamora Mestre, J.L.; Niampira, A. Lightweight ventilated façade: Acoustic performance in laboratory conditions, analysing the impact of controlled ventilation variations on airborne sound insulation. *Build. Acoust.* **2020**, *27*, 367–379. [CrossRef]
3. Heindl, W.; Sigmund, A. Influence of air-backed bracket-mounted cladding on the thermal insulation of external walls. *Bauphysik* **1984**, *6*, 137–141.
4. Anderson, B. *Conventions for U-Value Calculations*; The Building Research Establishment: Watford, UK, 2006.
5. Kolesnyk, I. Determination of Linear and Point Thermal Transmittance of the Most Usual Thermal Bridges in External Walls. *Bud. O Zoptymalizowanym Potencjale Energetycznym* **2013**, *2*, 47–54.
6. Šadauskiene, J.; Ramanauskas, J.; Šeduikyte, L.; Daukšys, M.; Vasylius, A. A simplified methodology for evaluating the impact of point thermal bridges on the high-energy performance of a Passive House. *Sustainability* **2015**, *7*, 16687–16702. [CrossRef]

7. García, B.Z.; Goikolea, B.A.; Martín, J.M.G.; Hernández García, J.L. Comparison of theoretical heat transfer model with results from experimental monitoring installed in a refurbishment with ventilated façade. *IOP Conf. Ser. Earth Environ. Sci.* **2020**, *410*, 012104. [CrossRef]

8. Levinskytė, A.; Bliūdžius, R.; Kapačiūnas, R. The comparison of a numerical and empirical calculation of thermal transmittance of ventilated facade with different heat-conductive connections. *J. Sustain. Archit. Civ. Eng.* **2018**, *23*, 39–48. [CrossRef]

9. *ISO 6946:2017*; Building Components and Building Elements. Thermal Resistance and Thermal Transmittance. International Organization for Standardization: Geneve, Switzerland, 2017.

10. *ISO 10211:2017*; Thermal Bridges in Building Construction—Heat Flows and Surface Temperatures—Detailed Calculations. International Organization for Standardization: Geneve, Switzerland, 2017.

11. Alghamdi, A.; Alharthi, A.; Alanazi, A.; Halawani, M. Effects of Metal Fasteners of Ventilated Building Facade on the Thermal Performances of Building Envelopes. *Buildings* **2021**, *11*, 267. [CrossRef]

12. Nowak, K.; Byrdy, A. Effect of mounting brackets on thermal performance of buildings with ventilated facades. *J. Build. Phys.* **2019**, *43*, 46–56. [CrossRef]

13. Doda, A. Thermal Bridging in Rainscreen Cladding. Available online: https://www.patrickryanassociates.com/media/files/Cold-Bridging-in-Rainscreen-Cladding-System.pdf (accessed on 30 June 2022).

14. Sadauskiene, J.; Ramanauskas, J.; Vasylius, A. Impact of Point Thermal Bridges on Thermal Properties of Buidling Envelopes. *Therm. Sci.* **2020**, *24*, 2181–2188. [CrossRef]

15. Hilti Ventilated Facades Technical Manual V1.0. Available online: https://www.hilti.hu/content/dam/documents/pdf/e4/engineering/manuals/MFT_Technical%20Manual%20PRINTING%20V1.0%20low.pdf (accessed on 30 June 2022).

16. Ujma, A.; Pomada, M. Analysis of the temperature distribution in the place of fixing the ventilated facade. *E3S Web Conf.* **2019**, *97*, 01041. [CrossRef]

17. Theodosiou, T.G.; Tsikaloudaki, A.G.; Kontoleon, K.J.; Bikas, D.K. Thermal bridging analysis on cladding systems for building facades. *Energy Build.* **2015**, *109*, 377–384. [CrossRef]

18. Theodosiou, T.; Tsikaloudaki, K.; Bikas, D. Analysis of the Thermal Bridging Effect on Ventilated Facades. *Procedia Environ. Sci.* **2017**, *38*, 397–404. [CrossRef]

19. Theodosiou, T.; Tsikaloudaki, K.; Tsoka, S.; Chastas, P. Thermal bridging problems on advanced cladding systems and smart building facades. *J. Clean. Prod.* **2019**, *214*, 62–69. [CrossRef]

20. Ingeli, R.; Gašparík, J.; Paulovičová, L. Impact of an innovative solution for the interruption of 3-D point thermal bridges in buildings on sustainability. *Sustainability* **2021**, *13*, 11561. [CrossRef]

21. Šadauskienė, J.; Ramanauskas, J.; Šeduikytė, L.; Buska, A. Assessment of buildings with ventilated facade systems and evaluation of point thermal bridges. *J. Sustain. Archit. Civ. Eng.* **2015**, *11*, 59–71. [CrossRef]

22. Grauer, M. Ventilated rainscreen cladding system subframe contribution to annual source energy use in mid-size office buildings. *Energy Build.* **2019**, *187*, 269–280. [CrossRef]

23. Salvalai, G.; Sesana, M.M. Experimental Analysis of Different Insulated Façade Technologies in Summer Condition. *J. Green Build.* **2019**, *14*, 77–91. [CrossRef]

24. Olshevskyi, V.; Statsenko, E.; Musorina, T.; Nemova, D.; Ostrovaia, A. Moisture Transfer in Ventilated Facade Structures. *MATEC Web Conf.* **2016**, *53*, 01010. [CrossRef]

25. Feris, P.; Strachan, P.; Dimoudi, A.; Androutsopoulos, A. Investigation of the performance of a ventilated wall. *Energy Build.* **2011**, *43*, 2167–2178.

26. Sharma, A.; Mishra, K.B. Experimental investigations on the influence of 'chimney-effect' on fire response of rainscreen façades in high-rise buildings. *J. Build. Eng.* **2021**, *44*, 103257. [CrossRef]

27. Kosínski, P.; Brzyski, P.; Suchorab, Z.; Łagód, G. Heat Losses Caused by the Temporary Influence of Wind in Timber Frame Walls Insulated with Fibrous Materials. *Materials* **2020**, *13*, 5514. [CrossRef]

28. Rahiminejad, M.; Kholvalyg, D. Thermal resistance of ventilated air-spaces behind external claddings; definitions and challenges (ASHRAE 1759-RP). *Sci. Technol. Built Environ.* **2021**, *27*, 788–805. [CrossRef]

29. Domínguez-Torres, C.A.; León-Rodríguez, Á.L.; Suárez, R.; Dominguez-Delgado, A. Empirical and Numerical Analysis of an Opaque Ventilated Facade with Windows Openings under Mediterranean Climate Conditions. *Mathematics* **2022**, *10*, 163. [CrossRef]

30. Ibañez-Puy, M.; Vidaurre-Arbizu, M.; Sacristán-Fernández, J.A.; Martín-Gómez, C. Opaque Ventilated Façades: Thermal and energy performance review. *Renew. Sustain. Energy Rev.* **2017**, *79*, 180–191. [CrossRef]

31. Gagliano, A.; Aneli, S. Analysis of the energy performance of an Opaque Ventilated Façade under winter and summer weather conditions. *Sol. Energy* **2020**, *205*, 531–544. [CrossRef]

32. *Heat Transfer Module User's Guide*; COMSOL Multiphysics®v. 5.6; COMSOL AB: Stockholm, Sweden, 2021.

33. TNM Decree 7/2006. On the Determination of the Energy Performance of Buildings. Available online: https://njt.hu/jogszabaly/2006--7-20--6F (accessed on 30 June 2022).

34. Hallik, J.; Kalamees, T. A new method to estimate point thermal transmittance based on combined two-dimensional heat flow calculation. *E3S Web Conf.* **2020**, *172*, 08005. [CrossRef]

35. Yang, W.; Šadauskien, J.; Ramanauskas, J.; Krawczyk, D.A.; Klumbyt, E.E.; Fokaides, P.A. Investigation of Thermal Bridges of a New High-Performance Window Installation Using 2-D and 3-D Methodology. *Buildings* **2022**, *12*, 572.

36. *MSZ 24140:2015*; Power Engineering Dimensioning Calculuses of Buildings and Building Envelope Structures. Hungarian Standards Institution: Budapest, Hungary, 2015.
37. Arregi, B.; Garay, R.; Garrido-Marijuan, A. Assessment of thermal performance and surface moisture risk for a rear-ventilated cladding system for façade renovation. *IOP Conf. Ser. Earth Environ. Sci.* **2020**, *410*, 012102. [CrossRef]
38. Nagy, B.; Stocker, G. Numerical Analysis of Thermal and Moisture Bridges in Insulation Filled Masonry Walls and Corner Joints. *Period. Polytech. Civ. Eng.* **2019**, *63*, 446–455. [CrossRef]
39. Arregi Goikolea, B.; Garay Martinez, R.; Riverola Lacasta, A.; Chemisana Villegas, D. Heat transfer through anchoring elements in a rear-ventilated rainscreen insulation system for façade retrofit. In Proceedings of the Construction Pathology, Rehabilitation Technology and Heritage Management (REHABEND 2018 Congress), Cáceres, Spain, 15–18 May 2018.

 *buildings*

Article

BECOP: A New Metric for Measuring the Energy Performance of the Building Envelope

Samir E. Chidiac * and Ghassan E. Marjaba

Department of Civil Engineering, McMaster University, Hamilton, ON L8S 4L7, Canada; marjabg@mcmaster.ca
* Correspondence: chidiac@mcmaster.ca

Abstract: Thermal properties of the building envelope (BE) prescribed by codes and standards do not provide a consistent and comprehensive measure of its performance. Qualitative comparative analysis employed by the codes to assess energy savings is deterrent to technology development as the potential energy savings are never realized. A new metric, referred to as the building envelope coefficient of performance (BECOP), is proposed, which compares the BE performance to an ideal system. BECOP, which is invariant to calculation methods and applicable to all building types and climate zones, is a comprehensive metric for assessing the thermal performance of building envelopes while accounting for the various building characteristics. The sensitivity and range of BECOP were assessed for Canadian climate and construction methods. Using case studies, BECOP results revealed that current practices and regulations pertaining to the building envelope are inconsistent and fail to provide any measure of efficiency. It was also found that current building envelope technologies are not energy efficient. A max BECOP value of 35% is obtained for the best building envelope technology, revealing inefficiencies and energy saving potentials.

Keywords: energy efficiency; building envelope; coefficient of performance; BECOP; housing; codes

Citation: Chidiac, S.E.; Marjaba, G.E. BECOP: A New Metric for Measuring the Energy Performance of the Building Envelope. *Buildings* **2022**, *12*, 553. https://doi.org/10.3390/buildings12050553

Academic Editors: Gianpiero Evola and Elena Lucchi

Received: 18 March 2022
Accepted: 22 April 2022
Published: 26 April 2022

Publisher's Note: MDPI stays neutral with regard to jurisdictional claims in published maps and institutional affiliations.

1. Introduction

The need to balance between growing global demands for energy and sustainability is paramount; however, its realization is stunted by today's technologies, knowledge, and policies [1–3]. For reference, the built environment in Canada produces about 17% of greenhouse gas (GHG) emissions [4], with 14% attributed to existing residential buildings corresponding to 17% of all combined energy [5]. In contrast, CAD 12 billion in energy saving were realized in Canada in 2013 through residential energy conservation measures [4]. These statistics are supported by an econometric multivariate analysis, where inefficient thermal envelopes and heating systems have been identified as the dominant energy inefficiency problems in households [6].

Buildings' energy consumption depends on the climate, orientation, size, occupants, building envelope specifications, HVAC system specifications, lighting specifications, available controls and equipment, etc. [7]. Regulating buildings' energy efficiencies in codes and standards, which is increasingly being sought, [8,9], is starting to be recognized as one of the most cost-effective tools for achieving energy efficiency in buildings. In 2017, the Building Code of British Columbia (BCBC) prescribed the highest level of performance as annual net zero energy (NZE) consumption during occupancy [10]. Towards this objective, thermal energy demand intensity (TEDI) and mechanical energy use intensity (MEUI) are utilized to regulate the energy efficiency of buildings [10]. An examination of the premise of these two metrics reveals that TEDI and MEUI are not compatible and that only MEUI has been standardized. The MEUI includes absolute measures of efficiency, whereas the TEDI includes relative measures of thermal performance being climate condition and building archetype. Accordingly, building envelope energy efficiency as an absolute measure is missing.

Typical metrics employed to assess buildings energy efficiency are through qualitative comparative analysis (QCA), where the causal effects of the building characteristics on the energy consumption are measured. The representative metrics include (1) energy consumption per unit time, usually annual totals [11]; (2) energy use intensity (EUI) [12]; (3) relative energy consumption, as opposed to absolute energy consumption [13]; (4) hybrid or combinations of the first three; and (5) other metrics, mostly calculated or deduced from regression type models [14–18]. Moreover, economic justifications, which have been an inherent rationale for the incremental changes in codes and standards, have been substantiated by QCA results. The implications have been detrimental to sustainability, being ecological and economical, and most critically, the inability to quantify actual energy saving potentials has stunted the development of new and innovative energy efficient building envelope components and systems. This postulation is confirmed by findings of scientific studies reported in the literature:

(1)   A multi-objective optimization study was carried out to assess the EU prescribed cost-optimal approach of a balance of energy and economic targets [19]. The study showed that a zero-energy target is possible with current technology, provided a lower indoor thermal comfort is allowed [19]. Accordingly, energy efficiencies of the current building envelope technologies are not adequate to meet zero-energy target.

(2)   Results of life cycle cost implications of energy efficiency measures in new residential buildings reveal that higher levels of energy efficiency requirements via building regulations are justified based on both economic and environmental grounds [20]. Findings demonstrate that the current energy efficiency requirements, particularly for the building envelope, are too low for new residential buildings to meet sustainability requirements.

(3)   Results from a net zero energy buildings (NZEB) study show that increasing the thermal energy efficiency of the building envelope is a step towards fulfilling all of the NZEB balances [15]. Others have reported that increasing the building envelopes insulative properties is more economically and ecologically effective in colder climates and less effective in warmer climates, depending on the internal heat loads [21]. Findings on NZEB, which agree with the previous studies' findings, confirm that the energy efficiency of the building envelope needs significant improvement and that the design requirements are climate dependent.

(4)   A review of building envelope components for passive buildings concluded that the additional cost of an energy efficient building envelope can be recouped by the reduced size of mechanical systems [22]. The results confirm that improving the thermal resistance of the building envelope is both economically and ecologically viable.

(5)   Results from case studies conducted on a house located in Toronto Canada show that a 70% reduction in energy consumption of code minimum requirements by improving the building envelope's thermal properties is achievable with a less than 7% increase in the construction budget [23].

In brief, the results from several scientific studies reveal with certainty that higher energy efficiencies for the building envelope are needed to meet the NZEB target and that the most cost-effective and ecologically sound house design is always more energy efficient than the current energy code requirements. Accordingly, this study was undertaken to review the progress of Canada's National Codes pertaining to the thermal performance of building envelopes for residential buildings with specific focus on housing, to discuss the implications of the codes on the development of new technologies, and to postulate a new metric for assessing the thermal efficiency of building envelopes. Case studies are then presented to demonstrate the range, sensitivity and applications of the proposed thermal efficiency metric for the building envelope.

2. Historical Development of Building Envelope Energy Efficiency Measures for Canadian Housing

2.1. Chronological Review

The National Building Code of Canada (NBC) first issued in 1941, is the model building code of Canada. Since 1960, NBC was revised every 5 years except for the change from prescriptive to objective-based codes between the 1995 and 2005 editions. Review of the NBC 1985 to 2010 pertaining to Housing and Small Buildings reveals that energy efficiency was not part of the code requirements. The prescribed requirements were for thermal insulation, air leakage and vapor barriers to prevent moisture condensation and to ensure comfortable conditions for the occupants [Article 9.26.2.1, [24]; Article 9.25 [25]]. In 2012 and through a special amendment for Part 9 of NBC, energy efficiency requirements were added to Section 9.36 in a 2012 Amendment [25]. "The Environment" was added as an NBC objective in 2012 to mitigate the probability of harming the environment due to excessive use of energy [OE1.1] [25]. The corresponding minimum requirements aimed at energy efficiency were prescribed using three paths to compliance: prescriptive, tradeoff, and performance. For the prescriptive path, the effective thermal resistances were specified for the building envelope and are reproduced in Tables 1 and 2 for reference.

Table 1. (a) Effective thermal resistance of above ground opaque assemblies in building without heat recovery ventilator [26]. (b) Effective thermal resistance of above ground opaque assemblies in building with heat recovery ventilator [26].

Above-Ground Opaque Building Assembly	Heating Degree Days of Building Location, in Celsius Degree Days					
	Zone 4	Zone 5	Zone 6	Zone 7A	Zone 7B	Zone 8
	<3000	3000 to 3999	4000 to 4999	5000 to 5999	6000 to 6999	≥7000
	Minimum Effective Thermal Resistance (RSI), (m^2 K)/W					
(a)						
Ceiling below attics	6.91	8.67	8.67	10.43	10.43	10.43
Cathedral ceilings and flat roofs	4.67	4.67	4.67	5.02	5.02	5.02
Walls	2.78	3.08	3.08	3.08	3.85	3.85
Floors over unheated spaces	4.67	4.67	4.67	5.02	5.02	5.02
(b)						
Ceiling below attics	6.91	6.91	8.67	8.67	10.43	10.43
Cathedral ceilings and flat roofs	4.67	4.67	4.67	5.02	5.02	5.02
Walls	2.78	2.97	2.97	2.97	3.08	3.08
Floors over unheated spaces	4.67	4.67	4.67	5.02	5.02	5.02

Table 2. Thermal conductance of fenestration and doors [26].

Components	Heating Degree Days of Building Location, in Celsius Degree Days					
	Zone 4	Zone 5	Zone 6	Zone 7A	Zone 7B	Zone 8
	<3000	3000 to 3999	4000 to 4999	5000 to 5999	6000 to 6999	≥7000
	Maximum U-Value, W/(m^2 K), Minimum Energy Rating in Brackets (if Available)					
Fenestration and doors	1.80 (21)	1.80 (21)	1.60 (25)	1.60 (25)	1.40 (29)	1.40 (29)
Skylights	2.90	2.90	2.70	2.70	2.40	2.40

The Model National Energy Code of Canada for Buildings, introduced in 1997 [27], was Canada's first national standard for building energy performance that was updated in 2011 and renamed the National Energy Code of Canada for Buildings (NECB) [28] and further updated in 2015 and 2017 [29,30] to ensure a high level of energy efficiency in new Canadian buildings. An objective of NECB is sustainable and energy efficient buildings with a focus on five key building elements: building envelope, lighting, HVAC, water heating, and electrical power systems and motors. Building envelope, which is the objective of this study, includes floors, walls, windows, doors and roofing, and air infiltration rates. Like the NBC, the NECB offers three compliance paths: prescriptive, trade-off, and performance. The prescriptive requirements for the building envelope thermal properties are reproduced from 1970 to 2017 per climate zone for wall, roof, ground floor and window in Tables 3–6, respectively [28–32]. Comparing NBC 2015 Part 9 and NECB 2017 building envelope's thermal resistance requirements reveals that the latter prescribes higher energy efficiency requirements. Accordingly, NECB data are analyzed to critically assess the code's approach vis-a-vie energy efficiency.

Table 3. Thermal conductance of wall (W/m^2 K) [28–32].

Climate Zone	1970		2007				2011	2015	2017
	Electric Heating	Non-Electric Heating	Steel Frame		Wood Frame				
			Continuous Insulation	Cavity Insulation	Continuous Insulation	Cavity Insulation			
Zone 4	0.40	0.62	0.75	0.38	1.14	0.38	0.315	0.315	0.315
Zone 5	0.34	0.51	0.57	0.27	-	0.27	0.278	0.278	0.278
Zone 6	0.34	0.51	0.57	0.27	0.57	0.38	0.247	0.247	0.247
Zone 7A	0.34	0.51	0.57	0.27	0.57	0.27	0.210	0.210	0.210
Zone 7B	0.34	0.51	0.57	0.27	0.57	0.27	0.210	0.210	0.210
Zone 8	0.28	0.45	0.57	0.27	0.57	0.27	0.183	0.183	0.183

Table 4. Thermal conductance of roof (W/m^2 K) [28–32].

Climate Zone	1970		2007				2011	2015	2017
	Electric Heating	Non-Electric Heating	Attic Space		Without Attic Space				
			Wood Frame	Steel Frame	Wood Frame	Steel Frame			
Zone 4	0.51	0.68	0.19	0.19	0.26	0.26	0.227	0.227	0.193
Zone 5	0.45	0.62	0.13	0.13	0.22	0.19	0.183	0.183	0.156
Zone 6	0.45	0.62	0.12	0.12	0.15	0.15	0.183	0.183	0.156
Zone 7A	0.45	0.62	0.12	0.12	0.15	0.15	0.162	0.162	0.138
Zone 7B	0.45	0.62	0.12	0.12	0.15	0.15	0.227	0.162	0.138
Zone 8	0.45	0.57	0.11	0.11	0.15	0.15	0.183	0.142	0.121

Table 5. Thermal conductance of ground floor (W/m^2 K) [28–32].

Climate Zone	1970		2007		2011	2015	2017
	Heating Source		Wood Frame	Steel Frame			
	Electric Heating	Non-Electric Heating	Cavity Insulation				
Zone 4	0.51	0.68	0.27	0.15	0.227	0.227	0.227
Zone 5	0.45	0.62	0.23	0.15	0.183	0.183	0.183
Zone 6	0.45	0.62	0.23	0.15	0.183	0.183	0.183
Zone 7A	0.45	0.62	0.15	0.15	0.162	0.162	0.162
Zone 7B	0.45	0.62	0.15	0.15	0.162	0.162	0.162
Zone 8	0.45	0.57	0.15	0.15	0.142	0.142	0.142

Table 6. Thermal conductance of window (W/m^2 K) [28–32].

Climate Zone	2007	2011	2015	2017
Zone 4	2	2.4	2.4	2.1
Zone 5	2	2.2	2.2	1.9
Zone 6	2	2.2	2.2	1.9
Zone 7A	2	2.2	2.2	1.9

Table 6. *Cont.*

Climate Zone	2007	2011	2015	2017
Zone 7B	2	2.2	2.2	1.9
Zone 8	2	2.6	1.6	1.4

2.2. Analytical Review of the Building Envelope Minimum Requirements

Historically, the progression of the building envelope thermal resistance requirements is through a percentage increase in thermal resistance. Comparative analysis of the NECB wall properties shows that, on average, thermal resistance increased by 72% and 29% for the year 2007 and 2011, and 0% thereafter. For the NECB roof properties requirement, the thermal resistance increased by 260%, −7%, 11% and 17% for the year 2007, 2011, 2015 and 2017, respectively. For the ground floor, thermal resistance increased by 230% and 10% for the year 2007 and 2011, and 0% thereafter. As for the windows, thermal resistance decreased, on average, by 13% for the 2011 edition and remained the same, except for Zone 8 in the 2015 edition, and then increased, on average, by 15% for the 2017 edition. Although the motivation is energy efficiency, the logic supporting the changes is not consistent and appears to be arbitrary. The normalized heat transfer rate through the building envelope, calculated according to Equation (1), is employed to assess the impact of the thermal properties requirements on the building energy consumption.

$$\dot{q} = U \cdot \text{HDD}_{\text{Avg}} \tag{1}$$

In which U and HDD_{Avg} are the conductance (W/m^2 K) and average heating degree days corresponding to the climate zone, respectively. HDD for Canadian climate are given in Table 7. The results, plotted in Figures 1–4, corresponding to wall, roof, ground floor and window, respectively, reveal that (a) the heating energy is designed to increase with HDD, (b) the heat transfer rate differs for the different building envelope systems, and (c) the requirements which are incremental hardly changed for the past 10 years. Accordingly, and focusing solely on the building envelope, NECB requirements are designed to accept higher heating energy with increased HDD, which is counter intuitive from an economic and ecological perspective. Moreover, NECB assesses improvements through comparison with the preceding thermal resistance requirements. This methodology, which is adopted by most, if not all, codes. However, although it is sound mathematically, it is misleading as it measures improvements with the worst case and not the best or perfect case and is deterrent to technology development, as the potential energy savings are never realized. As such, a consistent and comprehensive metric for measuring the energy performance of the building envelope is needed.

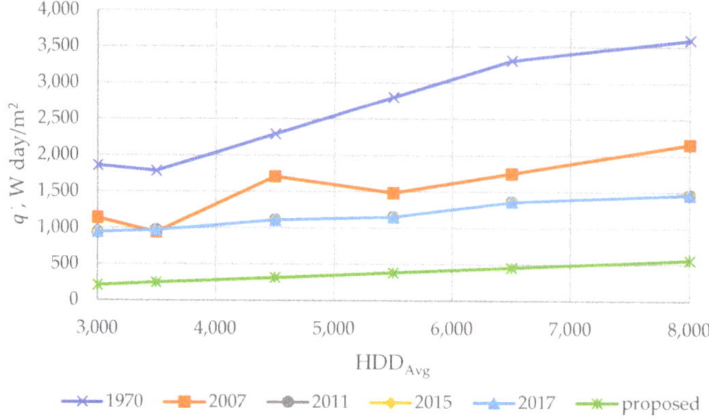

Figure 1. Normalized heat transfer through the walls per code specified properties.

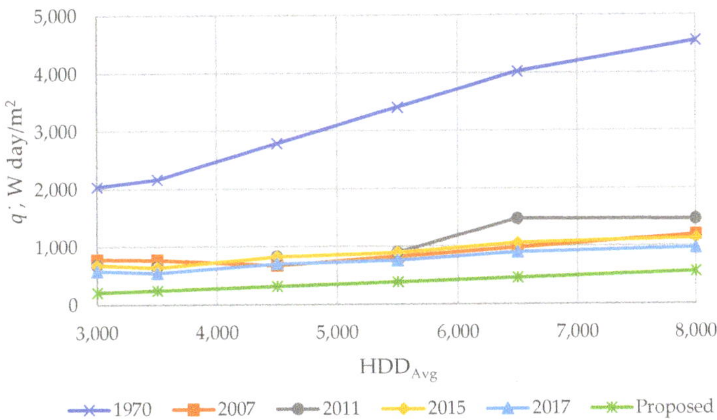

Figure 2. Normalized heat transfer through the roof per code specified properties.

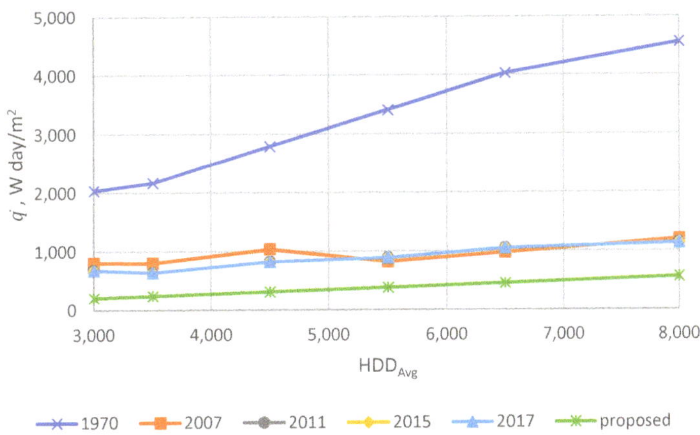

Figure 3. Normalized heat transfer through the ground floor per code specified properties.

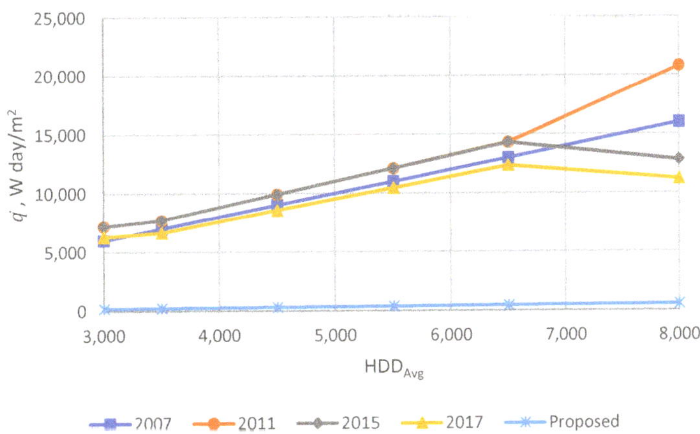

Figure 4. Normalized heat transfer through the windows per code specified properties.

Table 7. Heating degree days for Canadian climate zones.

Zone	HDD Below 1 °C	HDD$_{\text{Avg}}$
4	<3000	3000
5	3000–3999	3500
6	4000–4999	4500
7A	5000–5999	5500
7B	6000–6999	6500
8	≥7000	8000

3. Building Envelope Energy Efficiency Measure

Building energy efficiency metrics are assessment measures prescribed to compare the buildings energy consumption. For metrics that include mechanical equipment energy consumption, they include datums in the form of equipment efficiencies and COP that are standardized and embedded in the assessment measures. These datums ensure that the assessment employs the same yardstick and consistency for all equipment and systems. Whereas for metrics specifically designed for capturing the building envelope performance, the datums resemble more of moving averages in the form of compliance targets, which render the metrics inconsistent and irregular. To overcome this deficiency, an efficiency measure analogous to the mechanical equipment is adapted by postulating a "perfect" building envelope system as a datum. A "perfect" thermal insulating medium with a zero-thermal transmittance would be ideal for this application, except that a zero value for thermal property is problematic in energy modelling and mathematically. Accordingly, an equivalent "ideal" system for the building envelope is proposed that comprises the following properties:

- Thermal conductance: 7.0×10^{-2} W/m^2 K;
- Thermal diffusivity: 1.05×10^{-5} m^2/h;
- Absorptance: 0.2;
- Air leakage: 0.1 ACH at atmospheric pressure.

The proposed "ideal" building envelope system provides a datum for assessing the thermal performance, similar or analogous to the COP of equipment, and to be used as the "ideal" building envelope (walls, roof, slabs, windows, and doors). As such, this metric is referred to as the Building Envelope Coefficient of Performance (BECOP), where

$$\text{BECOP} = \frac{q_{\text{Ideal}}}{q_{\text{BE}}} \tag{2}$$

In which q_{Ideal} and q_{BE} are the heat losses through the building envelope, while employing the ideal system and the subject building, respectively. BECOP provides a consistent and relative measure of the building envelope thermal performance while keeping all other building variables the same. The metric is independent of the calculation method, being energy modelling tools, hand calculations or any other statistical or hybrid tools, and requires that the same analysis method be used throughout. The proposed BECOP is applicable to all building types and Climate Zones, and is designed to provide a measure that is compatible and comparable to existing systems within the building, such as HVAC, lighting, etc. Accordingly, the energy saving potential of the building envelope shifts from a passive to an active approach, where large energy savings can be realized [22]. In this study, the energy modelling tool EnergyPlus [33] is used to demonstrate the applicability, versatility and sensitivity of BECOP.

3.1. Range of BECOP

The range of BECOP for Canada is gauged by employing extreme Climate Zones 4 and 8 along with three levels of design specifications of the building envelope thermal performance, referred to as "low", "typical" and "high" thermally efficient, relative to North American construction practices. The corresponding building characteristics and envelope properties are given in Tables 8 and 9. The building is a single dwelling house, two-storeys high, with a basement, rectangular in shape, and long face oriented in the E-W direction (90° to North), and has the same window areas on all four sides. The plug loads, lighting loads, occupancy loads, domestic hot water load, etc., and their respective schedules are taken from the National Building Code of Canada Section 9.36 [24] and/or the National Energy Code for Buildings [30] and are given in Table 10.

Table 8. Properties and Characteristics of a House Located in Abbotsford, BC (Zone 4).

Building Envelope Properties	Building Envelope Thermal Efficiency		
	Low	Typical	High
HDD18	2920	2920	2920
CDD18	74	74	74
Floor Footprint Area (m^2)	118.45	118.45	118.45
Aspect Ratio	1.5	1.5	1.5
Window-to-wall Ratio	60%	40%	20%
Wall Height (m)	2.74	2.74	2.74
Overall Wall U-value (W/m^2 K) [R-value]	0.32 [R18]	0.159 [R36]	0.103 [R55]
Attic U-value (W/m^2 K) [R-value]	0.189 [R30]	0.095 [R60]	0.072 [R79]
Foundation Wall Overall U-value (W/m^2 K) [R-value]	0.322 [R18]	0.169 [R34]	0.172 [R33]
Area Weighted Average Window U-value (W/m^2 K)	3.166 [R1.8]	1.704 [R3.3]	0.836 [R6.8]
Area Weighted Average Window SHGC	0.493	0.267	0.25
Air Tightness (ACH at atm.)	0.75	0.35	0.1

Table 9. Properties and Characteristics of a House Located in Iqaluit, NU (Zone 8).

Building Envelope Properties	Building Envelope Thermal Efficiency		
	Low	Typical	High
HDD18	9924	9924	9924
CDD18	0	0	0
Floor Footprint Area (m^2)	118.45	118.45	118.45
Number of Stories	2	2	2
Aspect Ratio	1.5	1.5	1.5
Window-to-wall Ratio	60%	40%	20%
Orientation (degrees)—90° is south facing	90	90	90
Number of Basements	1	1	1
Wall Height (m)	2.74	2.74	2.74
Overall Wall U-value (W/m^2 K) [R-value]	0.32 [R18]	0.159 [R36]	0.103 [R55]
Attic U-value (W/m^2 K) [R-value]	0.189 [R30]	0.095 [R60]	0.072 [R79]
Foundation Wall Overall U-value (W/m^2 K) [R-value]	0.322 [R18]	0.169 [R34]	0.172 [R33]
Area Weighted Average Window U-value (W/m^2 K)	3.166 [R1.8]	1.704 [R3.3]	0.836 [R6.8]
Area Weighted Average Window SHGC	0.493	0.267	0.25
Air Tightness (ACH at atm.)	0.75	0.35	0.1

Table 10. Operational and electrical specifications of a single dwelling house.

Occupants	
Number of occupants	4
Occupancy Schedule	NECB 2017 Schedule G Table A-8.4.3.2.(1)-G
Setpoints	
Heating	20 °C
Cooling	25 °C
Setbacks	None

Table 10. *Cont.*

Lighting	
Target Illuminance	150 lux
Normalized Power Density	6.25 W/m^2
Lighting Schedule	Simplified: 12 am to 4 pm—0 4 pm to 11 pm—1 11 pm to 12 am—0
Equipment	
Power Density	4.25 W/m^2
Equipment Schedule	NBC 2015 Table 9.36.5.4
Domestic Hot Water	
Peak Flow Rate	0.0000167 m^3/s
Usage Schedule	NBC 2015 Table 9.36.5.8
Daily usage	225 L / house
Natural Ventilation	
Ventilation Rate	0.24 ACH
Schedule	Simplified: 12 am to 4 pm: 0 4 pm to 6 pm: 0.5 6 pm to 10 pm: 1 10 pm to 11 pm: 0.6667 11 pm to 12 am: 0

The thermal properties are derived from current codes, past codes, and expected future codes. The past codes represent older and low energy efficient construction practices, and the expected future codes represent anticipated future technologies and construction practices with higher energy efficiency, as compiled in Tables 8 and 9. The results, presented in Tables 11 and 12 show the energy consumption due to heat loss through the building envelope for the three archetypes and two climate zones. Abbotsford, BC and Iqaluit, NU represent Climate Zones 4 and 8, respectively. The first review of the BECOP reveals the extreme inefficiency of the building envelope from an energy perspective, where the best BECOP is below 35% compared to above 90% for furnaces and other electrical equipment. For low performance building envelopes, the calculated BECOP for Zone 4 is 0.1% and 4.0% for Zone 8. For the typical construction, the BECOP for Zone 4 is 0.2% and 9.2% for Zone 8. For the high-performance envelope, the BECOP for Zone 4 is 0.7% and 32.2% for Zone 8. These values indicate that the current BECOP ranges between 0.1% and 35%. A BECOP of 100% implies that the building envelope thermal performance is equivalent to that of the idealized building envelope.

Table 11. BECOP Values for a House Located in Abbotsford, BC (Climate Zone 4).

Energy Consumption	Building Envelope Thermal Efficiency		
	Low	Typical	High
Total Energy [kWh]	70,658	39,494	22,779
Energy Per Total Building Area [kWh/m^2]	199	116	70
Heating Energy [kWh]	51,650	21,934	5816
Cooling Energy [kWh]	1019	128	7780
BECOP-Heating Energy	0.1%	0.2%	0.7%

Table 12. BECOP Values for a House Located in Iqaluit, NU (Climate Zone 8).

Energy Consumption	Building Envelope Thermal Efficiency		
	Low	Typical	High
Total Energy [kWh]	165,024	80,958	34,586
Energy Per Total Building Area [kWh/m^2]	464	238	106
Heating Energy [kWh]	146,738	63,426	17,604
Cooling Energy [kWh]	144	9	0
BECOP-Heating Energy	4.0%	9.2%	33.2%

Closer examination of the BECOP values reveals that the metric captures the coupled effect of the thermal resistance, Climate Zone, and internal heat gains. With the latter being constant, as the HDD increases, the impact of an efficient building envelope is captured and reflected with an increase in BECOP value. Moreover, the significance is most visible for the high-performance construction, where the BECOP value goes from less than 1% to 33%. If the values are compared across the levels of construction, a clear upward trend is observed from Figure 5. For Zone 8, the impact of the building envelope properties on the BECOP is significant, with the value increasing exponentially to 33%.

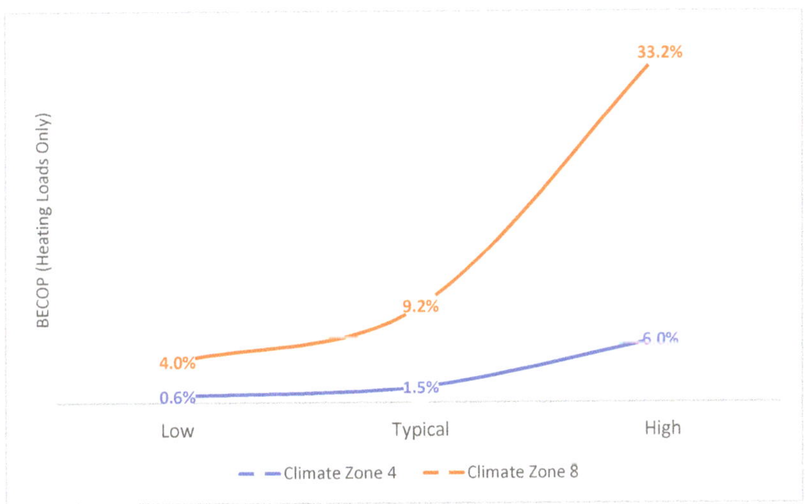

Figure 5. BECOP Values versus Level of Construction.

For further context, the target heating energy consumption of 15 kWh/m^2.year is the current limit prescribed by Passive House [34]. Accordingly, the house annual heating energy consumption would be 4872 kWh for all climate zones. The corresponding BECOP values for Climate Zones 4 and 8 are 0.8% and 120%, respectively. The results clearly show the deficiency and inconsistency in the approach currently followed by codes and standards pertaining to building envelope. The target needs to be a measure of efficiency or performance relative to a datum if energy efficiency is in fact the intended measure.

3.2. Sensitivity of BECOP

The results given in Tables 11 and 12 are further studied to determine the sensitivity of the metric to the Climate Zone, design specification level, and both. For Zone 4, one observes that the BECOP values go from 0.1% to 0.7% and for Zone 8 from 4.0% to 33.2%. A ratio of approximately 9 is observed between the low and high construction, regardless of the Climate Zone. By examining the BECOP values across the Climate Zones, one observes a ratio of approximately 60 between Zone 8 and Zone 4, regardless of the design specification level. The fact that the same building envelope has a higher BECOP in Zone 8 relative to Zone 4 may be counterintuitive if the traditional logic that the envelope in Zone 8 would lose more energy than in Zone 4, simply due to the temperature difference. Instead, BECOP yields the improvement/opportunity potential by quantifying the relative performance of the design to the ideal design while accounting for all the building properties, including internal heat gains. The values indicate that BECOP is sensitive to the climate and that the measure is uniform when the properties of the building envelope are the same. When both the climate and the properties change, the ratio of BECOP is no longer the same, as the impact is amplified by the changes in both the climate and building envelope properties. The ratio of BECOP of a highly efficient building envelope in Zone 8 to a poorly efficient building envelope in Zone 4 is about 475, whereas the ratio of BECOP of a highly efficient building envelope in Zone 4 to a highly efficient building envelope in Zone 8 is about 6. This response is reflected in Figure 5, where the increase in BECOP as a result of an improved building envelope performance is significantly higher in Climate Zone 8, where the improvement is more impactful. On the other hand, in Climate Zone 4, where the improved envelope has a more modest and linear impact, the BECOP displays that effect. These results clearly show the sensitivity of BECOP to the coupled effect of building envelope performance parameters in combination with the Climate Zone, along with the building's properties and characteristics.

4. Application of BECOP

Three case studies are presented to demonstrate the applicability of the proposed metric. The case studies were selected to demonstrate the strength of the metric and the associated benefits/potentials, as well as identify potential weaknesses.

4.1. Case Study 1–Design of a New House

The first case study illustrates how a design professional could employ the BECOP to take inventory of the design decisions. For reference, the house is to be constructed in Toronto, ON, with a total living space specified by the owner to be approximately 240 m^2 without the basement and a ceiling height of 2.74 m. A 20% window to wall ratio (WWR) is selected contingent on the energy consumption. The city of Toronto, ON is in Zone 5 with a corresponding HDD18 and CDD18 of 3892 and 292, respectively [35]. Accordingly, the variables to evaluate are the building orientation and the building envelope specifications. Firstly, the effect of WWR is investigated by considering three possibilities: 20%, 40% and 60%. The properties and characteristics of the house's pre-design are given in Table 13. The corresponding energy consumption and BECOP are given in Table 14. Given the relative thermal properties of the building envelope, the energy consumption due to heating is expected to increase as WWR goes from 20% to 40%. This impact of doubling WWR is captured by BECOP as it drops from 5.4% to 4.4% reflecting a 19% relative loss in efficiency, which can be misleading as the actual loss in efficiency is only 1%. By further increasing the WWR to 60%, BECOP decreases to 3.7%, representing a 30% and 1.7% drop in relative and absolute loss of efficiency, respectively. These results show the significant difference between absolute and relative measure. The actual loss in efficiency is 1% and 1.7% as the WWR increases from 20% to 40% and from 20% to 60%, respectively. Therefore, reporting a relative loss in efficiency of 19% and 30%, which has been the norm for building envelope, can mislead the designer and lead to an erroneous design.

The heating energy consumption increases by 23% and 47% as WWR goes from 20% to 40% and from 20% to 60%, respectively. These results show that a percent increase in energy consumption is linearly proportional to WWR. Comparing the heating energy consumptions with those of BECOP, one observes that the former yields a linear trend, whereas the latter a non-linear one. Moreover, the information in the form of percent change in energy consumption can be misleading, as the results imply that the energy efficiency of the building decreased by 23% when WWR is increased by 20%. A non-apparent and critical implication is the sensitivity of the relative change in energy consumption to the building envelope properties, i.e., as the building envelope properties change the increase in heating energy consumption will be significantly different for the same WWR increases. In contrast, the change in the BECOP reflects the impact of WWR as it is a measure of the overall building envelope efficiency relative to a fixed ideal system.

Table 13. Effect of Window-to-Wall Ratio.

Building Characteristics and Building Envelope Properties	Window-to-Wall Ratio (%)		
	20	**40**	**60**
Floor Footprint Area (m^2)	118.45	118.45	118.45
Number of Stories	2	2	2
Aspect Ratio	1.5	1.5	1.5
Orientation (degrees)—90° is south facing	90	90	90
Number of Basements	1	1	1
Overall Wall U-value (W/m^2 K) [R-value]	0.159 [R36]	0.159 [R36]	0.159 [R36]
Attic U-value (W/m^2 K) [R-value]	0.086 [R66]	0.086 [R66]	0.086 [R66]
Foundation Wall Overall U-value (W/m^2 K) [R-value]	0.322 [R18]	0.322 [R18]	0.322 [R18]
Area Weighted Average Window U-value (W/m^2 K)	1.704 [R3.3]	1.704 [R3.3]	1.704 [R3.3]
Area Weighted Average Window SHGC	0.267	0.267	0.267
Air Tightness (ACH at atm.)	0.25	0.25	0.25

Table 14. Effect of Window-to-Wall Ratio on BECOP.

Energy Consumption	Window-to-Wall-Ratio (%)		
	20	**40**	**60**
Total Energy [kWh]	37,756	42,619	47,735
Energy Per Total Building Area [kWh/m^2]	109	123	138
Heating Energy [kWh]	20,197	24,899	29,752
Cooling Energy [kWh]	29	172	400
BECOP-Heating Energy	5.38%	4.36%	3.65%

In brief, BECOP provides an efficiency pattern that can be used to optimize the design. The loss in BECOP is indicative and intuitive for a designer to understand a loss in efficiency as opposed to increased energy consumption.

Secondly, the orientation of the building using a 40% WWR is investigated by varying the east west orientation (0° to north) to the north–south orientation (90° to north). The results in the form of energy consumption and BECOP are summarized in Table 15. They reveal that a change in orientation has no effect on the BECOP value as it goes from 4.38% to 4.36%. Although the change in BECOP value is considered negligible, it nonetheless shows the sensitivity of the metric to small changes in energy consumption. The heating energy consumptions give the same results. In brief, the minor change in BECOP and heating energy implies that the orientation has no impact on the house energy consumption for this configuration.

Table 15. Effect of Building Orientation on BECOP.

Energy Consumption	Orientation	
	0° to North (Facing E-W)	**90° to North (Facing N-S)**
Total Energy [kWh]	42,717	42,619
Energy Per Total Building Area [kWh/m^2]	124	123
Heating Energy [kWh]	24,918	24,899
Cooling Energy [kWh]	234	172
BECOP-Heating Energy	4.38%	4.36%

Further examination of the results provides an important insight into how the BECOP can provide additional information. With the slight drop in the heating energy from 24,917 kWh to 24,899 kWh, it implies a small benefit can be realized with the house oriented in the N-S direction. In contrast, the BECOP value drops from 4.38% to 4.36% indicating a decrease in the efficiency. This implies that the ideal building experienced a more significant drop in heating energy than did the investigated house, indicating that there are more potentials to improve the building envelope in the orientation facing N-S than in the E-W direction. This information, which is not intuitive from the heating energy consumption alone, is valuable and can lead the designer down the path of seeking further improvements.

4.2. Case Study 2–Retrofit Design for an Existing House

Upgrading the thermal resistance of an existing two-storey single family detached dwelling with a basement located in Toronto, ON is sought. The house is rectangular in shape with an aspect ratio of 1.5, floor area of 118.5 m^2, wall height of 2.74 m, 40% WWR, and long side facing south. The energy renovation measures (ERMs) include thermal upgrading of windows, walls, walls and windows, roof, or walls, windows, and roof. Assuming some budgetary constraint, the designer could estimate what improvements for each option could be achieved within the constraints. If BECOP is calculated for each of the options, the designer would have sufficient information to select the most impactful and cost-effective option. Tables 16 and 17 show the effect of several energy retrofit measures (ERMs), improving the overall wall U-value from 0.159 W/m^2 K (R36) to 0.142 W/m^2 K (R40), improve window U-value from 1.704 W/m^2 K (R3.3) to 0.921 W/m^2 K (R6), both improvements, improving the overall U-value of the attic from 0.086 W/m^2 K (R66) to 0.071 W/m^2 K (R80), and all three improvements combined.

Table 16. ERM Designs.

Building Envelope Properties	ERMs					
	Base	1 (Walls)	2 (Windows)	3 (1 & 2)	4 (Attic)	5 (1, 2 & 4)
Overall Wall U-value (W/m^2 K) [R-value]	0.159 [R36]	0.142 [R40]	0.159 [R36]	0.142 [R40]	0.159 [R36]	0.142 [R40]
Attic U-value (W/m^2 K) [R-value]	0.086 [R66]	0.086 [R66]	0.086 [R66]	0.086 [R66]	0.071 [R80]	0.071 [R80]
Foundation Wall Overall U-value (W/m^2 K) [R-value]	0.322 [R18]	0.322 [R18]	0.322 [R18]	0.322 [R18]	0.322 [R18]	0.322 [R18]
Area Weighted Average Window U-value (W/m^2 K)	1.704 [R3.3]	1.704 [R3.3]	0.921 [R6]	0.921 [R6]	1.704 [R3.3]	0.921 [R6]
Area Weighted Average Window SHGC	0.267	0.267	0.240	0.240	0.267	0.240
Air Tightness (ACH at atm.)	0.25	0.25	0.25	0.25	0.25	0.25

Table 17. Effect of ERMs on BECOP Values.

Energy Consumption	ERMs					
	Base	1	2	3	4	5
Total Energy [kWh]	42,619	42,316	35,199	34,897	42,477	34,756
Energy/Total Building Area [kWh/m^2]	123	123	102	102	123	101
Heating Energy [kWh]	24,899	24,650	17,497	17,250	24,758	17,111
Cooling Energy [kWh]	172	171	168	167	171	165
BECOP-Heating Energy	4.36%	4.41%	6.21%	6.29%	4.39%	6.35%

From Table 17, it is evident that improving the windows provides the most savings in terms of energy consumption and improving the wall and attic U-values provides minimal benefit. Considering that the wall and attic insulation levels of the base-house are quite high relative to current codes and construction practices, this indicates that they have reached the point of diminishing returns with today's technology. This knowledge is useful to the designer to make an informed decision. Moreover, the BECOP provides insight that is not evident from the energy consumption data, i.e., increasing the wall and attic thermal resistance using today's technology would not improve the efficiency of the building envelope without improving other aspects of the building such as orientation, geometry, air tightness, WWR, etc. This insight allows the designer to investigate other options such as the ones presented in Case Study 1, as well as air tightness improvement, window shading, etc. As such, the path and design decisions will be guided by the information embedded in the relative changes in the BECOP. The absolute value of the BECOP also provides insight into the fact that there is still an opportunity to improve the performance with novel and advanced materials and systems, that may guide the designer to further explore. Moreover, the BECOP value accounts for the whole building envelope and rewards for having a compatible thermal resistance envelope, something which cannot be discerned directly from energy consumption data.

The changes in the BECOP value for each ERM relative to the base case are compared to the corresponding changes in energy consumptions, Table 18. The BECOP values indicate that Option 3 provides a 1.93% increase in the building envelope efficiency, which is significantly less than the 31% reduction in the heating energy consumption. Moreover, changes in BECOP values of 0.05% and 0.03% are obtained for upgrading the wall and attic with a corresponding 1% and 0.6% reduction in heating energy. BECOP provides a measure of the building envelope efficiency which is different from energy savings. Although it is more appealing to report a saving of 31% in heating energy consumption, albeit it is a real measure, it dissuades from realizing that the heating energy saving potential for the house is significantly greater than the one obtained.

Table 18. Percent Reduction in Energy Consumption due to ERMs.

Energy Consumption	**ERM–1**	**ERM–2**	**ERM–3**	**ERM–4**	**ERM–5**
Total Energy	0.7	17.4	18.1	0.3	18.5
Energy/Total Building Area	0.1	17.4	17.6	0.3	18.0
Heating Energy	1.0	29.7	30.7	0.6	31.3
Cooling Energy	0.9	2.5	3.4	0.7	4.1
BECOP-Heating Energy	0.05	1.85	1.93	0.03	1.99

4.3. Case Study 3–Regulatory Compliance

Efficiency and COP are measures used to assess the absolute performance of equipment or systems. For buildings, these absolute measures are only prescribed for the electrical and mechanical equipment and systems such as the lighting, HVAC, pumps, fans, etc. For the building envelope thermal performance, there are no absolute measures for efficiency or COP. Present practice is to either specify a minimum thermal resistance for each sub-system (wall, window, roof, and floor) based on the climate zone, or an annual energy use and/or intensity. The regulators supporting rationale stems from comparative energy consumption, statistical analysis, or both. Case study 2 is a prime example where a 31% reduction in the heating energy consumption would resonate well with regulators not knowing that the savings correspond to 1.93% increase in the building envelope efficiency. In brief, the current approach does not provide an absolute measure of the thermal efficiency of the building envelope or account for the thermal compatibility of the various sub-systems that form the building envelope system.

BECOP is a simple, practical, and performance-based metric for regulating the energy efficiency of building envelopes. For illustration purpose, a BECOP of 4.50% as a minimum requirement for the City of Toronto is prescribed. This approach specifies an absolute efficiency measure as well as allows flexibility in the design to achieve the desired BECOP. From the results of the previous case studies, one can establish that a 20% WWR would meet the requirement, Table 14. From Table 15, the thermal resistance of the base house needs to be upgraded enough to improve its BECOP from 4.4% to 4.5%. Alternatively, upgrading the windows would provide more than sufficient improvement to comply with the regulation. Different approaches, designs or combinations are possible to achieve the same BECOP target, which is directly related to actual energy performance. Moreover, BECOP can be used early in the design process to account for the orientation and geometry, among other properties at no cost.

Replacing a target heating energy consumption with a BECOP value would revolutionize the regulatory compliance requirements. It would transform a deficient and inconsistent approach currently followed by codes and standards pertaining to building envelope to a measure of its performance relative to a well-defined datum. Moreover, BECOP measures the efficiency of the building envelope while accounting for the entire building properties, characteristics, climate-zone, occupancy, and operation.

5. Discussion

BECOP was developed to measure the thermal performance of the building envelope in a useful, consistent, and systematic manner. The benefits and strengths of BECOP were noted while analyzing the results of the case studies. A noted weakness is the range of the BECOP given the low efficiency of the building envelope compared to the ideal system. Nonetheless, this weakness can become a catalyst for designing a more efficient building envelope. Furthermore, BECOP values showcase the energy saving potentials that can be realized with newer and innovative building envelope systems.

The ideal system was inspired by a perfect opaque vacuum (zero conductivity, convection, and radiation properties), which was then translated into practical values for BECOP. BECOP can accommodate future advances in the building envelope technologies as it is a measure of performance and not a direct measure of efficiency.

For this study, the focus was on BECOP due to heating given Canada's climate and was measured while accounting for all internal gains. An alternative approach is to exclude the internal gains as they depend on the occupancy and use of the building and not on the thermal performance of the building envelope. Accordingly, BECOP will provide an impartial measure of the building envelope thermal efficiency for both heating and cooling.

As demonstrated through the case studies, a significant capability of the metric is its ability to capture the performance and compatibility of the system as a whole, which traditional metrics currently used (total energy, TEUI, TEDI, etc.) are not able to capture, since there is no built-in datum. The BECOP achieves this target by ensuring that a compatible design is rewarded relative to an incompatible design. This results in a manual optimization of certain parameters, which is not possible to achieve without having an optimal design (or "idealized" design). Another advantage of the BECOP is its ability to penalize missed opportunities in the system design and reward the captured opportunities. For example, the BECOP will capture and inform the designer if, for a particular orientation, improving the thermal resistance of particular components does not offer a benefit. On the other hand, the same properties for a different orientation would further penalize the savings not achieved by that configuration. This is further demonstrated when comparing a fixed and absolute space heating EUI target (e.g., 15 kWh/m^2.year). This target does not address the increased difficulty in achieving a set level of performance in colder climates, with the argument that colder climate requires a thermally efficient building envelope among other energy systems. The corresponding BECOP values identify which building envelope still has room for improvement (BECOP 0.8% in Climate Zone 4), and which one requires to be 20% more efficient than the idealized system (BECOP 120% or 1.2 in Climate Zone 8). These BECOP values show that it is not possible to build to PassivHaus in climate Zone 8 and demonstrate the impracticality of imposing a space heating EUI target for all climate zones.

6. Conclusions & Recommendations

The results from this study have revealed the following conclusions:

- Current practices and regulations pertaining to the building envelope appears arbitrary and do not provide a defined measure of efficiency.
- BECOP, which provides a performance measure, captures any deviations from an idealized system, and yields a measure of efficiency and thermal compatibility of the building envelope.
- BECOP, which is invariant to the calculation methods and applicable to all building types and climate zones, exposes the difference between energy savings and building envelope efficiency.
- QCA for the purpose of energy efficiency design can be misleading in establishing the optimal design and is deterrent to technology development as the potential energy savings are never realized.
- Maximum BECOP values of 35% reveal the inefficiencies in the current building envelope technologies and the building envelope energy saving potentials.
- BECOP provides a measure of the distance away from an optimized/idealized design.
- The BECOP, or similar metric that utilizes a fixed ideal datum is a step in the right direction to revolutionize the regulatory methodology and philosophy and, subsequently, demand innovation from the construction industry, ensuring a positive economic and ecological impact.

This study is a first step in highlighting the differences between heating energy savings and efficiency of the building envelope. Accordingly, it is recommended that further studies be carried out to refine the properties of the ideal datum, as well as carry out extensive sensitivity analyses to guide in the interpretation of the BECOP values.

Author Contributions: Conceptualization, S.E.C. and G.E.M.; Formal analysis, G.E.M.; Investigation, G.E.M.; Methodology, S.E.C.; Resources, S.E.C.; Supervision, S.E.C.; Writing—original draft, G.E.M.; Writing—review and editing, S.E.C. All authors have read and agreed to the published version of the manuscript.

Funding: This research received no external funding.

Institutional Review Board Statement: Not applicable.

Informed Consent Statement: Not applicable.

Data Availability Statement: Not applicable.

Acknowledgments: This research was funded by McMaster University Centre for Effective Design of Structures.

Conflicts of Interest: The authors declare no conflict of interest.

References

1. Marjaba, G.E.; Chidiac, S.E. Sustainability and resiliency metrics for buildings—Critical review. *Build. Environ.* **2016**, *101*, 116–125. [CrossRef]
2. Nelms, C.; Russell, A.D.; Lence, B.J. Assessing the performance of sustainable technologies for building projects. *Can. J. Civ. Eng.* **2005**, *32*, 114–128. [CrossRef]
3. Dadzie, J.; Runeson, G.; Ding, G. Sustainable technologies as determinants of energy efficient upgrade of existing buildings. In Proceedings of the 2018 International Conference on Smart Grid and Clean Energy Technologies (ICSGCE), Kajang, Malaysia, 29 May–1 June 2018; pp. 145–149.
4. Government of Canada. *Pan-Canadian Framework on Clean Growth and Climate Change: Canada's Plan to Address Climate Change and Grow the Economy*; Government of Canada: Ottawa, ON, Canada, 2016.
5. Natural Resources Canada. *Energy Use Data Handbook (1990 to 2016)*; Natural Resources Canada: Ottawa, ON, Canada, 2019.
6. Gamtessa, S.F. An explanation of residential energy-efficiency retrofit behavior in Canada. *Energy Build.* **2013**, *57*, 155–164. [CrossRef]
7. Babaei, T.; Abdi, H.; Lim, C.P.; Nahavandi, S. A study and a directory of energy consumption data sets of buildings. *Energy Build.* **2015**, *94*, 91–99. [CrossRef]
8. Evans, M.; Yu, S.; Staniszewski, A.; Jin, L.; Denysenko, A. The international implications of national and local coordination on building energy codes: Case studies in six cities. *J. Clean. Prod.* **2018**, *191*, 127–134. [CrossRef]
9. Rosenberg, M.; Jonlin, D.; Nadel, S. A Perspective of Energy Codes and Regulations for the Buildings of the Future. *J. Sol. Energy Eng.* **2016**, *139*, 010801. [CrossRef]

10. Province of British Columbia. *Regulation of the Minister of Natural Gas Development and Minister Responsible for Housing and Deputy Premier Ministerial Order No. 158*; Province of British Columbia: Victoria, BC, Canada, 2017.

11. Thomsen, K.E.; Rose, J.; Morck, O.; Jensen, S.Ø.; Østergaard, I. Energy consumption in an old residential building before and after deep energy renovation. *Energy Procedia* **2015**, *78*, 2358–2365. [CrossRef]

12. Wang, E. Benchmarking whole-building energy performance with multi-criteria technique for order preference by similarity to ideal solution using a selective objective-weighting approach. *Appl. Energy* **2015**, *146*, 92–103. [CrossRef]

13. Wang, E.; Shen, Z.; Alp, N.; Barry, N. Benchmarking energy performance of residential buildings using two-stage multifactor data envelopment analysis with degree-day based simple-normalization approach. *Energy Convers. Manag.* **2015**, *106*, 530–542. [CrossRef]

14. Mathew, P.; Greenberg, S.; Sartor, D.; Rumsey, P.; Weale, J. Metrics for energy efficiency. *ASHRAE J.* **2008**, *50*, 40–47.

15. Mohamed, A.; Hasan, A.; Sirén, K. Fulfillment of net-zero energy building (NZEB) with four metrics in a single family house with different heating alternatives. *Appl. Energy* **2014**, *114*, 385–399. [CrossRef]

16. Arregi, B.; Garay, R. Regression analysis of the energy consumption of tertiary buildings. *Energy Procedia* **2017**, *122*, 9–14. [CrossRef]

17. Chidiac, S.E.; Catania, E.J.C.; Morofsky, E.; Foo, S. Effectiveness of single and multiple energy retrofit measures on the energy consumption of office buildings. *Energy* **2011**, *36*, 5037–5052. [CrossRef]

18. Chidiac, S.E.; Catania, E.J.C.; Morofsky, E.; Foo, S. A screening methodology for implementing cost effective energy retrofit measures in Canadian office buildings. *Energy Build.* **2011**, *43*, 614–620. [CrossRef]

19. Penna, P.; Prada, A.; Cappelletti, F.; Gasparella, A. Multi-objectives optimization of Energy Efficiency Measures in existing buildings. *Energy Build.* **2015**, *95*, 57–69. [CrossRef]

20. Morrissey, J.; Horne, R.E. Life cycle cost implications of energy efficiency measures in new residential buildings. *Energy Build.* **2011**, *43*, 915–924. [CrossRef]

21. Li, D.H.W.; Yang, L.; Lam, J.C. Zero energy buildings and sustainable development implications—A review. *Energy* **2013**, *54*, 1–10. [CrossRef]

22. Sadineni, S.B.; Madala, S.; Boehm, R.F. Passive building energy savings: A review of building envelope components. *Renew. Sustain. Energy Rev.* **2011**, *15*, 3617–3631. [CrossRef]

23. Saunders, K. *The Toronto Method of Construction—Re-Thinking How We Insulate Our Homes*; Sustainable Building: Toronto, ON, Canada, 2016.

24. Canadian Commission on Buildings and Fire Codes. *National Building Code of Canada*, 2015th ed.; National Research Council of Canada: Ottawa, ON, Canada, 1985. Available online: https://nrc-publications.canada.ca/eng/view/object/?id=453db3d5-7d64-4d00-8c6b-fd1fe342992f (accessed on 10 March 2022).

25. Canadian Commission on Buildings and Fire Codes. *National Building Code of Canada*; National Research Council of Canada: Ottawa, ON, Canada, 2010. Available online: https://nrc-publications.canada.ca/eng/view/object/?id=cbd245df-bc91-4033-a538-fb20fcf536a1 (accessed on 10 March 2022).

26. Canadian Commission on Buildings and Fire Codes. *National Building Code of Canada*; National Research Council of Canada: Ottawa, ON, Canada, 2015. Available online: https://nrc-publications.canada.ca/eng/view/object/?id=c8876272-9028-4358-9b42-6974ba258d99 (accessed on 10 March 2022).

27. Canadian Commission on Buildings and Fire Codes. *Model National Energy Code Canada for Buildings*; National Research Council of Canada: Ottawa, ON, Canada, 1997. Available online: https://nrc-publications.canada.ca/eng/view/object/?id=3513c10b-54db-49c8-8d1e-6bd49e9ac597 (accessed on 10 March 2022).

28. Canadian Commission on Buildings and Fire Codes. *National Energy Code of Canada for Buildings*; National Research Council of Canada: Ottawa, ON, Canada, 2011. Available online: https://nrc-publications.canada.ca/eng/view/object/?id=a34734e9-6f66-404f-883c-2361a5a08549 (accessed on 10 March 2022).

29. Canadian Commission on Buildings and Fire Codes. *National Energy Code for Buildings*; National Research Council of Canada: Ottawa, ON, Canada, 2015. Available online: https://nrc-publications.canada.ca/eng/view/object/?id=ef02ee0b-e374-47ed-a0dd-c7bd63c45499 (accessed on 10 March 2022).

30. Canadian Commission on Buildings and Fire Codes. *National Energy Code for Buildings*, 2017th ed.; National Research Council of Canada: Ottawa, ON, Canada, 2017. Available online: https://nrc-publications.canada.ca/eng/view/object/?id=3eea8f31-47ef-4280-86b0-1c148744f8f1 (accessed on 10 March 2022).

31. Associate Committee on the National Building Code and National Research Council of Canada. *Canadian Code for Residential Construction*; National Research Council of Canada: Ottawa, ON, Canada, 1972.

32. *ASHRAE Standard 90.2-2007*; Energy-Efficient Design of Low-Rise Residential Buildings. ASHRAE: Atlanta, BC, Canada, 2007.

33. US Department of Energy. *Getting Started with EnergyPlus: Basic Concepts Manual—Essential Information You Need about Running Energy Plus*; US Department of Energy: Washington, DC, USA, 2010; p. 67.

34. Passive House Canada. *A Developer's Guide to Passive House Buildings*; Passive House Canada: Victoria, BC, Canada, 2017.

35. ASHRAE. Climatic Design Information. In *ASHRAE Fundamentals Handbook*; ASHRAE: Atlanta, GA, USA, 2013.

 *buildings*

Article

Optimal Insulation Assessment, Emission Analysis, and Correlation Formulation for Indian Region

Mehmet Ali Kallioğlu [1], Ahmet Yılmaz [2], Ashutosh Sharma [3], Ahmed Mohamed [4], Dan Dobrotă [5], Tabish Alam [6], Rohit Khargotra [7,*] and Tej Singh [8]

[1] Faculty of Technology, Batman University, Batman 72100, Turkey
[2] Computer Engineering Department of Engineering Faculty, Karamanoğlu Mehmetbey University, Karaman 70200, Turkey
[3] Department of Mechanical Engineering, Dr B R Ambedkar National Institute of Technology, Jalandhar 144011, India
[4] Structural Engineering and Construction Management Department, Future University in Egypt, Cairo 11835, Egypt
[5] Faculty of Engineering, Department of Industrial Engineering and Management, Lucian Blaga University of Sibiu, 550024 Sibiu, Romania
[6] CSIR-Central Building Research Institute, Roorkee 247667, India
[7] Institute of Materials Engineering, Faculty of Engineering, University of Pannonia, 8200 Veszprem, Hungary
[8] Savaria Institute of Technology, Faculty of Informatics, ELTE Eötvös Loránd University, 9700 Szombathely, Hungary
* Correspondence: rohitkhargotra@phd.mk.uni-pannon.hu

Abstract: The current study depicts the effects of different insulation materials and fuel types on the cooling and heating performance of buildings situated in hot and dry, warm and humid, composite, and cold climatic conditions in India. Ten different locations chosen from diverse climatic regions were selected, and various potential parameters for expanded polystyrene and extruded polystyrene insulation materials were evaluated. Potential parameters, such as optimal insulation thickness, annual savings, and payback period, were computed for cooling and heating requirements and were found in the ranges of 0.0428–0.891 m, 10.83–19.19 \$/m^2, and 1.49–2.36 years for cooling, as well as 0.0063–0.1522 m, 0.29–55.92 \$/m^2, and 0.95–6.52 years for heating, respectively. An emission analysis was also carried out for the estimation of greenhouse gas (GHG) emissions by the engagement of optimal insulation thickness for heating. The GHG emissions from natural gas, coal, and diesel by the employment of various insulating materials were found in the ranges of 5.39–11.28, 9.47–32.68, and 2.26–4.51 kg/m^2-year, respectively. A correlation formulation (power) for optimal insulation thickness was also carried out. For checking the preciseness of the developed mathematical models, statistical tools were utilized, and their obtained values in the satisfactory range signified the accurateness of the developed models.

Keywords: optimal insulation; environmental analysis; modeling; degree days; correlation

Citation: Kallioğlu, M.A.; Yılmaz, A.; Sharma, A.; Mohamed, A.; Dobrotă, D.; Alam, T.; Khargotra, R.; Singh, T. Optimal Insulation Assessment, Emission Analysis, and Correlation Formulation for Indian Region. *Buildings* 2023, 13, 569. https://doi.org/10.3390/buildings13020569

Academic Editors: Elena Lucchi and Gianpiero Evola

Received: 14 January 2023
Revised: 5 February 2023
Accepted: 13 February 2023
Published: 20 February 2023

1. Introduction

Energy is vital to human existence and advancement. The increasing demand for energy is driven by various factors, such as the development of new technologies, the impact of climate change, and population growth. However, the majority of the world's energy needs are currently met by fossil fuels, which are nonrenewable and finite resources. As a result, these energy sources are becoming increasingly scarce, with depletion happening at an alarming rate. This has led to the urgent need to find sustainable and renewable alternatives to fossil fuels in order to meet the growing energy demands and preserve the planet for future generations [1–3]. The use of thermal energy systems, such as those found in homes and businesses for temperature regulation (heating and cooling), can often result in a loss in thermal energy to the surrounding environment. To counter this, insulation materials are commonly applied to the exterior walls, windows, ceilings, and floors of these buildings. The use of insulation can greatly reduce energy loss and lead to significant energy savings. By minimizing energy loss in this way, the overall carbon footprint of a building can also be reduced [4,5]. The optimum insulation thickness for the exterior walls of a building can depend on several factors, including the climate, location, and intended use of the building. The R-value, which is a measure of thermal resistance, is typically used to determine appropriate insulation thickness for exterior walls. As a general rule, the R-value required for exterior walls in a building depends on the insulation materials, climate zone, overall cost, and building type [6,7]. Nyers et al. [8] developed an investment-saving method for determining the optimal insulation thickness of external walls in Serbia. A mathematical model was developed considering the steady-state one-dimensional heat conduction of walls made up of bricks and polystyrene. The overall savings and payback terms were evaluated at optimal insulation thickness and found to be 8.5 $/m^2 and 1.22 years, respectively. Liu et al. [9] computed the optimal thickness of insulation for external walls in China's cold winter and hot summer zones. A mathematical model that incorporated heat and moisture transfer was developed and used to estimate the annual energy consumption. Two different types of insulation materials, viz., expanded polystyrene (EPS) and extruded polystyrene (XPS), were compared with lifecycle savings and payback period. The optimal insulation thicknesses for XPS and EPS were found at 0.053–0.069 m and 0.081–0.105 m, respectively. The maximum lifecycle saving and the payback period were found at 16.6–28.5 $/m^2 and 1.89–2.56 years, respectively. The results also revealed that the EPS material performed better than XPS, with more lifecycle savings and a shorter payback period. In another study, Yuan et al. [10] considered the effects of both insulation thickness and reflectivity on outer walls in different climatic zones of Japan. The reflectivity and the insulation thickness were supposed to be varied between 0.1 and 0.8 mm and 10 and 100 mm, respectively. The results showed that a low value of reflectivity and thick insulation were preferable for cold regions and vice versa for hot areas. The effect of insulation thickness on enjoinment was studied by Dombayci [11] in Turkey for cold seasons. A one-dimensional steady-state heat conduction model was developed for a sandwiched wall made up of the materials in the following order: plaster, brick, insulation, brick, and plaster. The results revealed that, at optimum insulation thickness, the energy consumption decreased by 46.6%, and the CO_2 and SO_2 emissions were reduced by 41.53%. Kaynakli et al. [12] evaluated the optimum insulation thickness for different applications of insulation material (internal, sandwiched, and external) by taking into account the indoor and outdoor conditions. The key factors on which the optimum thickness of insulation depended were found to be indoor temperature and indoor and outdoor relative humidity. The results indicated that, for low indoor and outdoor relative humidity ($\leq$0.6), the type of insulation application did not affect insulation thickness. However, insulation application over the outer edge of an external wall performed better than all the other cases in the studied conditions. In a different study, Bolattürk [13] evaluated the optimum thickness of foam for sixteen cities from four different climate zones of Turkey. The fuels considered for heating and cooling in their study were coal, natural gas, diesel, LPG, and electricity. The results revealed that the optimum insulation thickness, energy saving, and life cost were in the ranges of 2–17 cm, 22–79%, and 1.3–4.5 years, respectively, depending on the different climate zones. The results also indicated that high values of optimum insulation thickness were evaluated for cold climatic regions with significantly less of a payback period. Mahlia et al. [14] evaluated the optimum insulation thickness for different materials in the climatic conditions of Malaysia and formed correlations as a function of the thermal conductivities of different insulating materials. The result revealed that a nonlinear mathematical model (polynomial) predicted the results most precisely. This correlation was $x_{opt} = a + bk + ck^2$, where a = 0.0818, b = −2.973, and c = 64.6, and "k" in the correlation was the thermal conductivity (W/m $°$C) of the insulating material. Sisman et al. [15] conducted a study and determined optimal insulation thickness for cities in four different climatic zones of Turkey for stone wool material for heating. These properties constituted equations that could be used to specify the optimum point based on the NDD variable for both external walls and roofs. The external wall equation was $x_{opt} = a \times NDD^b$, where $a = 0.001$ and $b = 0.7533$.

India has diversified climatic zones, including hot and dry, warm and humid, composite, cold, and temperate. A few studies have been reported to estimate optimal insulation thicknesses in different climatic zones in India. Sundaram and Bhaskaran [16] optimized insulation thickness for five cities in India, selecting the warm and humid and composite climates. Three different insulating materials attached at the inner sides of outer walls were considered for the study, and various thermo-economic parameters, such as optimum insulation thickness, annual energy cost, annual electric energy consumption, and payback period, were determined. The results revealed that EPS insulation material performed more effectively than the other studied materials in terms of energy saving and payback period. Mishra et al. [17] computed the optimal insulation thickness for Dehradun, located in a cold climatic region of India. Two dissimilar insulation materials, EPS and XPS, were considered for three different walls of dissimilar materials. The heat loss was calculated by the degree day method, and the fuel assumed for heating was natural gas. The results revealed that the optimum insulation thickness and energy saving varied between 5.2 and 7.4 cm and 31.41 and 67.59 $/m^2, respectively. Raza and Aggarwal [18] determined optimum insulation thickness using two different methods, viz., the degree day method and annual full load cooling hours operation. The insulation material was considered to be sandwiched between bricks, and three different insulation materials were investigated. The results revealed that the EPS insulation materials using LPG for heating were the most effective combination. In another study, Singh et al. [19] computed the optimal XPS thickness for various climates of India. The study concluded that, at a 25 °C indoor temperature for all the selected zones, the optimum insulation, annual savings, and payback period were found in the ranges of 0.015–0.031 m, 0.33–2.21 $/m^2, and 3.9–6.7 years, respectively.

The literature survey clearly indicates the necessity of evaluating the optimal insulation thickness to help reduce the amount of energy needed to heat and cool buildings, thus decreasing the amount of greenhouse gas emissions associated with energy production. Therefore, in this study, optimal insulation thickness is evaluated based on the degree day method for ten locations selected from different climatic zones in India. EPS and XPS insulation materials are investigated, and various fuels, such as natural gas, coal, diesel, and electricity, are considered for heating and cooling. In 2015, the United Nations (UN) defined 17 sustainable development goals that address the most pressing global issues [20]. This study is directly related to Target 12 (Responsible Consumption and Production). It is also indirectly related to Target 7, Target 11, and Target 13. These targets frame pressing issues in the construction, energy, and manufacturing sectors that need to be addressed in the next quarter century.

2. Materials and Methods

Calculations were performed for cities in 10 different regions (Jaisalmer, Kota, Mumbai, Chennai, Bhubaneswar, New Delhi, Lucknow, Patna, Srinagar, and Shillong) with different climatic characteristics in Indian climate geography, as shown in Figure 1 [21,22]. Köppen–Geiger climatic classification was utilized to understand climatic attributes on a global scale. These climate groups were expressed in three letters. For Köppen classification, the first letter of these climate groups was as follows: equatorial region (A), arid region (B), hot temperate zone (C), snowy region (D), or polar region (E). The second letter in classification was the rainfall position of the region; the third letter was the region's temperature [21]. Jaisalmer was classified as a tropical and subtropical desert climate (Bwh); Mumbai, Chennai, and Bhubaneswar were classified as tropical savanna climates (Aw); Kota and New Delhi were classified as mid-latitude steppe and desert climates (Bsh); Lucknow and Patna belonged to the monsoon-influenced humid subtropical climate classification (Cwa); Srinagar belonged to the cold desert climate classification (Bwk); and Shillong was classified as a tropical monsoon climate (Am). Aw-Cwa was mid-rainy, and Bwh-Bsh-Bwk was arid and less rainy [22]. Energy need was computed by accepting

T $\leq$ 15 °C as the degree day value for heating (HDD) and T > 24 °C as the degree day value for cooling (CDD). Annual average values between 2007 and 2016 were used as heating and cooling day degree data; Table 1 shows the annual values of the regions. The data on HDD, CDD, and average temperature for the different locations were collected through the database of the JRC Photovoltaic Geographical Information System (PVGIS) [23]. Coal, natural gas, and diesel were utilized for heating in the lifecycle cost analysis (LCA), while electricity was used for cooling.

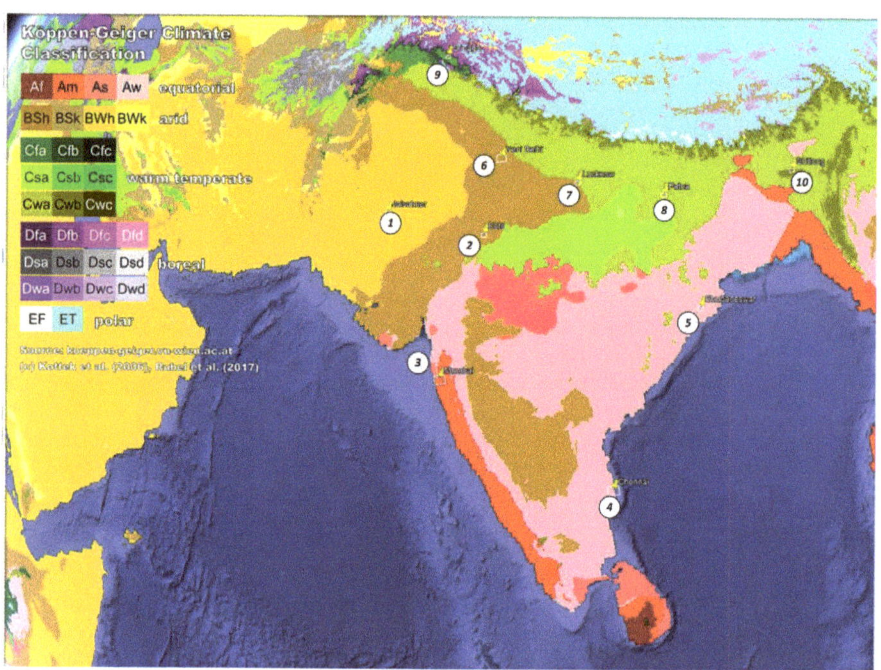

Figure 1. Selected Indian cities' geographic locations.

Table 1. Cooling and heating values of the selected cities [23].

No.	City	Latitude (Degree)	Altitude (m)	HDD (°C/Year)	CDD (°C/Year)	Average Temp. (°C/Year)	Process	Climate
1	Jaisalmer	29.90 N	225	43	2565	27.38	Cooling	Hot dry
2	Kota	25.20 N	271	87	2123	26	Cooling	Hot dry
3	Mumbai	19.07 N	14	0	2092	26.82	Cooling	Warm humid
4	Chennai	13.08 N	6.7	0	2485	27.82	Cooling	Warm humid
5	Bhubaneswar	20.29 N	58	1	1895	26.24	Cooling	Warm humid
6	New Delhi	28.61 N	216	267	1791	24.34	Cooling	Composite
7	Lucknow	26.85 N	123	144	1842	24.89	Cooling	Composite
8	Patna	25.59 N	53	75	1791	25.02	Cooling	Composite
9	Srinagar	34.08 N	1585	2109	26	13.14	Heating	Cold
10	Shillong	25.57 N	1525	815	0	16.04	Heating	Cold

2.1. Building Wall Model

The largest heat losses in buildings are reported through walls, floors, roofs, and windows, with a total energy loss of nearly 15–35% [24]. Heat losses and gains arising from buildings vary by architecture, location, and structural materials. The highest loss rate in a building is due to exterior walls and, if optimum insulation is provided, 50% to 60% of the energy can be saved [25]. Therefore, insulation for external walls is an investment for maximum energy gain. Walls are made up of a single layer nowadays, but they may also be considered as a construction component made up of two or more layers and elements [26].

The optimal insulation thickness was found in this investigation by considering that heat losses only occurred through outer walls. The wall model, as shown in Figure 2, was composed of interior plaster, insulating material, concrete, and exterior plaster. Table 2 displays the properties of these wall components [27,28].

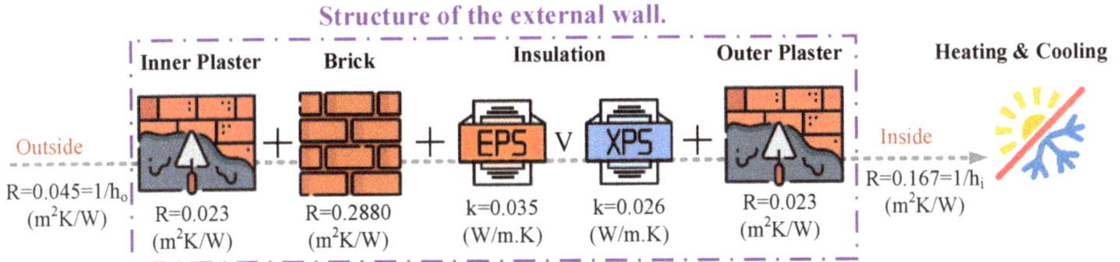

Figure 2. The exterior wall model in the study and the characteristics of its components.

Table 2. Construction of walls used in the analysis [27,28].

Wall Type	Density (kg·m^{-3})	Thickness (m)	Thermal Conductivity (W/m·K)	Thermal Resistance (m^{2}K/W)	Thermal Resistance R$_{TW}$ (Total Wall) (m^{2}K/W)
Inner plaster	1200	0.02	0.85	0.0230	
Hollow brick	750	0.13	0.45	0.2880	
Outer plaster	1100	0.02	0.85	0.0230	0.548
R inside	-	-		0.1670	
R outside	-	-	-	0.0450	

The overall heat transfer coefficient (U) was determined using Equation (1) [29]:

$$U = \frac{1}{R_i + R_w + R_{izo} + R_o} \quad \left(W/m^2K\right) \tag{1}$$

In Equation (1), R_i represents the thermal resistance of the inner surface, and R_o represents the thermal resistance of the outer surface. R_w is the thermal resistance of the wall layers that are not insulated, and R_{izo} is the insulation material's thermal resistance computed with Equation (2). In this equation, x is the insulation material's thickness, and k is the insulation material's thermal conductivity coefficient. The values of k were considered as 0.026 W/m·K and 0.035 W/m·K for XPS and EPS with costs (C_i) of 46.35 \$/m^3 and 105.94 \$/m^3, respectively [16].

$$R_{izo} = \frac{x}{k} \tag{2}$$

2.2. Building Wall Heating Load

Windows, ceilings, floors, exterior walls, and air infiltration result in heat losses arising in buildings. Calculations were revised in this analysis by assuming that they came only from the outer wall. For the outer wall, the heat loss was computed using Equation (3) [30]:

$$q = U.\Delta T \tag{3}$$

where ΔT is the temperature difference, and U is the total heat transfer coefficient. The number of degree days (NDD) and U were implemented to compute the yearly heat loss q_A (W/m^2) of the unit surface:

$$q_A = (3600 \times 24).NDD.U \tag{4}$$

Equation (5) provides the E_A (J/m^2-year) yearly energy required for heating and was computed by dividing q_A by the system efficiency:

$$E_A = \frac{86400.NDD.U}{\eta} = \frac{86400.NDD}{(R_{TW} + R_{izo}).\eta} \tag{5}$$

Equation (6) calculates the m_{fA} (kg/m^2-year) fuel amount consumed in a year:

$$m_{fA} = \frac{86400.NDD}{(R_{TW} + R_{izo}).H_u..\eta} \tag{6}$$

Equation (7) calculates the annual energy cost $C_{A,H}$ ($/m^2-year) utilized for heating a unit area:

$$C_{A,H} = \frac{86400.HDD.C_f}{(R_{TW} + R_{izo}).\eta.H_u.} \tag{7}$$

As shown in Table 3 [31], H_u is the fuel lower heat value, η is the the fuel efficiency, and C_f is the fuel price. Equation (8) computes the cost of cooling with respect to the amount of energy utilized:

$$C_{A,C} = \frac{86400.CDD.C_f}{(R_{TW} + R_{izo}).\ COP} \tag{8}$$

Here, COP is the cooling system performance coefficient taken as 2.5 [32]. The values of C_f, H_u, and η are listed in Table 3.

Table 3. Considered fuel characteristics [31,33].

Fuel	Chemical Equation	C_f	η	H_u
Coal	$C_{5.85}H_{5.26}O_{1.13}S_{0.008}N_{0.077}$	0.16610 $/kg	0.65	21.113×10^6 J/kg
Natural gas	$C_{1.05}H_4O_{0.034}N_{0.022}$	0.1305 $/m^3	0.93	34.526×10^6 J/m^3
Diesel	$C_{7.3125}H_{10.407}O_{0.04}S_{0.026}N_{0.02}$	0.69981 $/kg	0.80	42.911×10^6 J/kg
Electricity	–	0.08 $/kWh	0.99	3.5990×10^6 J/kWh

2.3. Optimal Insulation Analysis

The purpose was to reduce insulation costs by analyzing the optimal thickness of insulation desired for a building. The building's overall heating cost was calculated using LCCA by combining the energy price, any insulation price, the future value factor (PWF), and time (N) [34,35]. For the PWF price, the real interest rate (r) was determined based on two different terms using Equation (9), depending on the interest rate (φ) and the inflation rate (i):

$$\text{If } \varphi > i \text{ then } r = \frac{\varphi - i}{1 + i}; \text{ If } i > \varphi \text{ then } r = \frac{i - \varphi}{1 + \varphi} \tag{9}$$

For a ten-year process, current values were calculated using Equation (10) with inflation and interest rates of 4% and 8%, respectively [36,37]:

$$PWF = \frac{(r + 1)^N}{(r + 1)^N.r} \tag{10}$$

Equation (11) calculates the total C_T (\$) cost of an isolated building, where x and C_i are the used insulation material's thickness (m) and unit price (\$/m^3), respectively:

$$C_T = C_A.PWF + C_i.x \tag{11}$$

For optimal insulation thickness x_{opt} (m), the total cost must be minimal and was determined as follows:

$$\frac{dC_T}{dx} = \frac{d}{dx}.(C_A.PWF + C_i.\,x) \tag{12}$$

$$\frac{dC_T}{dx} = 0 \tag{13}$$

$$x_{opt} = \sqrt{86400}\,.\left(\frac{NDD.\,C_f.PWF.k}{H_u.C_i.\eta}\right)^{1/2} - k.R_{TW} \tag{14}$$

Equation (15) calculates the cost of the lifecycle (p_b) following the profit from the investment made. In this equation, the S_A represents annual savings:

$$p_b = \frac{C_T}{S_A} \tag{15}$$

2.4. Environmental Analysis

The steadily rising global population needs greater and greater quantities of energy every day. The related increasing energy demand has been generally utilized to heat houses. This demand is mainly supplied by fossil fuels, which are the widest and cheapest sources of energy, leading to the emission of greenhouse gases and harmful pollutants. Heating costs can be reduced by a certain degree with a certain increment in insulation thickness. The general chemical formula of combustion for fuel is expressed by Equation (16) [38]:

$$C_x + H_z + O_w + S_y + N_t + \alpha.A(O_2 + 3.76N_2) \rightarrow xCO_2 + \left(\frac{z}{2}\right)H_2O + ySO_2 + B.O_2 + E.N_2 \tag{16}$$

Here, the constants (A, B, and E) are calculated as follows:

$$A = \left(x + y + \frac{z}{4} - \frac{w}{2}\right) \tag{17}$$

$$B = \left(x + y + \frac{z}{4} - \frac{w}{2}\right).(\alpha - 1) \tag{18}$$

$$E = 3.76\alpha\,.\left(x + y + \frac{z}{4} - \frac{w}{4}\right) + \frac{t}{2} \tag{19}$$

SO_X and CO emissions are neglected in Equation (16). The emission rates of combustion products resulting from the burning of 1 kg of fuel can be calculated in Equations (20) and (21) [39]:

$$MCO_2 = \frac{x\,CO_2}{M} \equiv kg\ CO_2/kg\,fuel \tag{20}$$

$$MSO_2 = \frac{y\,SO_2}{M} \equiv kg\ SO_2/kg\,fuel \tag{21}$$

Here, M (kg/kmol) is the molar weight of the fuel determined as follows:

$$M = 12x + z + 16w + 32y + 14t \tag{22}$$

Depending on the total fuel consumption, emission values can be calculated using Equations (23) and (24):

$$mCO_2 = \frac{44.x}{M}m_{fA} \tag{23}$$

$$mSO_2 = \frac{64.y}{M}m_{fA} \tag{24}$$

3. Results and Discussion

3.1. Optimal Insulation Thickness

Optimal insulation thickness varied with fuel type, insulating material, and climate. Heat losses in buildings displayed a downtrend. Therefore, the heat losses and heat loads of buildings decreased. As a result, the overall amounts of fuel and emissions were reduced all at once. However, the overall cost (fuel and insulation) reduced initially and then climbed again after the minimal figure was reached. The ideal insulation thickness was determined by the point at which the overall cost was the lowest. Investment cost and total cost both increased due to unnecessarily increased insulation thickness. Figure 3 illustrates the relationship between the total cost of insulation thickness and externally insulated wall applications.

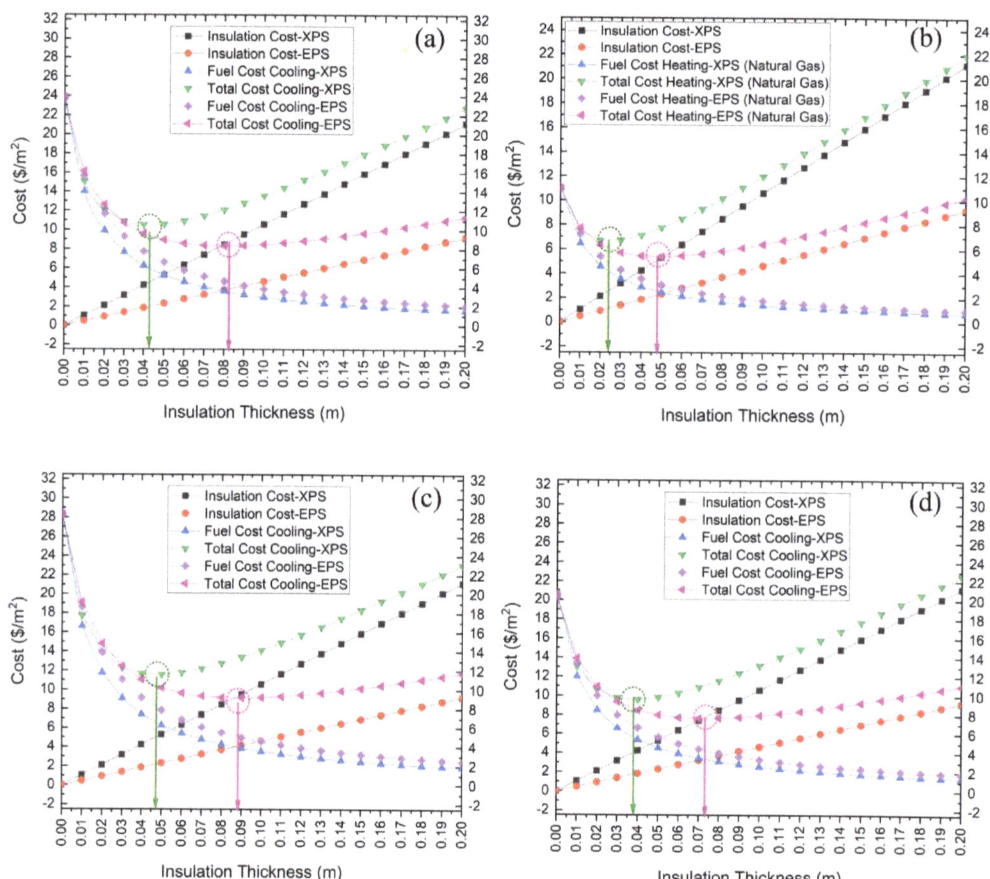

Figure 3. Cost–insulation thickness relationship for EPS and XPS insulation materials: (**a**) Mumbai, (**b**) Srinigar, (**c**) Chennai, and (**d**) Patna.

Optimal insulation thickness was computed for various fuels and insulating materials using Equation (14). The results can be seen in Tables 4 and 5 for externally insulated walls in Jaisalmer, Kota, Mumbai, Chennai, Bhubaneswar, New Delhi, Lucknow, and Patna for cooling, as well as in Srinagar and Shillong for heating. Insulation became meaningless because of the initial investment cost and economic parameters in solutions for low numbers of degree days. It can be seen when Table 4 is reviewed that XPS had differences compared to EPS in using electricity for cooling when analyzing the optimal insulation thickness. The lowest thermal insulation thickness values related to using XPS material in building insulation were in New Delhi (Bsh) and Patna (Cwa), which were in the composite climatic zone, by 0.0382 (m). The highest value was in Jaiselmer by 0.0484 (m), which was in a hot and dry region.

Table 4. Results of various insulation materials for cooling.

Materials	X_{opt} (m)	Annual Savings ($/m^2)	Payback Period (Years)	Annual Savings Rate (%)
Jaisalmer				
XPS	0.0484	17.47	1.96	40.28
EPS	0.0908	19.94	1.49	31.83
Kota				
XPS	0.0428	13.63	2.16	43.71
EPS	0.0809	15.82	1.64	34.66
Mumbai				
XPS	0.0424	13.36	2.18	43.99
EPS	0.0802	15.54	1.65	34.89
Chennai				
XPS	0.0475	16.77	1.99	40.84
EPS	0.0891	19.19	1.51	32.29
Bhubaneswar				
XPS	0.0401	11.70	2.29	45.88
EPS	0.0754	13.73	1.74	36.46
New Delhi				
XPS	0.0382	10.83	2.36	46.99
EPS	0.0728	12.79	1.79	37.38
Lucknow				
XPS	0.0389	11.25	2.33	46.44
EPS	0.0741	13.25	1.77	36.92
Patna				
XPS	0.0382	10.83	2.36	46.99
EPS	0.0728	12.79	1.79	37.38

Table 5. Results of various insulation materials for heating.

Fuel	Materials	X_{opt} (m)	Annual Savings ($/m^2)	Payback Period (Years)	Annual Savings Rate (%)
			Srinagar		
Natural gas	XPS	0.0242	4.36	3.27	60.39
	EPS	0.0483	5.63	2.47	48.80
Coal	XPS	0.0188	2.62	3.86	67.70
	EPS	0.0387	3.62	2.91	55.29
Diesel	XPS	0.0834	51.73	1.24	27.06
	EPS	0.1522	55.92	0.95	21.15
			Shillong		
Natural gas	XPS	0.0097	0.69	5.49	83.69
	EPS	0.0227	1.25	4.11	70.57
Coal	XPS	0.0063	0.29	6.52	90.66
	EPS	0.0168	0.68	4.85	78.19
Diesel	XPS	0.0465	16.05	2.03	41.44
	EPS	0.0873	18.42	1.54	32.79

New Delhi, Patna, and Jaisalmer had the highest values by 0.0908 (m) for cooling when EPS material was used in building insulation. These values proportionally varied by degree day number. The results for XPS material under a cooling load were 2.20 years on average (the payback period was 1.96 (years) at least and 2.36 (years) at best). The annual savings were 10.83 ($/m^2) at least, 17.47 ($/m^2) at best, and 13.23 ($/m^2) on average. The annual savings rate was 40.28 (%) at least, 46.99 (%) at best, and 44.39 (%) on average. According to the results for the EPS material, the payback period was 1.49 (years) at least, 1.79 (years) at best, and 1.67 years on average. The annual savings were 12.79 ($/m^2) at least, 19.94 ($/m^2) at best, and 15.38 ($/m^2) on average. The annual savings rate was 31.83 (%) at least, 37.38 (%) at best, and 35.22 (%) on average. According to the average results for both materials, 0.0607 (m) was the optimal insulation thickness, 1.93 (years) was the payback period, 14.30 was the annual savings ($/m^2), and 39.80 (%) was the annual savings rate. Cities located in a hot climate zone only consumed energy for cooling. It can be seen that serious energy savings could be achieved with products selected in the optimum thickness values from insulation materials ranging from 0.1 cm to 0.12 cm in the current market.

Insulating properties of XPS and EPS materials in the cases of using natural gas, coal, and diesel fuel under a heating load can be seen in Table 5.

The lowest insulation thickness for XPS material was 0.0063 (m) for coal (Shillong); the highest value was 0.0834 (m) for diesel fuel (Srinagar). The lowest and highest values for EPS material for the same regions and types of fuels, respectively, were 0.0168 (m) and 0.1522. Regarding the results for XPS material under a heating load, the payback period was 1.24 (years) at least, 6.52 (years) at best, and 3.74 years on average. The annual savings were 0.29 ($/m^2) at least, 51.73 ($/m^2) at best, and 12.62 ($/m^2) on average. The annual savings rate was 27.06 (%) at least, 90.66 (%) at best, and 61.82 (%) on average. According to the results for EPS material, the payback period was 0.95 (years) at least, 4.85 (years) at best, and 2.80 years on average. The annual savings were 0.68 ($/m^2) at least, 55.92 ($/m^2) at best, and 14.25 ($/m^2) on average. The annual savings rate was 21.15 (%) at least, 78.19 (%) at best, and 51.13 (%) on average.

Regarding the average results in both regions for both materials, 0.0462 (m) was the optimal insulation thickness, the payback period was 3.27 (years), 13.43 was the annual savings ($/m^2), and 56.47 (%) was the annual savings rate. The three main parameters that affected the results were fuel components, insulating properties, and the number of degree days. In general terms, variables were close to each other for India; good results for both heating and cooling were obtained.

3.2. Environmental Analysis

Figure 4 shows changes based on insulation thickness in the annual CO_2 and SO_2 gases. The CO_2 emissions at different values of insulation thickness and for different types of fuels are presented in Figure 4a. For both EPS and XPS materials, the highest and lowest CO_2 emissions were observed corresponding to coal and diesel, respectively. Fuel quantity for heating unit volume decreased with increasing insulation thickness. This is because there was an observed decrease in the emission of deleterious gases. According to the average results in two different cities, XPS and EPS respectively became 66% and 52% in the case of providing insulation and optimal insulation thickness. In addition, a decrease was seen in SO_2 and CO_2 emission values, with more sustainable ecological environmental structures.

3.3. Degree Days and Correlations

There have been some studies that have proposed methods to optimize the thicknesses of insulation materials. For example, Mahlia et al. [14] used the thermal conductivity (k) of different insulation materials to compute the optimal values. However, they did not consider the needed annual energy of buildings. On the contrary, Sisman et al. [15] computed optimal values using NDD but without using the k value of the insulation material. The results have shown that these two proposed methods are promising. However, a question arises here: what if both parameters, the k value of the insulation material and NDD, were used to compute optimal thickness? To investigate this problem, we tried to develop a mathematical model that used both parameters (k and *NDD*). The model was developed using a multiple linear regression method, which is a machine-learning technique [40,41]. Equation (25) is a general form of the model:

$$\mathbf{x_{opt} = a + b.k + c.NDD} \tag{25}$$

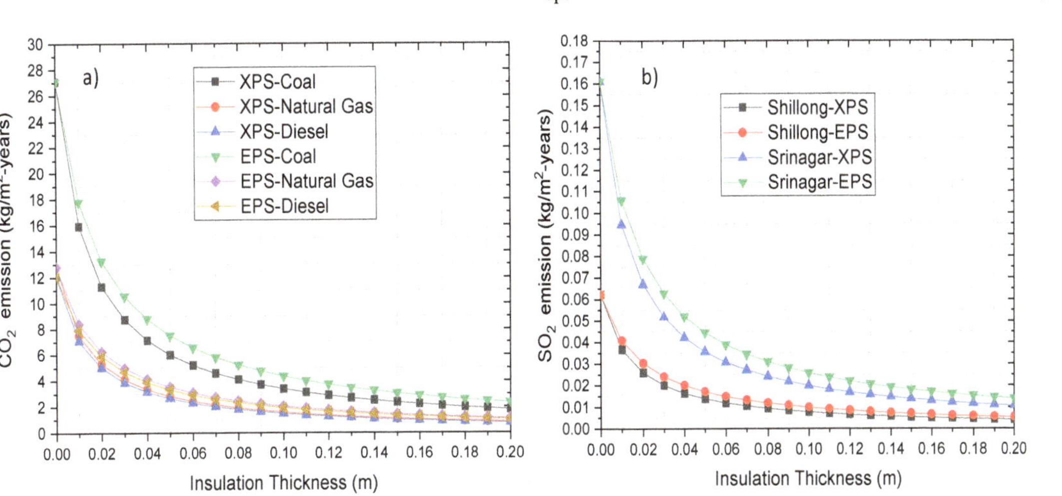

Figure 4. Fuel gas emission–insulation thickness: (**a**) CO_2 (Shillong) (**b**) SO_2-Coal (Shillong and Srinagar).

The optimal values of a, b, and c in Equation (25) were configured using the multiple curve-fitting tool in MATLAB for electricity, natural gas, coal, and diesel separately. For simplicity, we called these configurations Model (26), Model (27), Model (28), and Model (29), respectively.

Table 6 shows these configurations and their results. The results were validated by common statistical methods [42]. The R^2 values of all these correlations statistically varied from 0.9943 to 0.9598; related values were close to ($R \leq 1$). The root mean square error (RMSE) value was 0.0016 at best and 0.0140 at least because of the high harmony between the estimated and computed values. The sum of squares error (SSE) value was 2.209×10^{-5} at best and 1.96×10^{-4} at least. The ideal was zero in these two statistical methods (RMSE ≥ 0) -(SSE ≥ 0). Both the RMSE and SSE values were close to zero, and they gave consistent results. Figure 5 illustrates graphs of the optimal insulation thicknesses of the used insulating (XPS and EPS) materials, as computed by Equation (14) for different fuels.

Table 6. Developed model regression constants.

Fuel	Model	a	b	c	R^2	SSE	RMSE
Electricity	26	-0.1039	4.161	1.823×10^{-5}	0.9943	3.443×10^{-5}	0.0016
Natural gas	27	-0.05929	2.061	1.549×10^{-5}	0.9604	3.08×10^{-5}	0.0055
Coal	28	-0.05079	1.689	1.329×10^{-5}	0.9598	2.209×10^{-5}	0.0047
Diesel	29	-0.1509	6.089	3.934×10^{-5}	0.9661	1.96×10^{-4}	0.0140

The results obtained with the configurations of Equation (16) for electricity, natural gas, coal, and diesel (electricity: Model (26); natural gas: Model (27); coal: Model (28); diesel: Model (29)) were compared with previous works [14,15] and the reference values computed for different climate regions [17,19,33,43–45]. The comparison results are reported in Table 7. The heat map in Table 7 shows how close the computed thicknesses in each row were to the optimal thickness (reference value). These values in each row are the absolute values of the differences between the computed thicknesses and the reference value, i.e., errors of the model. Red shows the lowest error, green shows the highest error, and yellow shows the average error. Sisman's model always had the highest error, while the others had similar errors in most cases. The mean errors of our model, Mahlia's model [14], and Sisman's model [15] for the instances in which we trained our model (instances 1–28 in Table 7) were 0.0713, 0.807, and 6.3, respectively. The mean errors of these models on the other cases in which we tested our model (cases 29–60 in Table 7) were 1.136, 2.023, and 10.868, respectively. When we considered all the instances in the table, the mean errors were 0.639, 1.456, and 8.737, respectively. The results show that Sisman's model performs poorly in all the cases we tested, while others (Mahlia's and our model) performed relatively well. When one compares our model with Mahlia's model considering their mean errors, one can see that our model is superior to Mahlia's model. Furthermore, this superiority was not only in the training instances but also in the testing instances.

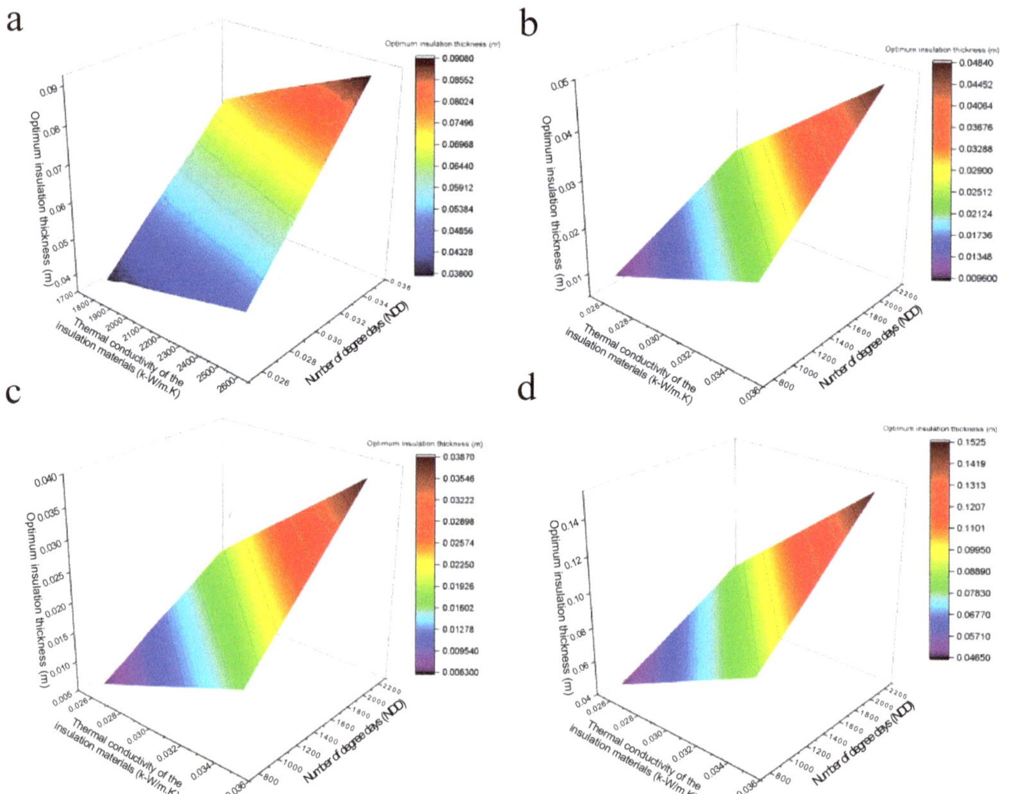

Figure 5. Optimal insulation thickness and k-NDD values: (**a**) electricity; (**b**) natural gas; (**c**) coal; (**d**) diesel.

Buildings **2023**, 13, 569

Table 7. Relative error statistical results of reference results with estimated models.

												Low	Mid	High
												Relative Error (Er) %		
	Location			Reference							Ref. Value	Developed Model	Previous Models	
Case	Country	City	Reference	Climate	NDD	Process	Fuel	Material	k (W/m·K)	x_{opt} (m)	Developed Eq. for Corresponding Fuel	Mahlia et al. [14]	Sisman et al. [15]
1	India	Jaisalmer	Present Study	Bwh	2565	Cooling	Electricity	XPS	0.0260	0.0484	0.0536	0.0047	6.6422
2	India	Jaisalmer	Present Study	Bwh	2565	Cooling	Electricity	EPS	0.035	0.0908	0.0261	0.3736	3.0736
3	India	Kota	Present Study	Bsh	2123	Cooling	Electricity	XPS	0.026	0.0428	0.0033	0.1255	6.4946
4	India	Kota	Present Study	Bsh	2123	Cooling	Electricity	EPS	0.035	0.0809	0.0064	0.2969	2.965
5	India	Mumbai	Present Study	Aw	2092	Cooling	Electricity	XPS	0.026	0.0424	0.0006	0.1361	6.4819
6	India	Mumbai	Present Study	Aw	2092	Cooling	Electricity	EPS	0.035	0.0802	0.0048	0.2908	2.9555
7	India	Chennai	Present Study	Aw	2485	Cooling	Electricity	XPS	0.026	0.0475	0.0429	0.0141	6.6034
8	India	Chennai	Present Study	Aw	2485	Cooling	Electricity	EPS	0.035	0.0891	0.0238	0.3616	3.0534
9	India	Bhubaneswar	Present Study	Aw	1895	Cooling	Electricity	XPS	0.026	0.0401	0.0327	0.2013	6.3431
10	India	Bhubaneswar	Present Study	Aw	1895	Cooling	Electricity	EPS	0.035	0.0754	0.011	0.2456	2.9053
11	India	New Delhi	Present Study	Aw	1791	Cooling	Electricity	XPS	0.026	0.0382	0.0342	0.261	6.3874
12	India	New Delhi	Present Study	Aw	1791	Cooling	Electricity	EPS	0.035	0.0728	0.021	0.2187	2.8764
13	India	Lucknow	Present Study	Cwa	1842	Cooling	Electricity	XPS	0.026	0.0389	0.0277	0.2383	6.4096
14	India	Lucknow	Present Study	Cwa	1842	Cooling	Electricity	EPS	0.035	0.0741	0.0157	0.2324	2.8898
15	India	Patna	Present Study	Cwa	1791	Cooling	Electricity	XPS	0.026	0.0382	0.0342	0.261	6.3874
16	India	Patna	Present Study	Cwa	1791	Cooling	Electricity	EPS	0.035	0.0728	0.021	0.2187	2.8764
17	India	Srinagar	Present Study	Bwk	2109	Heating	Natural gas	XPS	0.026	0.0242	0.1142	0.9906	12.189
18	India	Srinagar	Present Study	Bwk	2109	Heating	Natural gas	EPS	0.035	0.0483	0.0577	0.1776	5.6081
19	India	Srinagar	Present Study	Bwk	2109	Heating	Coal	XPS	0.026	0.0188	0.1251	1.5623	15.9773
20	India	Srinagar	Present Study	Bwk	2109	Heating	Coal	EPS	0.035	0.0387	0.0606	0.4698	7.2474
21	India	Srinagar	Present Study	Bwk	2109	Heating	Diesel	XPS	0.026	0.0834	0.0837	0.4224	2.827
22	India	Srinagar	Present Study	Bwk	2109	Heating	Diesel	EPS	0.035	0.1522	0.0461	0.6263	1.0971
23	India	Shillong	Present Study	Am	815	Heating	Natural gas	XPS	0.026	0.0097	0.2866	3.9661	15.0769
24	India	Shillong	Present Study	Am	815	Heating	Natural gas	EPS	0.035	0.0227	0.122	1.5057	5.8699
25	India	Shillong	Present Study	Am	815	Heating	Coal	XPS	0.026	0.0063	0.3722	6.6463	23.7534
26	India	Shillong	Present Study	Am	815	Heating	Coal	EPS	0.035	0.0168	0.1403	2.3857	8.2825
27	India	Shillong	Present Study	Am	815	Heating	Diesel	XPS	0.026	0.0465	0.1511	0.0359	2.3537
28	India	Shillong	Present Study	Am	815	Heating	Diesel	EPS	0.035	0.0873	0.0799	0.3485	0.7863
29	Libya	Tripoli	[43]	Csa	492	Cooling	Electricity	EPS	0.037	0.069	0.1452	0.127	0.5453
30	Turkey	Adana	[33]	Csa	874	Heating	Natural gas	XPS	0.024	0.069	0.9462	0.3093	1.3823
31	Turkey	Adana	[33]	Csa	874	Heating	Natural gas	EPS	0.035	0.069	0.6176	0.1757	1.3823
32	Turkey	Adana	[33]	Csa	874	Heating	Natural gas	Glass wool	0.05	0.024	1.3874	2.9438	5.849
33	Turkey	Adana	[33]	Csa	874	Heating	Natural gas	Rock wool	0.048	0.035	0.5193	1.5124	3.6965
34	Turkey	Adana	[33]	Csa	874	Heating	Natural gas	Polyurethane	0.017	0.05	1.2143	0.0014	2.2875
35	Turkey	Adana	[33]	Csa	874	Heating	Coal	XPS	0.031	0.048	0.7253	0.0774	2.4245
36	Turkey	Adana	[33]	Csa	874	Heating	Coal	EPS	0.044	0.017	1.0671	3.4737	8.6692
37	Turkey	Adana	[33]	Csa	874	Heating	Coal	Glass wool	0.061	0.031	1.0598	3.5427	4.3025

Table 7. *Cont.*

Case	Location		Reference							Ref. Value	Developed Model	Previous Models	
											Relative Error (Er) %		
	Country	City	Reference	Climate	NDD	Process	Fuel	Material	k (W/m·K)	x_{opt} (m)	Developed Eq. for Corresponding Fuel	Mahlia et al. [14]	Sisman et al. [15]
38	Turkey	Adana	[33]	Csa	874	Heating	Coal	Rock wool	0.059	0.044	0.3745	1.9833	2.7358
39	Turkey	Adana	[33]	Csa	874	Heating	Coal	Polyurethane	0.022	0.061	1.0331	0.2187	1.6947
40	Palestine	Jericho	[44]	Csa	1989	Cooling	Electricity	Polystyrene	0.038	0.059	0.5325	0.0527	4.1762
41	Palestine	Hebron	[44]	Dfb	456	Cooling	Electricity	Polystyrene	0.038	0.022	1.8404	1.8231	3.5769
42	Palestine	Jerusalem	[44]	Csa	768	Cooling	Electricity	Polystyrene	0.038	0.049	0.3913	0.2675	2.0433
43	Palestine	Tulkarem	[44]	Csa	1066	Cooling	Electricity	Polystyrene	0.038	0.057	0.2913	0.0896	2.3492
44	Palestine	Gaza	[44]	Bsh	1097	Cooling	Electricity	Polystyrene	0.038	0.062	0.1962	0.0017	2.1463
45	Palestine	Bethelem	[44]	Csa	971	Cooling	Electricity	Polystyrene	0.038	0.053	0.3561	0.1719	2.3573
46	Palestine	Jenin	[44]	Csa	1399	Cooling	Electricity	Polystyrene	0.038	0.068	0.1716	0.0866	2.4454
47	Palestine	Nablus	[44]	Csa	854	Cooling	Electricity	Polystyrene	0.038	0.052	0.3412	0.1944	2.1064
48	Turkey	Ağrı	[45]	Dsb	4423	Heating	Coal	XPS	0.031	0.0261	1.3123	0.9815	20.3638
49	Turkey	Ağrı	[45]	Dsb	4423	Heating	Natural gas	XPS	0.031	0.0314	1.3284	0.6471	16.7578
50	Turkey	Aydın	[45]	Csa	1213	Heating	Coal	XPS	0.031	0.0022	7.0408	22.508	94.6422
51	Turkey	Aydın	[45]	Csa	1213	Heating	Natural gas	XPS	0.031	0.005	3.6781	9.3435	41.0826
52	Turkey	Elazığ	[45]	Dsa	2653	Heating	Coal	XPS	0.031	0.019	0.9383	1.722	18.9686
53	Turkey	Elazığ	[45]	Dsa	2653	Heating	Natural gas	XPS	0.031	0.0182	1.5108	1.8416	19.8463
54	Turkey	Kocaeli	[45]	Cfa	1786	Heating	Coal	XPS	0.031	0.0113	1.2394	3.5768	23.9209
55	Turkey	Kocaeli	[45]	Cfa	1786	Heating	Natural gas	XPS	0.031	0.0106	2.044	3.879	25.5666
56	India	-	[19]	Hot and humid	1288	Cooling	Electricity	XPS	0.036	0.028	1.476	1.0891	6.8622
57	India	-	[19]	Hot and dry	1111	Cooling	Electricity	XPS	0.036	0.031	1.1323	0.8869	5.3529
58	India	-	[19]	Composite	1121	Cooling	Electricity	XPS	0.036	0.031	1.1382	0.8869	5.396
59	India	Dehradun	[17]	Cfa	3587	Heating	Natural gas	XPS	0.033	0.057	0.1278	0.0519	7.3542
60	India	Dehradun	[17]	Cfa	3587	Heating	Natural gas	EPS	0.031	0.073	0.1758	0.2915	5.5231

4. Conclusions and Recommendations

The current study examined how different insulation materials and fuel types affected commercial and domestic building cooling and heating performances in different climatic zones of India. The following conclusions can be drawn from the current study:

- For sites situated in hot and dry, warm and humid, and composite climatic regions, the ranges of XPS insulation thickness, annual savings, and payback period for these regions were found as 0.0382–0.0484 m, 10.83–17.47 \$/m^2, and 1.93–2.36 years, respectively. Similarly, these ranges for EPS insulation thickness, annual savings, and payback period were 0.0728–0.0908 m, 12.79–19.9347 \$/m^2, and 1.49–1.79 years, respectively.

- For sites situated in cold climatic regions, the ranges of XPS insulation thickness, annual savings, and payback period were 0.0097–0.0834 m, 0.29–51.73 \$/m^2, and 1.24–6.52 years, respectively. The ranges for EPS insulation thickness, annual savings, and payback period were 0.0168–0.1522 m, 0.68–55.92 \$/m^2, and 0.95–4.85 years, respectively.

- The ranges of GHG emissions for XPS material with natural gas, coal, and diesel fuels were 9.67–11.28 kg/m^2-year, 27.12–32.68 kg/m^2-year, and 2.89–4.51 kg/m^2-year, respectively. Similarly, the ranges of GHG emissions for EPS material with natural gas, coal, and diesel fuels were 5.39–9.17 kg/m^2-year, 9.47–23.75 kg/m^2-year, and 2.26–3.43 kg/m^2-year, respectively.

- The optimal insulation thickness for XPS material was lower than that of EPS material, while the payback period and annual savings for XPS material were greater and lower, respectively, than those of EPS material in all the circumstances.

- XPS was proved to be more effective than EPS, and the correlations obtained could aid in the determination of optimal insulation thickness for a specific location based on the number of degree days.

The study showed that using insulation on building exterior walls led to significant annual savings and had a payback period of less than five years, making it economically feasible. The study also found that increasing insulation thickness reduced GHG emissions from heating fuels. Future research can determine the best insulation thicknesses for different climate zones and materials.

Author Contributions: Conceptualization, M.A.K., A.Y. and A.S.; Data curation, M.A.K., A.Y. and A.S.; Investigation, M.A.K., A.Y., A.S. and R.K. Methodology, M.A.K., A.Y. and R.K., Software, M.A.K., A.Y., A.S. and R.K., Visualization, M.A.K., A.S. and T.S., Writing—original draft, M.A.K., A.Y. and A.S. Writing—review and editing, M.A.K., A.M., D.D., T.A. and T.S., Validation, A.Y. and A.S., Resources, A.M., D.D., T.A. and T.S. All authors have read and agreed to the published version of the manuscript.

Funding: This research received no external funding.

Institutional Review Board Statement: Not Applicable.

Informed Consent Statement: Not Applicable.

Data Availability Statement: Data available on request.

Conflicts of Interest: The authors declare no conflict of interest.

Nomenclature

Am	tropical monsoon climate	q_A	annual heat loss in unit area (J/m^2-year)
Aw	tropical savanna climate	r	actual interest rate
Bsk	cold semi-arid (steppe) climate	R_i	inside air film thermal resistances (m^{2}K/W)
$C_{A.C}$	annual cooling energy cost ($/m^2-year)	R_{izo}	thermal resistance of insulation layer (m^{2}K/W)
$C_{A.H}$	annual heating energy cost ($/m^2-year)	R_o	outside air film thermal resistance (m^{2}K/W)
C_i	cost of insulation in ($/m^3)	R_{TW}	sum of Ri.Rw.Ro (m^{2}K/W)
CDD	cooling degree days (°C-days)	R_w	total thermal resistance of wall materials without insulation (m^{2}K/W)
C_f	price of fuel ($/kg; $/m^3)	S_A	annual savings ($/m^2)
Csa	hot summer mediterranean climate	SO_2	sulfur dioxide
Csb	warm summer mediterranean climate	U	overall heat transfer coefficient (W/m^{2}K)
C_T	total cost ($)	x	thickness of insulation material (m)
CO_2	carbon dioxide	x_{opt}	optimum insulation thickness (m)
Dbf	warm summer humid continental climate	XPS	extruded polystyrene
E_A	annual energy requirement (J/m^2-year)		
EPS	expanded polystyrene	Greek letters	
g	inflation rate	η	heating system efficiency
HDD	heating degree days (°C-days)	ΔT	temperature difference (°C)
H_u	heating value of fuel (J/kg; J/m^3; J/kwh)		
i	interest rate	Subscripts	
k	thermal conductivity of insulation material (W/m K)	A	annual
LCA	lifecycle cost analysis	C	cooling
M	molar weight of fuel	H	heating
m_{fA}	amount of fuel consumed per year (kg/m^2-year)	i	inside
N	lifetime (years)	izo	insulation
NDD	number of (degree days (°C-days)	o	outside
p_b	payback period (years)	opt	optimum
PWF	present worth factor	t	total
q	heat loss (MJ m^2 year^{-1})	w	wall

References

1. Kallioğlu, M.A.; Ercan, U.; Avcı, A.S.; Fidan, C.; Karakaya, H. Empirical Modeling between Degree Days and Optimum Insulation Thickness for External Wall. *Energy Sources Part A Recovery Util. Environ. Eff.* **2020**, *42*, 1314–1334. [CrossRef]
2. Altun, A.F. Determination of Optimum Building Envelope Parameters of a Room Concerning Window-to-Wall Ratio, Orientation, Insulation Thickness and Window Type. *Buildings* **2022**, *12*, 383. [CrossRef]
3. Wang, H.; Huang, Y.; Yang, L. Integrated Economic and Environmental Assessment-Based Optimization Design Method of Building Roof Thermal Insulation. *Buildings* **2022**, *12*, 916. [CrossRef]
4. Shaik, S.; Gorantla, K.; Ghosh, A.; Arumugam, C.; Maduru, V.R. Energy Savings and Carbon Emission Mitigation Prospective of Building's Glazing Variety, Window-to-Wall Ratio and Wall Thickness. *Energies* **2021**, *14*, 8020. [CrossRef]
5. Skaar, C.; Gaarder, J.-E.; Bunkholt, N.S.; Sletnes, M. Energy Upgrading of Basement Exterior Walls: The Good, the Bad and the Ugly. *Buildings* **2023**, *13*, 133. [CrossRef]
6. Kallioğlu, M.A.; Sharma, A.; Chinnasamy, V.; Chauhan, R.; Singh, T. Optimum Insulation Thickness Assessment of Different Insulation Materials for Mid-Latitude Steppe and Desert Climate (BSH) Region of India. *Mater. Today Proc.* **2020**, *44*, 4421–4424. [CrossRef]
7. Xu, H.; Ding, J.; Li, T.; Mu, C.; Gu, X.; Wang, R. A Study on Optimum Insulation Thickness in Walls of Chinese Solar Greenhouse for Energy Saving. *Agronomy* **2022**, *12*, 1104.
8. Nyers, J.; Kajtar, L.; Tomić, S.; Nyers, A. Investment-Savings Method for Energy-Economic Optimization of External Wall Thermal Insulation Thickness. *Energy Build.* **2015**, *86*, 268–274. [CrossRef]
9. Liu, X.; Chen, Y.; Ge, H.; Fazio, P.; Chen, G. Determination of Optimum Insulation Thickness of Exterior Wall with Moisture Transfer in Hot Summer and Cold Winter Zone of China. *Procedia Eng.* **2015**, *121*, 1008–1015. [CrossRef]
10. Yuan, J.; Farnham, C.; Emura, K.; Alam, M.A. Proposal for Optimum Combination of Reflectivity and Insulation Thickness of Building Exterior Walls for Annual Thermal Load in Japan. *Build. Environ.* **2016**, *103*, 228–237. [CrossRef]
11. Dombaycı, Ö.A. The Environmental Impact of Optimum Insulation Thickness for External Walls of Buildings. *Build. Environ.* **2007**, *42*, 3855–3859. [CrossRef]
12. Kaynakli, O.; Bademlioglu, A.H.; Ufat, H.T. Determination of Optimum Insulation Thickness for Different Insulation Applications Considering Condensation. *Teh. Vjesn.* **2018**, *25*, 32–42.
13. Bolattürk, A. Determination of Optimum Insulation Thickness for Building Walls with Respect to Various Fuels and Climate Zones in Turkey. *Appl. Therm. Eng.* **2006**, *26*, 1301–1309. [CrossRef]
14. Mahlia, T.M.I.; Taufiq, B.N.; Masjuki, H.H. Correlation between Thermal Conductivity and the Thickness of Selected Insulation Materials for Building Wall. *Energy Build.* **2007**, *39*, 182–187. [CrossRef]
15. Sisman, N.; Kahya, E.; Aras, N.; Aras, H. Determination of Optimum Insulation Thicknesses of the External Walls and Roof (Ceiling) for Turkey's Different Degree-Day Regions. *Energy Policy* **2007**, *35*, 5151–5155. [CrossRef]
16. Shanmuga Sundaram, A.; Bhaskaran, A. Optimum Insulation Thickness of Walls for Energy-Saving in Hot Regions of India. *Int. J. Sustain. Energy* **2014**, *33*, 213–226. [CrossRef]
17. Mishra, S.; Usmani, J.A.; Varshney, S. Energy Saving Analysis in Building Walls through Thermal Insulation System. *Int. J. Eng. Res. Appl.* **2012**, *2*, 128–135.
18. Raza, A.; Aggarwal, V. Evaluation of Optimum Insulation Thickness for Cost Saving of Wall Design in India. Available online: https://www.researchsquare.com/article/rs-1565499/v1.pdf (accessed on 13 October 2022).
19. Singh, H.K.; Prakash, R.; Shukla, K.K. Economic Insulation Thickness of External Walls in Hot and Composite Regions of India. In Proceedings of the Global Conference on Renewable Energy GCRE2016, Patna, India, 4–6 March 2016; p. 6.
20. Clark, H.; Wu, H. The Sustainable Development Goals: 17 Goals to Transform Our World. In *Furthering Work United Nations*; UN: New York, NY, USA, 2016; pp. 36–54.
21. Kottek, M.; Grieser, J.; Beck, C.; Rudolf, B.; Rubel, F. World Map of the Köppen-Geiger Climate Classification Updated. *Meteorol. Z.* **2006**, *15*, 259–263. [CrossRef]
22. Rubel, F.; Brugger, K.; Haslinger, K.; Auer, I. The climate of the European Alps: Shift of very high resolution Köppen-Geiger climate zones 1800–2100. *Meteorol. Z.* **2017**, *26*, 115–125. [CrossRef]
23. JRC Photovoltaic Geographical Information System (PVGIS)—European Commission. Available online: https://re.jrc.ec.europa.eu/pvg_tools/en/#MR (accessed on 15 February 2022).
24. Szokolay, S. *Introduction to Architectural Science*; Routledge: London, UK, 2012.
25. Shehadi, M. Energy Consumption Optimization Measures for Buildings in the Midwest Regions of USA. *Buildings* **2018**, *8*, 170. [CrossRef]
26. Kon, O.; Caner, İ. The Effect of External Wall Insulation on Mold and Moisture on the Buildings. *Buildings* **2022**, *12*, 521. [CrossRef]
27. Aditya, L.; Mahlia, T.M.I.; Rismanchi, B.; Ng, H.M.; Hasan, M.H.; Metselaar, H.S.C.; Muraza, O.; Aditiya, H.B. A Review on Insulation Materials for Energy Conservation in Buildings. *Renew. Sustain. Energy Rev.* **2017**, *73*, 1352–1365. [CrossRef]
28. Standard, T. 825 (TS 825). In *Thermal Insulation in Buildings*; Official Gazette (23725): Ankara, Turkey, 1999; Volume 14.
29. Nellis, G.; Klein, S. *Heat Transfer*; Cambridge University Press: Cambridge, UK, 2008; ISBN 1-139-47661-0.
30. Qian, X.; Lee, S.W.; Yang, Y. Heat Transfer Coefficient Estimation and Performance Evaluation of Shell and Tube Heat Exchanger Using Flue Gas. *Processes* **2021**, *9*, 939. [CrossRef]

31. Statistics, E.; Central Statistics Office, Ministry of Statistics and Programme Implementation. Govt. India New Delhi (22nd Issue) 2015. Available online: www.mospi.gov (accessed on 8 February 2022).
32. Cengel, Y. *Heat and Mass Transfer: Fundamentals and Applications*; McGraw-Hill Higher Education: Columbus, OH, USA, 2014.
33. Kurekci, N.A. Determination of Optimum Insulation Thickness for Building Walls by Using Heating and Cooling Degree-Day Values of All Turkey's Provincial Centers. *Energy Build.* **2016**, *118*, 197–213. [CrossRef]
34. Llantoy, N.; Chàfer, M.; Cabeza, L.F. A Comparative Life Cycle Assessment (LCA) of Different Insulation Materials for Buildings in the Continental Mediterranean Climate. *Energy Build.* **2020**, *225*, 110323. [CrossRef]
35. Hasan, A. Optimizing Insulation Thickness for Buildings Using Life Cycle Cost. *Appl. Energy* **1999**, *63*, 115–124. [CrossRef]
36. Inflation Rate in India 1984–2024. Available online: https://www.statista.com/statistics/271322/inflation-rate-in-india/ (accessed on 16 August 2020).
37. SBI—Loans, Accounts, Cards, Investment, Deposits, Net Banking—Personal Banking. Available online: https://sbi.co.in/ (accessed on 23 February 2021).
38. Yildiz, A.; Gurlek, G.; Erkek, M.; Ozbalta, N. Economical and Environmental Analyses of Thermal Insulation Thickness in Buildings. *J. Therm. Sci. Technol.* **2008**, *28*, 25–34.
39. Ozel, M. Cost Analysis for Optimum Thicknesses and Environmental Impacts of Different Insulation Materials. *Energy Build.* **2012**, *49*, 552–559. [CrossRef]
40. Asadi, S.; Amiri, S.S.; Mottahedi, M. On the Development of Multi-Linear Regression Analysis to Assess Energy Consumption in the Early Stages of Building Design. *Energy Build.* **2014**, *85*, 246–255. [CrossRef]
41. Chapra, S.C.; Canale, R.P. *Numerical Methods for Engineers*; McGraw-Hill Higher Education: Boston, MA, USA, 2010.
42. Qian, X.; Lee, S.; Soto, A.; Chen, G. Regression Model to Predict the Higher Heating Value of Poultry Waste from Proximate Analysis. *Resources* **2018**, *7*, 39. [CrossRef]
43. Alghoul, S.K.; Gwesha, A.O.; Naas, A.M. The Effect of Electricity Price on Saving Energy Transmitted from External Building Walls. *Energy Res. J.* **2016**, *7*, 1–9. [CrossRef]
44. Alsayed, M.F.; Tayeh, R.A. Life Cycle Cost Analysis for Determining Optimal Insulation Thickness in Palestinian Buildings. *J. Build. Eng.* **2019**, *22*, 101–112. [CrossRef]
45. Ucar, A.; Balo, F. Effect of Fuel Type on the Optimum Thickness of Selected Insulation Materials for the Four Different Climatic Regions of Turkey. *Appl. Energy* **2009**, *86*, 730–736. [CrossRef]

Perspective

Renewable Energies and Architectural Heritage: Advanced Solutions and Future Perspectives

Elena Lucchi

Department of Architecture, Built Environment and Construction Engineering (DABC), Politecnico di Milano, 20133 Milan, Italy; elena.lucchi@polimi.it

Abstract: The current legislative framework and the recent energy crisis ask for massive applications of renewable energy sources (RES) in the built environment to reduce energy demand, environmental emissions, and energy costs. The uncritical application of these policies, especially on architectural heritage, could generate serious conservation issues, compromising their heritage values, biodiversity, traditional appearance, and materiality. Thus, there is an urgent call to balance architectural heritage preservation with energy production using clear rules, policies, criteria, and heritage-compatible technologies. The present study aims at defining an updated overview of the application of solar, wind, geothermal energy, and bioenergy on architectural heritage. A deep literature review of the studies published in the years 2020–2023 has been performed, identifying main topics, challenges, advanced solutions, and future perspectives. Acceptability, design criteria, and cutting-edge technologies are also illustrated through case studies to better understand practical approaches.

Keywords: heritage; renewable energies; solar energy; photovoltaic; wind energy; geothermal energy; bioenergy

1. Introduction

The United Nations Educational, Scientific and Cultural Organization (UNESCO) World Heritage Convention defines *"cultural heritage"* as any *"monument"* (e.g., paintings, sculptures, architecture, inscriptions, cave dwellings), *"group of buildings"* (e.g., buildings with similar architectural value thanks to the presence of a continuous historical process of modification and transformation), and *"site"* (e.g., historic town, archeological site) *"(…) with an outstanding universal value that express history, art, or science of a specific culture"* [1]. Inside them, *"architectural heritage"* refers to buildings, ruins, or groups of them characterized by physical, intangible, historical, or emotional values that increase over the years, according to the International Council of Monuments and Sites (ICOMOS) [2]. These cultural values reflect and express human knowledge, beliefs, craftsmanship, and traditions [3]. *"Architectural heritage"* can be both a physical *"artifact"* or a *"cultural meaning"* that expresses constructive cultures or events that occurred during the life of the building [1]. Each object has a specific *"heritage significance"*, defined as the combination of the heritage values assigned to a building and its setting [4]. Architectural objects are classified into a protected (also called listed or historic) and not protected (also called not listed, traditional, or historical) group according to the presence of an *"architectural interest"*). The criteria for identifying and assessing the presence of an *"architectural interest"* are [3]: (i) age connected with the architectural history (e.g., pre-industrial, industrial, modernism, and post-war periods); (ii) aesthetic merits related to the visual appearance and materiality, as well as to significant technological innovation, engineering, or socio-economic distinction; (iii) selectivity or rarity connected with the unique architectural quality; and (iv) national interest that emphasizes distinctive regional elements, and vernacular features.

The preservation of architectural objects is faced with risks related to physical damage, environmental pollution, tourism pressure, climatic changes, and a lack of financial

Citation: Lucchi, E. Renewable Energies and Architectural Heritage: Advanced Solutions and Future Perspectives. *Buildings* **2023**, *13*, 631. https://doi.org/10.3390/buildings13030631

Academic Editor: Antonio Formisano

Received: 1 February 2023
Revised: 15 February 2023
Accepted: 23 February 2023
Published: 27 February 2023

funding [5,6]. The conservation of *"architectural heritage"* requires any operation that aims at preserving its physical matters, visual appearances, and heritage values for a long time [7]. This trans-disciplinary work is based on the interaction among different competencies, not only a dull summation of specialist skills. It concerns a deep knowledge of its historical–critical foundations through the study of original documents, the survey of physical structures, the analysis of historical marks, and the critical interpretation of actual works [7]. More recently, architectural conservation has shifted its paradigm from purely physical preservation to making buildings functionally relevant for the age through constant redevelopment and repurposing. Any intervention involves a dialogue between *"old"* and *"new"* parts with a complex activity that includes changes and extensions that reveal the hidden meanings of the architectural monument [8]. Different approaches are highlighted at the international level [9]:

- *"Critical-conservation"* aims at transferring the architectural heritage to the future in the best possible conditions, studying and conserving its original matters and values while also interpreting and facilitating *"its reading"* through reversible interventions [10].
- *"Pure conservation"* aims at the meticulous conservation of the architectural heritage in its environment, adding only new necessary elements as well as preserving layers and marks of time transformations, not subtracting original matters [11].
- *"Repair and maintenance"* [12] aim at designing, *"by analogy"*, forms and materials similar to the past through their reconstructions [11].

Carbonara [9] clarified that the operations affecting and transforming the *"architectural heritage"* with *"renovation"* or *"full redesign"* are not included in architectural conservation because they do not respect original matters and values (e.g., rehabilitation, functional repair, reinvention, or remaking of the entire building or an element). Moreover, *"building reuse"* and its ramifications (e.g., rejuvenation, improvement, recycling, recovery, regeneration, adaptive reuse) can be placed *"next to restoration"* as they preserve the existing property, giving new practical and economic functions [13], but preservation is not the main purpose of the intervention [9].

All these approaches emphasize the sustainability and circularity of cultural heritage [13]. The convergence between the 'culture of sustainability' and the 'culture of heritage protection' is revealed by their common primary intentions. The planet's resources, such as the natural environment and architectural heritage, are finite. Hence, they should be carefully protected and wisely used [14]. In the context of sustainable transitions, defined as *"[. . .] long-term, multidimensional, and fundamental transformation processed through which established socio-technical systems shift to more sustainable alternatives"* [15], any intervention on the *"architectural heritage"* requires a balance within the values and the constraints imposed by the historical matters and the criteria of environmental sustainability and affordability [14]. Thus, sustainable design options for cultural heritage must follow the same purposes, considering functional, structural, environmental, and energy adaptations as tools for conserving and transmitting the object to the future rather than a redevelopment process in opposition to conservation requirements [14,16]. Each design solution should follow the operative criteria suggested by the *"Restoration Charters"* [17], such as compatibility, minimum intervention, reversibility, distinguishability, expressive authenticity, durability, and respect for original materials [14,18]. Inside these new challenges, the attention to the issues of environmental sustainability and energy efficiency has progressively increased in recent years [19]. The COVID-19 pandemic and the current energy crisis have completely changed the worldwide energy situation, generating huge impacts on the *"architectural heritage"* [20]. On the one hand, pandemic lifestyles (e.g., smart working, home-schooling, online shopping) has improved energy consumption and costs, with a higher impact on old buildings [17,20]. On the other hand, the energy crisis and climate changes require cleaner energy production based on the use of renewable sources, adaptation, and mitigation activities for favoring energy autarky [21]. This opens the opportunity for the energy retrofit of buildings, integrating passive and active systems respecting their original materiality, meanings, and appearance [18]. This idea boosts the traditional concept of land and building reuse, embodied energy, and usage of raw materials. In parallel, the European legislative frameworks (and recently the worldwide legislations) ask for massive applications of renewable energy sources (RES) in the built environment to reduce the energy demand, the environmental emissions, and the costs for electricity, domestic hot water, heating, and cooling in the building sector [19,20]. Otherwise, RES targets in *"architectural heritage"* are hidden by the historic* constraints for preserving original and traditional values [16,17,20]. Additionally, the uncritical application of these policies could generate serious conservation issues, especially for heritage contexts (e.g., historic* buildings and towns, protected landscapes), compromising their heritage values, biodiversity, traditional visual appearance, and materiality. Thus, there is an urgent call to balance architectural heritage preservation with energy production using clear rules, policies, criteria, and heritage-compatible technologies [21].

Cabeza et al. reviewed the integration of RES into historical building envelopes, focusing on solar and geothermal energy [22]. This study showed several architectural applications at the material, system, and building levels, also discussing their energy potentiality and human wellbeing. The analyzed period is 2006–2017. Thus, the examples use mainly traditional technologies, such as conventional photovoltaic (PV) systems, thin films, and applied PV systems. On the contrary, the technological development of RES is very fast. Over the last 5 years, the renewable energy sector has undergone crucial expansions and evolutions, boosting the applicability of these systems also on the *" architectural heritage"* thanks to the customization of colors and textures, the geometric flexibility as well as the presence of compact shapes, mimetic design, low-rate reflection, and high-resolution printed images. Thus, the study aims at updating the knowledge of the state of the art of RES integration on *"architectural heritage"* to understand new possibilities, innovative developments, and future perspectives. After having defined the methodological approach (Section 2), a detailed discussion on the integration of active solar systems (Section 3), wind technologies (Section 4), geothermal energy (Section 5), and bioenergy (Section 7) in *"architectural heritage"* is presented. Here, main topics, challenges, advanced solutions, impacts, and future perspectives are delineated. In addition, integration criteria and cutting-edge technologies are illustrated through case studies to better understand cultural, climatic, environmental, and design specificities. In the end, conclusions on innovative developments and future perspectives are summarized (Section 7).

2. Materials and Methods

RESs are derived from natural sources that have a higher replenished rate than consumed. The United Nations (UN) classified RES into the following categories: (i) solar energy; (ii) wind energy; (iii) geothermal energy; (iv) bioenergy; (v) hydropower; and (vi) ocean energy [23]. As mentioned before, this study aims at updating the knowledge of RES application on the *"architectural heritage"*, analyzing scientific studies and applications for the years 2020–2023. To this purpose, only RES with a direct application to *"architectural heritage"* are analyzed, such as solar, wind, and geothermal energy as well as bioenergy. Otherwise, hydropower and ocean energy are not studied because they are applied at the territorial level, not at an architectural level. The study is structured in two phases:

- Phase 1: A literature review on renewable energy and *"architectural heritage"*.
- Phase 2: Definition and discussion of main topics, advanced solutions, and future perspectives.

First, the literature review was performed to identify and count the existing scientific studies published in the Scopus bibliometric database (Phase 1). The Scopus database was selected because it guarantees a more complete overview of the studies, thanks to its spectrum of publications that has 20% more coverage than Web of Science [24,25]. Additionally, Google Scholar and Researchgate were excluded for the low accuracy of the analysis that considers several overlapped manuscripts [24]. This bibliometric analysis allowed the determination of (i) the number of publications; (ii) their evolution during time; (iii) the provenience and geographic distribution of the publications; and (v) indexed and authors' keywords. To have the highest overview of the topic, the queries concern "titles, abstracts, and keywords" (TITLE-ABS-KEY). On the contrary, queries that consider only "keywords" (KEY) cut several important papers. Cultural heritage and technical keywords on solar energy, wind technologies, geothermal energy, and bioenergy have been analyzed through integrated queries to have the widest range possible of publications. The keywords used in the Scopus Database are shown below (Table 1).

Table 1. Keywords used in the Scopus database.

Keywords				
Cultural Heritage	**Solar Energy**	**Wind Energy**	**Geothermal Energy**	**Bioenergy**
Heritage	Solar energy *	Wind energy *	Geothermal energy *	Biomass *
Architectural heritage	Solar system *	Wind system *	Geothermal	Bioenergy *
Heritage building *	Solar technology *	Wind technology *	Heating, Ventilation, Air Conditioning (HVAC)	Wood energy *
Historic * building *	Photovoltaic * (PV)	Wind turbine *		Dung energy *
Built environment	Solar Thermal (ST)	Wind farm *		Charcoal energy *
	PVT		Heat pump *	

Note: * = plural and singular.

More specific heritage keywords (e.g., protected building*, listed building*, vernacular building*/architecture, traditional building*) did not produce any significant result. Conversely, the combination between heritage OR technical keywords was not focused on RES integration in architectural heritage but on energy retrofit of historic* buildings using internal insulation, windows, mechanical ventilation, etc. In the first step, the analyzed period was 1994–2023 to have wide results. Then, this period was reduced to the years 2020–2023 to update the knowledge and to understand future research perspectives. Scientific data have been cleansed after reading titles and abstracts to improve data relevance, eliminating duplications, etc. After this process, data were extracted and charted using database and filter services. First, a chronological view of the different periods was produced to show the evolution of the studies. Moreover, scientific studies were mapped and classified according to provenience, number, and indexed keywords. Authors and indexed keywords have been mapped with VOSviewer 1.6.18, the most widely open-source software for science mapping [26], to visualize data patterns and bibliometric networks. Associated keywords are clustered using the same colors. The popularity of a keyword is indicated by its size, while its proximity is interpreted as an indication of its similarity. In the second step, a detailed and critical discussion of the most relevant studies was carried out on the selected papers (Phase 2), focusing on the following questions: *"What are the main aspects considered?"*, *"What is the approach for RES integration on architectural heritage?"*; *"Is it possible to balance heritage preservation and energy production?"*; *"In which way?"*; *"What are the differences for integrating different RES?"*. Starting from these questions, a detailed discussion of main topics, advanced solutions, and future perspectives has been realized and presented.

3. Solar Energy

The integration of solar energy into architectural heritage refers to the use of photovoltaic (PV) and solar thermal (ST) systems. Fifty scientific documents have been found for the period 1994–2023, combining cultural heritage and solar energy keywords (Table 1). Between them, 23 papers have been published in the period 2020–2023. Thus, 46% of publications are from the last 3 years (Figure 1).

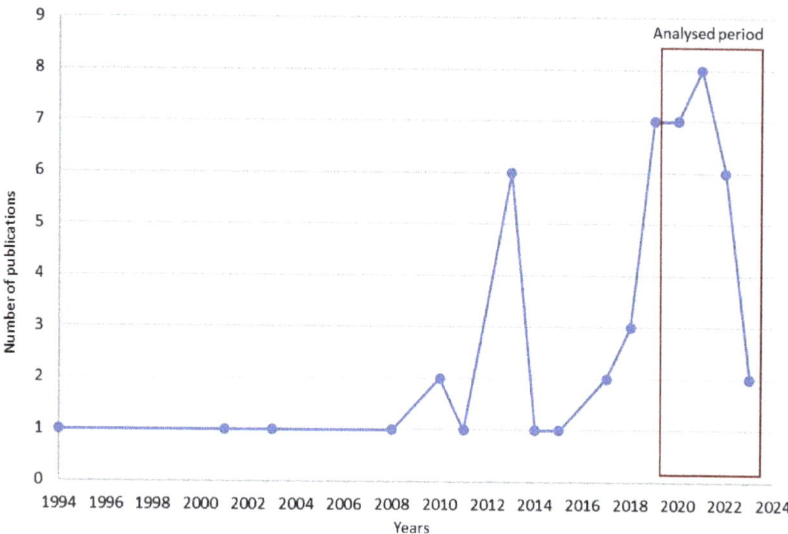

Figure 1. Chronological view of the studies on solar energies and cultural heritage: 50 scientific manuscripts have been realized from 1994 to 2023, 23 of them in the years 2020–2023 (Source: Author's elaboration using Scopus data).

The most active Countries in the analyzed period are Italy (9 papers), Switzerland (7 papers), and the United Kingdom (2 papers). Moreover, one paper on this topic was published in several Mediterranean Countries (e.g., Spain, Portugal, France, and Greece), Central Europe (e.g., Germany, Belgium, Poland), and Scandinavia (Sweden). Outside Europe, the active Countries are Peru, Iraq, Indonesia, and Egypt (Figure 2).

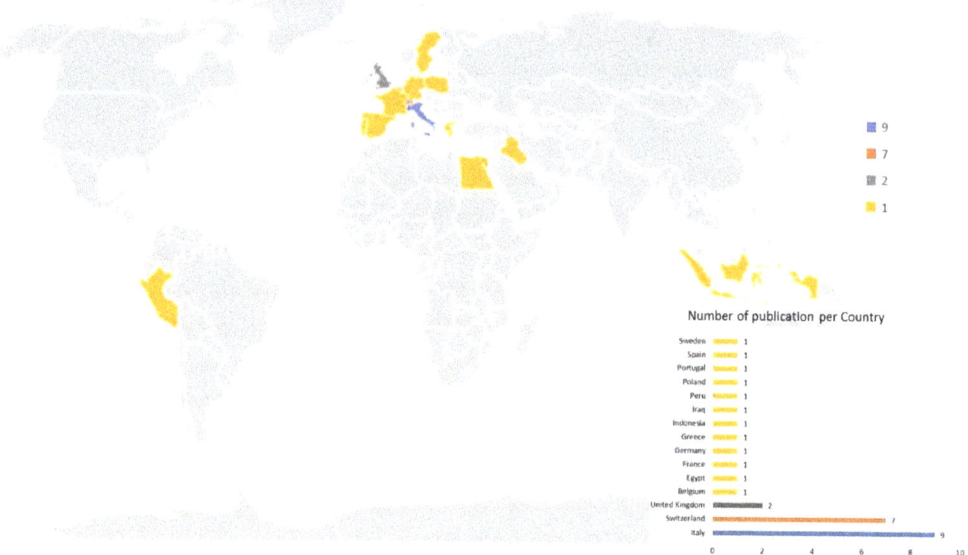

Figure 2. Map of scientific studies on solar energies and architectural heritage according to their provenience (Source: Author's elaboration using Scopus data).

The keywords of these studies have been analyzed. Authors' and indexed keywords produced a heterogenous cloud, difficult to be clustered for the overlapping of several keywords and concepts. Nine clusters are produced (Figure 3a): (i) solar energy retrofit; (ii) PV and building integrated PV (BIPV); (iii) sustainability; (iv) architectural conservation; (v) decision making; (vi) energy policies; (vii) energy production; (viii) climatic change; and (viii) award. On the contrary, indexed keywords can be divided into three clusters (Figure 3b): (i) solar energy production; (ii) energy efficiency and climate change; (iii) architectural conservation. This structure represents the three aims of the solar application on the architectural heritage that respond to energy, sustainability, and conservation purposes.

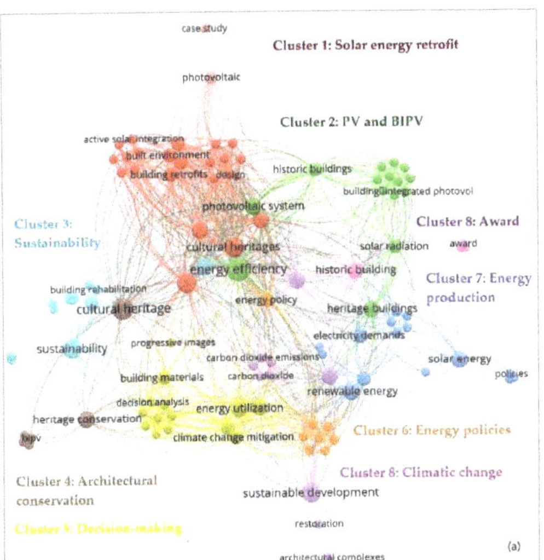

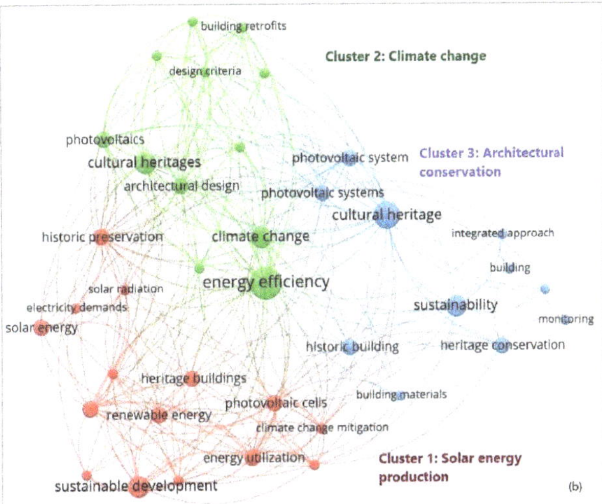

Figure 3. Scientific landscape of architectural heritage and solar energy and architectural heritage keywords: (**a**) 276 total keywords; (**b**) 186 indexed keywords (source: Author's elaboration using VOSviewer, based on Scopus data).

Solar energy and heritage keywords have been extracted from indexed keywords through detailed data mining to verify the main topics of these works. One hundred three indexed keywords have been selected, and four main topics can be defined (Figure 4): (i) solar acceptance; (ii) solar potential evaluation; (ii) visibility mapping; and (iv) solar integration criteria.

PV applications on architectural heritage are extensively investigated for their significant contribution to the reduction of energy requirements for electrical needs and thermal conditioning [19], as well as for their aesthetic appeal and multifunctionality [18]. Only one study investigates ST systems, while PVT is not studied. Initially, the studies focused on the acceptability of PV systems in the built environment [27]. Then, their technical advantages [28,29], energy performances [29,30], and economic benefits are demonstrated [29], also focusing on aesthetic design [27,28] and energy potentials for solar architecture [29,30]. Specific studies refer to historic buildings, with a section dedicated to RES integration in old [31,32], heritage [33,34], historical [35], and existing buildings [36], as well as in historical towns [37,38]. Here, the focus is on the criteria for ensuring the heritage compatibility of conventional technologies. Recently, attention has been focused mostly on innovative PV technologies [39,40], assessing their energy performance, risks, solutions, and design criteria. Lately, energy landscapes have been introduced [39–42]. Next, each cluster is deeply discussed.

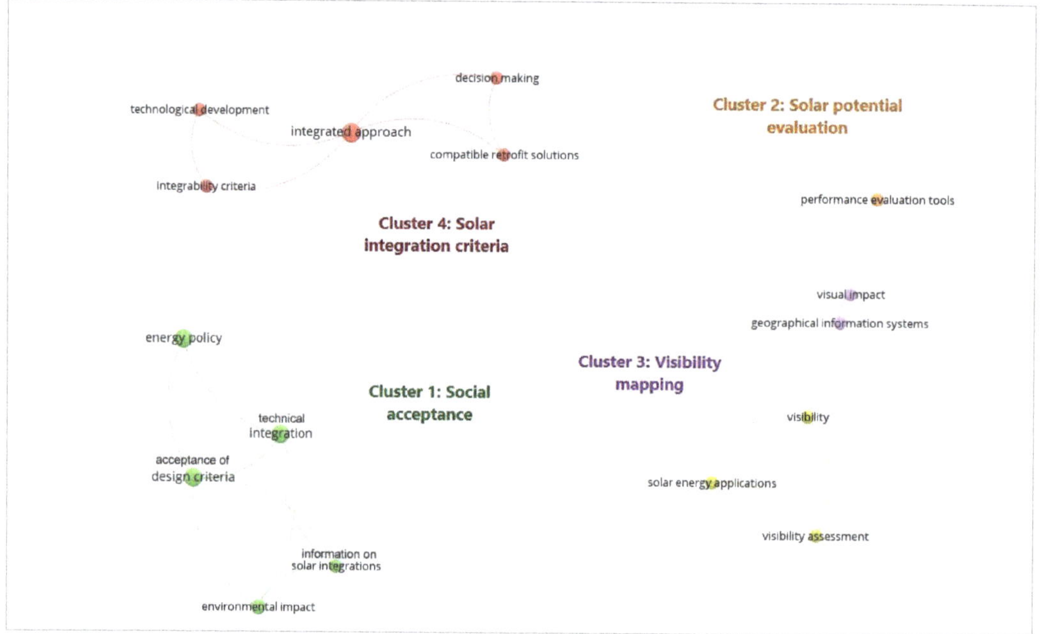

Figure 4. Main clusters of the studies based on co-occurrence network of the selected indexed keywords on solar energies and architectural heritage (source: Author's elaboration using VOSviewer, based on Scopus data).

3.1. Social Acceptance

Social acceptance and acceptability of active solar systems is a commonly debated topic, both on new and existing buildings [43,44]. Social acceptability is a mental representation (or a priori phenomenon) related to the use of a specific technology. On the contrary, social acceptance is a posteriori pragmatic evaluation of technology after knowing it. Active solar applications in architectural heritage are hindered by numerous barriers linked to the presence of outstanding values, traditional features, and materials [20,21]. Color ranges, high reflection, modularity, and geometric pattern of PV and ST systems have an impact on vernacular and historic buildings [45]. Thus, their application is not always compatible [45]. The literature mainly highlights the following barriers:

- Technical aspects.
- Costs.
- Policy.
- Information and knowledge.

In the past, the aesthetic aspect [27,28], technical knowledge [28], and economic issues [27,28] were underlined as key problems for the visual appearance of conventional technologies [27] and the economic crisis of the solar market [28]. More recently, these barriers have been less perceived thanks to the technological innovation of the solar sector, especially for the visual appearance and customization of innovative PV panels (e.g., thin films, hidden colored PV) [14,20,21]. Technical doubts affect the energy efficiency and the environmental impact linked to the production of innovative systems [46], especially for PV and PVT [21,47] (e.g., colored solar cells, thin films, solar concentrators). Technical doubts are strictly related to the economic barriers, which pertain mainly to large initial investments [43,46], long payback periods [46], and the absence of financial incentives [43]. In addition, the complexity and fragmentation of legislative frameworks and authorization processes are perceived as important elements for blocking the application of solar energies on cultural heritage [45]. The restrictions of local Heritage Authorities [43] and the absence of shared regulations [14,21] expand this problem [45]. Finally, information barriers concern the lack of information and confidence in innovative systems related to human expertise both for energy and heritage [20,45] as well as to training [18,43] and capacity building [43,46]. Recently, economic barriers have been decreasing progressively due to the increasing costs of fossil fuels [46]. Thus, economic aspects are perceived as the main benefits of solar energy applications. Positive aspects of PV integration in heritage buildings are connected to the enhancement of economic values [19], functionality [20], and human comfort [46]. Moreover, the creation of soft tourism and the multiplier economic effects are suggested as positive benefits related to heritage towns and buildings [27]. PV benefits are identified in scalability, reliability, versatility, low maintenance costs, on-site production, self-consumption coverage, and energy peak shaving [20,21,46]. A synthesis of barriers and benefits of active solar energies applied to architectural heritage is reported below (Table 2).

Table 2. Barriers and benefits for the social acceptance of active solar energy applications on architectural heritage (Source: Author's elaboration).

Aspect	Barrier	Benefit
Technical	Energy performance of innovative systems 🟩 Environmental impact of production 🟩	Innovative aesthetic appearance and versatility 🟩 Reliability and on-site production 🟩 Multifunctionality and scalability 🟩
Economic	Large initial investments 🟩 High costs 🟥 Long payback period 🟥 Lack of incentives 🟩	High energy costs 🟩 Appeal for soft tourism 🟥 Multiplier economic effects 🟥 Low maintenance costs 🟩
Policy	Complex legislation 🟥 Long authorization process 🟥	New local policies for solar applications 🟥
Information	Lack of knowledge of innovation 🟩 Lack of examples 🟥 Lack of training 🟥	New awareness after energy crisis and COVID-19 pandemic 🟩

Note: 🟩 = Common for the integration of active solar in buildings and architectural heritage; 🟥 = Specific for the integration of active solar systems in architectural heritage.

In general, people engagement and co-creating design are considered the correct approaches for improving the social acceptance of active solar technologies [14]. The development of tailored materials and solutions for building integration is always suggested as a possible measure for overcoming technical and information barriers [46].

3.2. Solar Potential Evaluation

The solar potential evaluation of heritage buildings is the starting point for decision-making purposes in urban planning. In the past, heritage buildings and towns were mainly excluded by these calculations for the presence of high urban and architectural constraints [46–48]. Recently, only a few studies investigated the impact of vernacular urban shapes, such as narrow streets, porches, and mutual shadows, on buildings. In all these cases, two deterministic approaches are used (Table 3):

- Bottom-up models.
- Solar cadasters.

Table 3. Approaches used for the solar potential evaluation of architectural heritage (Source: Author's elaboration).

Characteristics	Bottom-Up Models	Solar Cadaster
Object	Representative building typologies	Entire building stock
Time	Short	Long
Cost	Low	High
Heritage constraints	✓	✓
Urban constraints	✗	✓
Impact of surroundings	✗	✓
Impact of mutual shadows	✗	✓
Impact of urban geometric irregularities	✗	✓
Difficulties	Selection of representative buildings Cluster analysis of building differences	Detailed approach with high costs and long times

Note: ✗ = Neglected ✓ = Considered.

First, bottom-up models are mapping tools that cluster statistical and technological information for defining *"representative buildings"* characterized by similar dimensions, geometries, typologies, features, materials, and orientations for roofs and façades [48]. This approach is not appropriate for historic* features because the calculation of the solar potential of single representative buildings neglects heritage specificities, such as architectural constraints [49,50], urban geometric irregularities [51], surrounding structures, short-wave solar radiations [52], and mutual shadows from aggregated buildings [50–54]. Only a few studies investigate the impact of heritage features [49,50] and urban shapes [49–53] with the support of digital mapping. In the first case, only detailed investigations of urban, architectural, and historical values and constraints of roofs and façades permit the correct selection of building typologies and solar interventions [49]. The cluster analysis particularly demonstrates the difficulties of grouping heritage inhomogeneous building stocks due to the differences in constructive features, heritage values, utilization levels, and urban and building constraints [50]. In the second case, a study demonstrates that urban shadows are very important in historic towns, as the Urban Shading Ration (USR) can reach 60% of building façades and 25% of roofs [54]. The energy potential is significantly reduced by this aspect. Thus, the influence of mutual shadow on the energy potential is investigated, especially on building façades [51,52], ground [51], and roofs [51–53], also focusing on the influence of reflections [52], urban shadows [53,54], and complex geometries [55].

Second, solar cadasters are web-based mapping tools supported by mathematical models for determining the production capacity of active solar systems through two-dimensional (2D) maps or orthophotos [51]. Thus, the calculation is realized on the entire building stock. Examples of solar cadaster for heritage towns refer to the Swiss towns of Geneva (2018) [56] and Carouge (2018) [57] using 3D and 3D light detection and ranging (LiDAR) data, heritage and urban constraints, and building data. In the solar cadaster of Carouge, each building is analyzed in a detailed way, suggesting specific design criteria and installation procedures for PV and ST technologies.

In both cases, Geographic Information System (GIS) tools are matched with simulation software for data management, cluster analysis, and query interactions. The main models used for assessing the solar potential are divided according to the dimension of the urban areas [52]. In general, the higher the area, the lower the optical precision of reflection [52], and thus the calculation of USR, especially on building façades. A synthesis of these models is reported below (Table 4).

Table 4. Models used for assessing the solar potential at urban level (Source: Author's elaboration).

Aspect	Tool	Logo
Large scale analysis	CitySim [58]	
	Archelios Map [59]	
District level	Grasshopper (Honeybee, Daysim, Ladybug, and DIVA) [60]	
	Climate Studio [61]	

3.3. Visibility Mapping

Visibility mapping is strictly connected with the solar potential evaluation. The visibility of a solar installation can be assessed by [62]:

- Spatial modeling.
- Experts' inquiries.
- Simplified graphical methods.

In all cases, the visual impact is evaluated mainly from public spaces or significant views [21,60]. Thus, active solar systems can be located on hidden roofs, interior façades, behind parapets, outbuildings, or new additions [21,61].

LESO-QSV (Quality–Sensitivity–Visibility) is a cross-mapping tool for assessing the criticality of solar installations in heritage territories [63]. The "criticity" level of an installation combines the visibility of the solar system and the sensibility of the urban area. Heritage buildings are high-sensible in context, and thus, they require low-visibility technologies to reduce their impact. The evaluation of their visibility is based on the coherency of their geometry, materiality, and pattern (Figure 5).

This approach is combined with spatial modeling for assessing solar visibility in historic* towns. The cross-mapping between visual criticality and solar radiation maps of a specific surface evaluates the possibility/difficulty of solar installation [62,64]. This method advises decision-making on urban planning at different levels [62] (Table 5).

Table 5. Visual criticality and solar radiation maps to be used on heritage contexts (Source: Author's elaboration from [62]).

Planning Level	Visual Criticality Map	Solar Radiation Map
Strategic planning (1:100,000–1:30,000)	Photo shooting locations Relevant historical sightseeing	Aggregated solar radiation data over terrain models
Development planning (1:10,000–1:5000)	Roof visibility ratio Visual amplitude per surface	Calculated solar radiations on a roof surface
Detailed planning (1:2000–1:500)	Roof visibility ratio Façade visibility ratio	Calculated solar radiations on roof and façade surfaces

To this purpose, two new parameters have been defined: (i) *"roof visibility ratio"* and (ii) *"façade visibility ratio"*, respectively equal to the relationship between visible roof/façade areas and total roof/façade areas [62]. The combination of these maps and the potential energy consumption permits an understanding of the energy matching between production and consumption in historical areas [64].

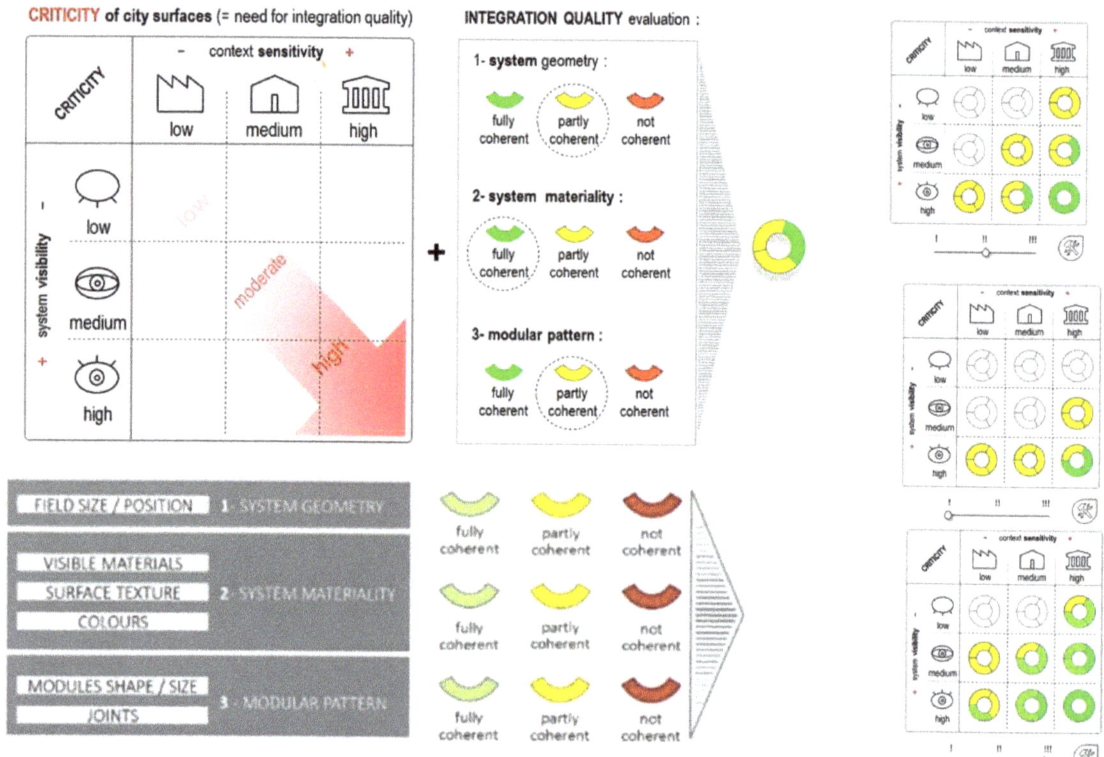

Figure 5. "Criticity map" considering visibility and sensitivity (Source: Authors' elaboration from LESO-QSV—Architectural integration tool, https://www.epfl.ch/labs/leso/research/domains/renewables_integration/leso-qsv, accessed on 14 February 2023).

Experts' inquiries involve experts, Heritage Authorities, and local and regional planning bodies for the evaluation process [21]. The assessment generally refers to specific buildings, considering their history, location, protection level, conservation states, and modifications during the years. This method is applied both to singular buildings and historical towns. First, at the building level, an approach [65,66] classified the architectural heritage in building elements according to the *"combinatory grouping approach"* proposed by the standard UNI 8290-1 [67]. Possible PV interventions and technologies are defined for each building element, evaluating their compatibility with the local Heritage Authority [65,66]. At the urban level, the *"target-based method"* evaluates *"target elements"* (e.g., the historic* building, the building envelope) rather than a set of significant points [68]. Thus, the visibility assessment is realized only for buildings that are of interest (e.g., listed, protected, or traditional buildings). Here, solar exposure (e.g., absence of shading, high irradiance) and heritage values (e.g., conservation state of the roof, absence of heritage constraints) are evaluated. A comparison between the cross-mapping and the target-based methods applied to the same historical center of Geneve in Switzerland shows a significant difference in the roof percentage that can be used for solar installations. Respectively, 50% and 64% of roofs can be used for solar installations using the two methods [68]. Thus, the target-based method respects heritage compatibility but also increases the energy potential of historic city centers.

Finally, simplified graphical methods check the visual impact of the solar installations considering the variation of the distance between the observer and significant views, the slope of the roof, and the building height [69]. Several examples have been produced.

3.4. Solar Integration Criteria

Design and evaluation criteria for the integration of active solar technologies into the historic* built environment are deeply investigated. Several countries developed national or local guidelines for balancing heritage preservation and energy production. These criteria refer to the architectural restoration theories that consider both physical and semantic issues, respectively linked to the preservation of original materiality and latent meanings [20,21]. There are no differences between the integration of PV and ST technologies [70], although PV systems are supported by a huge amount of the literature. Solar design criteria are *"universally recognized"* although their implementation has declined according to local climate, orography, morphologies, land features, resources as well as traditional features, building typologies, techniques, and materials. These criteria also differ according to the type of cultural heritage (building element, buildings, towns, landscape, site of historic resource) and the protection level (e.g., heritage protected or traditional features) [21]. Furthermore, the conservation level influences the heritage-compatibility: active solar installations are allowed in heritage contexts with lower conservation levels but avoided with high conservation levels for conserving original materials [71]. A taxonomy of international recommendations has been published, identifying recurring and transferable criteria, design suggestions, and a glossary for helping designers and Public Authorities in the selection and evaluation of appropriate design alternatives and products [21,70,72]. Additionally, new design solutions, shared criteria, positive local applications, and knowledge gaps on PV product innovation are identified through several focus groups with the Heritage Authorities [21]. The criteria are classified as aesthetic, technological, and energy integration [21,70]. Aesthetic criteria imply a compatible visual interaction with traditional characters, materials, and values [20,70]. Technological criteria are based on durability, reversibility, and detailed design [20,21]; energy integration entails an efficient coverage of the overall energy consumption [21,65,70]. The solar integration criteria can be summarized as follows:

- *"Visual compatibility"* maintains the original aesthetic appearance [21,65].
- *"Material compatibility"*: preserving original materials, construction techniques, and heritage significances as evidence of the *"material culture"* of a specific period and territory [20,21,68,70].
- *"Minimum intervention"*: thanks to the reduction of physical changes and material losses as well as to the preservation of the original visual appearance maintaining its geometries, proportions, shapes, sizes, colors, patterns, textures, and reflectance (Figure 6) [21,70].
- *"Reversibility"* of the solar interventions without damaging the original building (Figure 7) [20,21,70].
- *"Durability"* of the transformation preventing structural, electrical, hygrothermal, energy-efficiency risks, negative effects, or degradation process due to new solar installation [20,70].
- *"Balance between preservation and energy production"* dimensioning the active solar systems according to the real energy needs [21,46,70].
- *"Interdisciplinarity"* of different skills and competencies in architectural restoration, energy design, technology development, urban planning, and landscape design [18,71,73].

Figure 6. Positive example of the *"minimum intervention"* criteria on the PV roof of the La Certosa Island in Venice (Italy) with respect to the geometries, shapes, proportions, colors, and reflectance of traditional clay roofs [74] (Source: Elena Lucchi).

Figure 7. Positive example of the *"reversibility"* criteria on the solar intervention of Palazzo Leonori in Rome (Italy), where the transparent BIPV roof is detached from the original XIX Century Palace by metallic columns [75] (Source: Elena Lucchi).

Otherwise, the traditional restoration criteria of *"recognizability"* or *"distinguishability"* of the new intervention are contradictory and not accepted by all the recommendations. In some cases, the recognizability of the transformation is boosted to ensure a clear differentiation between new and existing elements, respecting original features and values [20,70]. This idea is correct, especially for modern buildings or industrial archaeology as well as for building extensions [21] (Figures 8 and 9).

Figure 8. Positive example of the *"recognizability"* criteria on the solar roof of the Winery Alois Lageder in Magrè (Italy) thanks to the presence of traditional PV panels oriented to the maximum solar exposure (Source: Elena Lucchi).

Figure 9. Negative examples of the *"recognizability"* criteria of the solar intervention in a XX Century School in Milan (Italy) and in traditional alpine buildings in Val Sarentino (Italy) where PV and ST panels are visible from public spaces for their irregular shapes, blue colors, and high reflectance (Source: Elena Lucchi).

In traditional or historic buildings, the *"concealment"* of the solar systems from public view or prominent visual assets is often suggested to reduce any potential visual impact of the new installation [20,21,70] (Figure 10). Hidden colored, thin films, semi-transparent, and textured PV systems resulted in promising visibility minimization [21,65]. Thus, the visibility of the solar system requires a deep analysis through a detailed mapping of architectural and environmental characteristics (Section 3.3).

Figure 10. Positive example of the *"concealment"* of the solar intervention in the Podere Case Lovara in Levanto (Italy) thanks to the presence of traditional PV and ST panels hidden behind a parapet and not visible from the protected landscape of the Cinque Terre Natural Park (Source: Fondo per l'Ambiente Italiano).

Tailored active solar design solutions can be supported by Building Information Modeling (BIM), which provides spatial and functional representations of architectural heritage elements using parametric objects. It permits early-stage visualization, data management, error correction, data sharing, and calculation. The studies focus mainly on PV optimization on rooftops without fulfilling specific integration criteria [76,77]. The main purposes of these studies are energy performance evaluation, shape and orientation investigation, layout and color preview, and cost reduction. Only one study highlights the theoretical benefits of Heritage BIM (HBIM) for PV installations on architectural heritage [46].

4. Wind Energy

The integration of wind energy into the *"architectural heritage"* refers mainly to the use of large wind turbines located on land (onshore) or water (offshore). Applications at the building level are neglected, probably for the strict heritage constraints that normally do not allow these applications on a historic building. Thus, the literature refers mainly to heritage sites. Nineteen scientific documents have been found for the period 1994–2023, combining cultural heritage AND wind energy keywords (Table 1). Between them, seven papers have been published in the period 2020–2023. Thus, 36% of publications are from the last 3 years (Figure 11).

Active Countries for the analyzed period are the UK, Australia, the US, Turkey, Portugal, Brazil, and Denmark, with one paper each (Figure 12).

The keywords of these studies have been analyzed. Authors' and indexed keywords produced a heterogenous cloud with five clusters (Figure 13a): (i) biodiversity; (ii) wind farms; (iii) risk assessment; (iv) decision-making; and (v) climatic change. Indexed keywords produced four more rational clusters (Figure 13b): (i) biodiversity; (ii) risk assessment; (iii) decision-making; and (iv) climate change. This classification represents the most important topics of the decision-making process on wind energy on heritage landscapes that require a detailed risk assessment for balancing the influence on biodiversity and climate change.

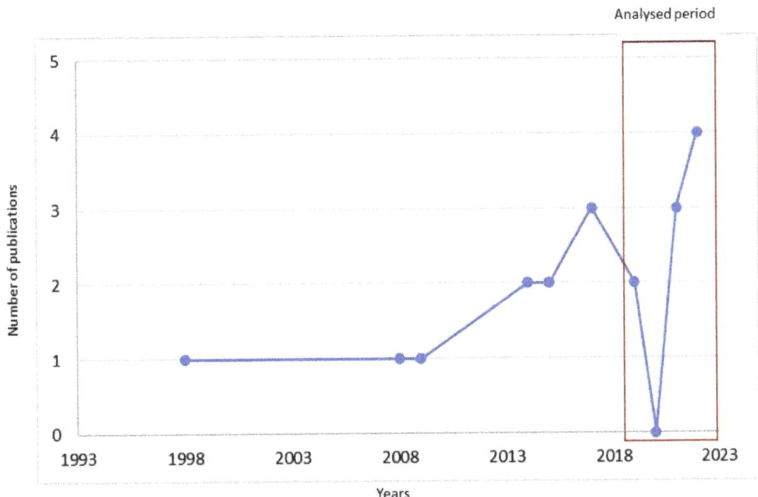

Figure 11. Chronological view of the studies on wind energy and architectural heritage: 19 scientific documents have been realized from 1998 to 2023, 7 of them in the years 2020–2023 (Source: Author's elaboration using Scopus data).

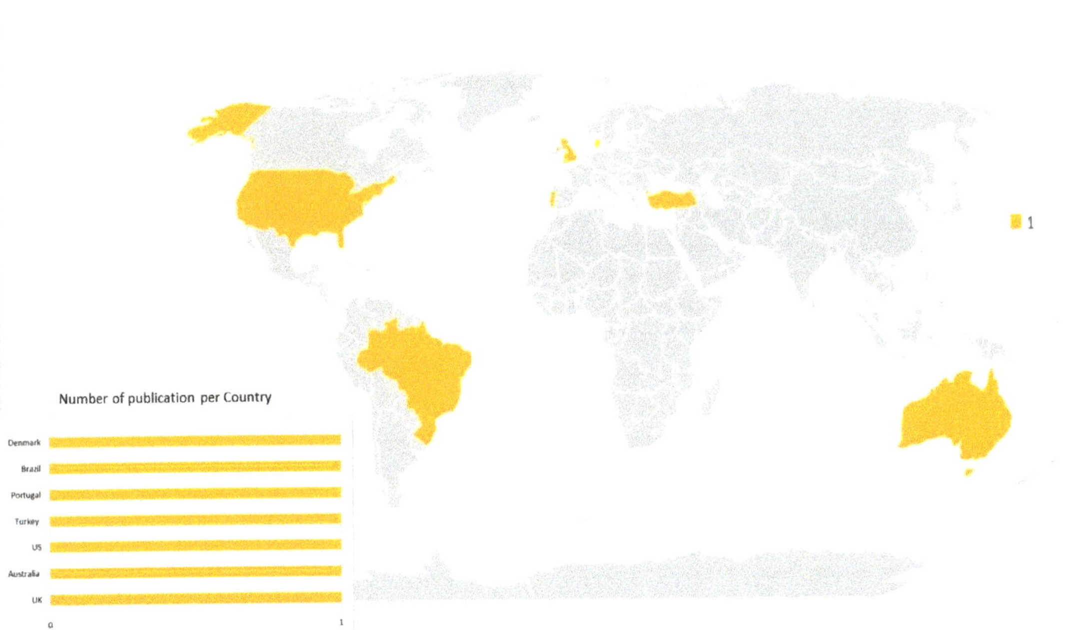

Figure 12. Map of scientific studies on wind energy and architectural heritage according to their provenience (Source: Author's elaboration using Scopus data).

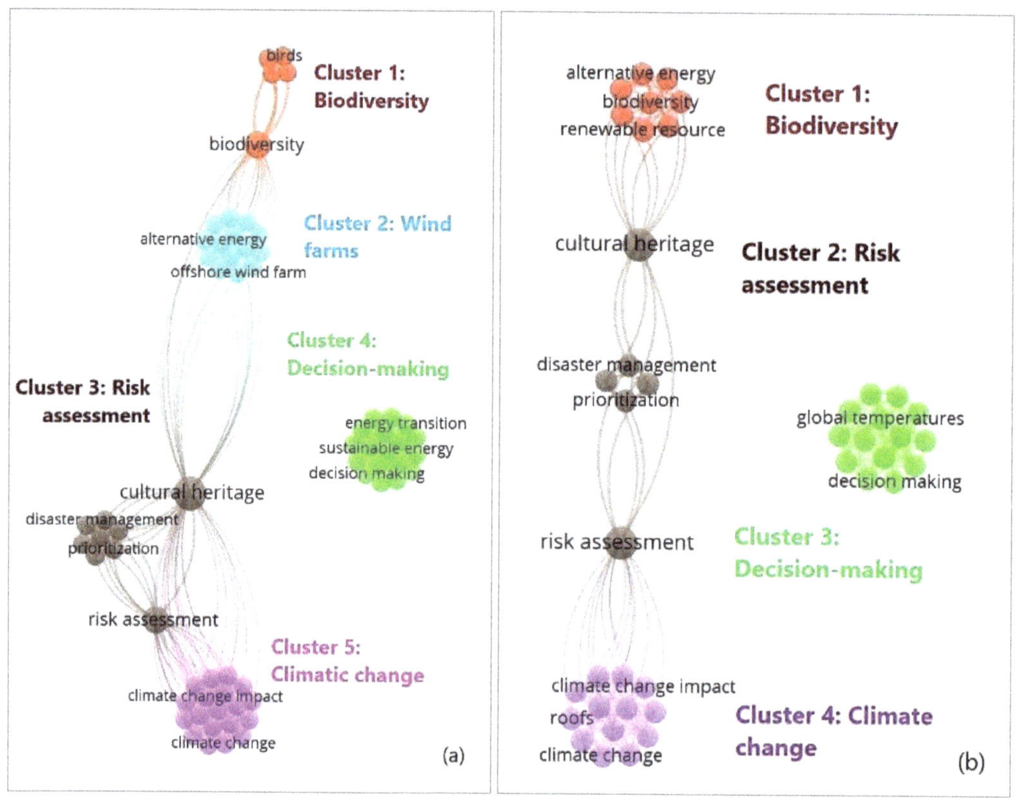

Figure 13. Scientific landscape of of wind energy and architectural heritage keywords: (**a**) 71 keywords; (**b**) 57 indexed keywords (source: Author's elaboration using VOSviewer, based on Scopus data).

Finally, wind and heritage keywords have been extracted from indexed keywords through detailed data mining to verify the main topics of these works. Twenty-eight indexed keywords have been recognized, and four main topics can be defined (Figure 14): (i) wind farm red; (ii) wind potential evaluation; (iii) social acceptance and visibility mapping; and (iv) wind integration criteria. Each topic is discussed deeply.

4.1. Wind Farms

The literature refers only to offshore wind farms. Restrictions for their implementation concern the protection of natural areas (e.g., archaeological monuments, shipwrecks, environmentally protected areas for biodiversity, refuges for wildlife) as well as the presence of technical constraints (e.g., marine pilot zones, underwater lines, pipelines), maritime uses (e.g., exploration or extraction of hydrocarbons and minerals), environmental risks (e.g., aquaculture and fishing banks), and military operations [78]. Moreover, maritime zones with mean wind velocity smaller than 4 m/s [78] and earthquake fault lines [79] are excluded, respectively, for their low energy potential and risks. The main impacts of floating wind farms concern the destruction or the disturbance of foraging or breeding habitats, the collision of marine species and seabirds [80], as well as the generation of negative social perceptions in local communities [80].

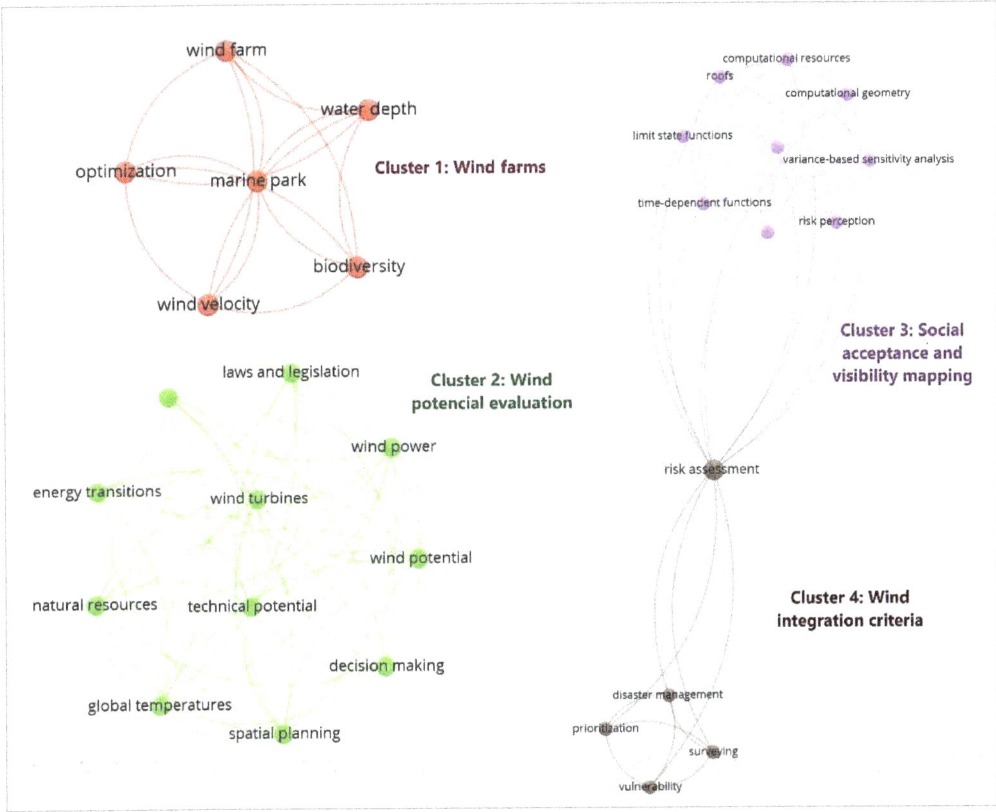

Figure 14. Main clusters of the studies on wind energy and architectural heritage based on co-occurrence network of the selected indexed keywords (source. Author's elaboration using VOSviewer, based on Scopus data).

4.2. Wind Potential Evaluation

The determination of wind potential and of optimized sites for utility-scale wind systems requires the integration of multiple factors that affect the conservation of landscapes and wild environments, as well as the costs of electricity generation and use [81]. Normally, multicriteria evaluation techniques (MCE) are coupled with GIS to compare environmental, economic, legal, social, and technical aspects [81,82]. The most common techniques are simple additive weighting (SAW), the analytical hierarchy process (AHP), the ideal point methods (e.g., TOPSIS), the elimination and choice expressing (ELECTRE), and the outranking techniques (e.g., PROMETHEE) [78]. AHP is the most used technique for the high complexity of wind studies. AHP is a semi-quantitative method that involves quantitative and qualitative criteria defined through public inquiries or expert working tables for guiding informed decision processes in a conscious way [82]. The parameters considered try to balance geo-resources and geo-hazards detection with land-use suitability to evaluate both wind potential and risks [78]. The evaluation is characterized by well-known steps [79]. First, areas not suitable for locating a wind farm are identified using general criteria (Section 4.1). These criteria are grouped into decision and exclusion criteria based on public inquiries [79] or general rules. These criteria are evaluated through the AHP method and scored according to their importance. Finally, the most suitable area is determined and georeferenced.

4.3. Social Acceptance and Visibility Mapping

Wind energy involves historically unprecedented changes to the visual integrity of a landscape [83]. Thus, wind integration criteria are strictly related to the assessment of the risk of wind turbines on the natural environment [84]. Wind turbines have a total high of 180–200 m. They are visible up to 80 km in lowlands and up to 30 km in the mountains [83]. Thus, they have an impact also on heritage towns, groups of isolated buildings, and heritage sites. The assessment of their location is very important to reduce a potential negative impact [83]. In many cases, wind technologies are excluded from conservation areas, sites of interest (e.g., heritage, archaeological, and paleontological sites), and agricultural lands with high fertility for the visual impact on high-sensitive heritage values [82]. They are not compatible not only with *"heritage core zones"* (the area with the listed property) but also with *"heritage buffer zones"* (the outside area of about 100 m that protects historic features from external influences) [83]. Otherwise, some visibility methodologies are provided for identifying suitable zones for their location. The most common is the production of a baseline with high-quality and georeferenced photographs from selected viewpoints to evaluate the impact and the risk connected with the installation of wind turbines [84,85]. Viewpoints must be selected according to the presence of heritage significance and historic testimony [85] (Figure 15).

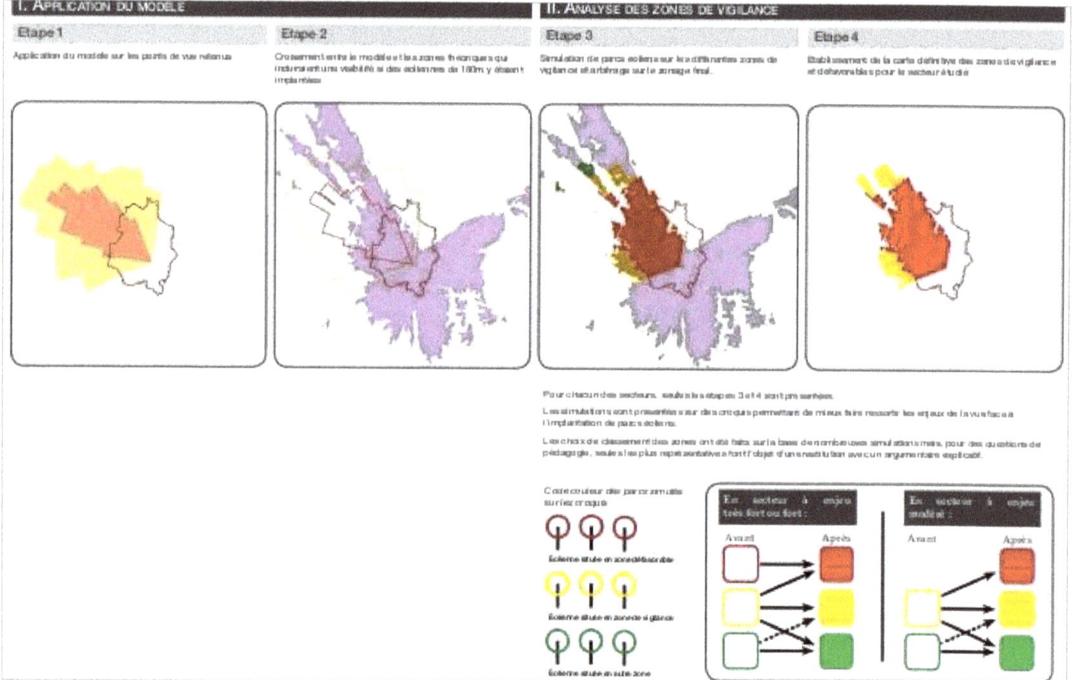

Figure 15. Zoning of wind turbines in the Vézelay landscape (France) based on angle of view and distance of the turbines: unfavorable (red), caution (yellow), and other zones (green) (Source: [84]).

The impact of these technologies may vary depending on their visibility, number, and distance from the property. Kloos et al. suggested the importance of maintaining the overall quality of the landscape by studying the interrelations between wind elements and surroundings [86]. The visibility of wind turbines should be assessed according to the following criteria: (i) technical dominance of the landscape image; (ii) visual dominance of the turbines; and (iii) distortion of the landscape scale. Noisy of wind turbines is not reported as a problem, probably because of the distance from inhabited areas.

4.4. Wind Integration Criteria

The main influencing factors for assessing the visual acceptability of wind turbines in a broader landscape perception are as follows:

- *"Distance"* from significant viewpoints and heritage property [76–84,87]. In general, a maximum distance of 10 km from conservation sites is required to reduce visibility while also considering changing weather and atmospheric conditions. Otherwise, suitable protection perimeters are within the range of 5–7 km from the property.
- *"Reduced visibility"* from important angles of view, maintaining an undisturbed horizon and defining correct color ranges and blade directions [85,86] (Figure 16).
- *Visual competition"* requires the conservation of an undisturbed visual setting, and quality of the landscape considering the interaction between wind turbines, a heritage property, and its surroundings [84,86].
- *"Environmental impacts"* on pre-existing elements express the need to reduce the disturbance on wildlife, seaside, biodiversity, and natural elements [78,82].
- *"Balance between preservation and wind potential"* consider the most suitable position also in terms of wind speed, wind direction, uniform directions, and absence of heritage values [76,81].

Figure 16. Positive example of the depiction of an outgoing view with panorama baseline on top and the view section with graphic accentuation, color code, and reference to evaluation scheme in Vézelay (France) (Source: [84]).

5. Geothermal Energy

The integration of geothermal energy refers mainly to ground source heat pumps in historic* buildings, and it is seldom investigated. Four scientific documents have been found for the period 2017–2023, combining cultural heritage and geothermal energy keywords (Table 1). Between them, one paper was published in the period 2020–2023. Thus, 25% of publications are from the last 3 years (Figure 17).

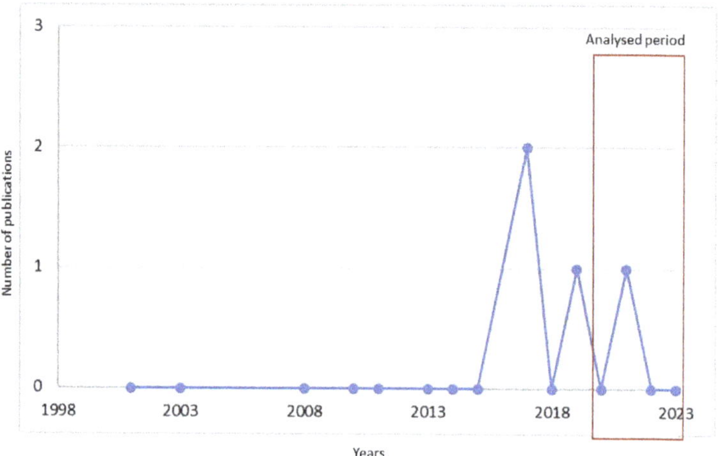

Figure 17. Chronological view of the studies on geothermal energy and architectural heritage: 4 scientific documents have been realized from 2017 to 2023, 1 of them in the years 2020–2023 (Source: Author's elaboration using Scopus data).

All documents described an Italian case study. The most recent publication presents the design and construction of a ground-coupled heat pump in a heritage building located in Naples (Italy) to demonstrate its feasibility in densely built historic towns [88]. The system is based on a low-enthalpy geothermal plant that uses groundwater temperature to produce both heating and cooling. The main problems for historic* towns concern the following difficulties:

- Access to mechanical devices for the construction of deep wells without altering the original architectural layout.
- Drilling and excavation due to the presence of narrow alleys.
- Interception of cavities that can reduce the effectiveness of the geothermal well.
- Long administrative and technical procedures.

In this case, the monitored coefficient of performance (COP) was better than the one certified by the manufacturer (5.65 instead of 4.32). Moreover, significant reductions in greenhouse gas emissions (CO_2) were found without causing alterations in the heritage values. Previous studies on historic buildings demonstrated the reduction of cooling and heating energy, respectively, of 30–50% and 20–40% [89,90]. Moreover, the construction of a geothermal heat pump showed lower payback periods compared with other heat pumps [90].

6. Bioenergy

Bioenergy is produced from several organic materials (also called biomass, e.g., wood, dung, charcoal) and agricultural crops for liquid biofuel. Only biomass can be integrated into the *"architectural heritage"*, while biofuels are used mainly for agricultural purposes. Thus, the present study considered only biomass and cultural heritage keywords (Table 1). Unfortunately, no studies have been found on this topic.

7. Conclusions and Research Perspectives

The present study aims at defining an updated overview of the application of RES on *"architectural heritage"*. RES analyzed are (i) solar energy (Section 3); (ii) wind energy (Section 4); (iii) geothermal energy (Section 5); and (iv) bioenergy (Section 6). A deep literature review of the studies published in the years 2020–2023 has been performed, identifying main topics, challenges, advanced solutions, impacts, and future perspectives. Acceptability, design criteria, and cutting-edge technologies are also illustrated through case studies to better understand practical approaches. Innovative aspects of the present study concern:

- Overcoming the knowledge fragmentation on RES integration in architectural heritage, updating the state of the art of this topic in the period 2020–2023.
- Bridging the traditional boundaries between architectural restoration, landscape design, urban planning, building physics, engineering, and social science.
- Description of current applications and future perspectives for the research sector on RES integration on architectural heritage.

In conclusion, some comments can be summarized considering current regulations, shortcomings, and challenges introduced by the updated state-of-the-art. The key findings are the following:

- The integration of active solar solutions into architectural heritage is more studied than wind technologies, geothermal energies, and bioenergy especially thanks to the presence of specific research projects.
- Among active solar solutions, photovoltaic systems are deeper studied thanks to the aesthetical and technical opportunities offered in the last years as well as to the publication of clear design criteria and recommendations.
- The integration of wind technologies is studied mainly at a territorial level for offshore and wind farms in natural areas, while the integration of wind technologies in historic* buildings is not considered due to the presence of strict regulation constraints.
- The integration of geothermal energy refers mainly to historic* buildings, delineating problems and opportunities for energy production.
- No publications have been found on bioenergy and *"architectural heritage"*.
- Acceptability (and acceptance) of solar and wind energy in heritage contexts is low for the presence of technical, economical, informative, and legislative barriers. This topic is internationally discussed, delineating recurring problems in all countries.
- Geothermal energy in historic* buildings is quite acceptable, and it did not produce specific literature.
- Visual and material compatibility are important criteria for maintaining the original appearance and minimizing the intervention on buildings and towns for all renewable energy technologies.
- Visibility mapping is at the basis of solar and wind energy integration, thanks to the elaboration of spatial modeling, experts' inquiries, and simplified graphical methods.
- Energy potential estimation of active solar and wind technologies in heritage contexts is less studied. In many cases, historic* city centers are excluded from the solar cadasters for the presence of heritage constraints. Otherwise, wind potential estimation refers mainly to offshore wind farms.
- HBIM constitutes a strong tool for balancing energy production and heritage protection, especially for the design of active solar and geothermal energies.

The detailed study of the state of the art of literature also highlights future perspectives for the research on RES applied to *"architectural heritage"*. These perspectives are different for active solar systems and other technologies because the state of the art of research on solar technologies is more advanced. In the first case, the following research perspective can be delineated:

- Practical applications and tests of energy, aesthetical, and sustainable performances (e.g., life cycle assessment) of innovative photovoltaic and solar thermal systems.
- Mapping of commercial products.
- Study on the real aesthetic and technological impact of these technologies on heritage and traditional buildings, supported by case study applications and interviews with the stakeholders.

- Economic analysis of direct and operational costs as well as of payback periods of innovative photovoltaic and solar thermal panels integrated into *"architectural heritage"*.
- Implementation of building integrated photovoltaics and solar thermal system on BIM and HBIM systems for implementing the visual appearance and for calculating the energy production from the early design to the construction phase.
- Prototyping of new products through focus groups with producers and Heritage Authorities.
- Otherwise, the research perspective for wind, geothermal, and bioenergy are the following:
- Implementation of research design projects on their application to cultural heritage, especially considering historic* buildings.
- Definition of the state of the art of the legislation in different countries to understand real barriers and constraints.
- Definition of clear rules, design criteria, and recommendations for *"architectural heritage"* and landscape applications to boost their applicability.
- Collection and mapping of positive and negative examples (products and case studies) to learn from the practice.

Funding: This research received no external funding.

Institutional Review Board Statement: Not applicable.

Informed Consent Statement: Not applicable.

Data Availability Statement: Not applicable.

Conflicts of Interest: The authors declare no conflict of interest.

Nomenclature

UNESCO	United Nations Educational: Scientific and Cultural Organization
EU	European
RES	Renewable Energy Sources
PV	Photovoltaic
BIPV	Building Integrate Photovoltaic
ST	Solar Thermal
USR	Urban Shading Ration
GIS	Geographic Information System
BIM	Building Information Modeling
HBIM	Heritage Building Information Modeling
HVAC	Heating Ventilation Air Conditioning

References

1. United Nations Educational Scientific and Cultural Organization (UNESCO). *Convention Concerning the Protection of the World Cultural and Natural Heritage*; UNESCO: Paris, France, 1972.
2. International Council of Monuments and Sites (ICOMOS), Glossary. Available online: https://www.icomos.org/en/2016-11-10-1 3-53-13/icomos-and-the-world-heritage-convention-4 (accessed on 8 February 2023).
3. Department for Digital, Culture, Media & Sport. *Principles of Selection for Listed Buildings*; Department for Digital, Culture, Media & Sport: London, UK, 2018.
4. *EN 16883:2017*; Conservation of Cultural Heritage—Guidelines for Improving the Energy Performance of Historic Buildings. Comitè Européen de Normalization (CEN): Brussels, Belgium, 2017.
5. Vecco, M. A definition of cultural heritage: From the tangible to the intangible. *J. Cult. Herit.* **2010**, *11*, 321–324. [CrossRef]
6. Lucchi, E. Review of preventive conservation in museum buildings. *J. Cult. Herit.* **2018**, *29*, 180–183. [CrossRef]
7. Carbonara, G. *Avvicinamento al Restauro*; Liguori Editore: Naples, Italy, 1997.
8. Goryacheva, A.V. Concerning Restoration Approaches in Italy at the Turn of the 20–21 Centuries. In Proceedings of the 2nd International Conference on Architecture: Heritage, Traditions and Innovations (AHTI 2020), Moscow, Russia, 26–27 February 2020; Volume 471, pp. 164–168.
9. Carbonara, G. An Italian contribution to architectural restoration. *Front. Archit. Res.* **2012**, *1*, 2–9. [CrossRef]
10. Cesare, B. *Teoria del Restauro*; Einaudi: Torino, Italy, 1963.
11. Dezzi Bardeschi, M. *Restauro: Punto e Da Capo. Frammenti per una (Impossibile) Teoria*; Franco Angeli: Milan, Italy, 2009.
12. Marconi, P. Materia e Significato. In *La Questione del Restauro Architettonico*; Laterza: Bari, Italy, 1999.

13. Plevoets, B.; Van Cleempoel, K. *Adaptive Reuse of the Built Heritage. Concepts and Cases of an Emerging Discipline*; Routledge: Oxfordshire, UK, 2019.
14. Baiani, S.; Altamura, P.; Lucchi, E.; Romano, G. Integration of Solar Technologies in Historical Buildings: Construction of an Evolutionary Framework of Good Practices. In Proceedings of the Mediterranean Green Forum, Florence, Italy, 20–22 July 2022.
15. Markard, J.; Raven, R.; Truffer, B. Sustainability transitions: An emerging field of research and its prospects. *Res. Policy* **2012**, *41*, 955–967. [CrossRef]
16. Carbonara, G. Energy efficiency as a protection tool. *Energy Build.* **2015**, *95*, 9–12. [CrossRef]
17. Franco, G.; Magrini, A. *Historical Buildings and Energy*; Springer: Cham, Switzerland, 2017.
18. Polo Lopez, C.S.; Lucchi, E.; Franco, G. Acceptance of Building Integrated Photovoltaic (Bipv) in Heritage Buildings and Landscapes: Potentials, Barriers and Assessments Criteria. In Proceedings of the 8th Euro-American Congress on Construction Pathology, Rehabilitation Technology and Heritage Management, Rehabend 2020, Granada, Spain, 28–30 September 2020.
19. Renewable Energy Policy Network for the 21st Century (REN21). *Renewables 2019 Global Status Report*; REN21 Secretariat: Paris, France, 2019.
20. De Medici, S. Italian Architectural Heritage and Photovoltaic Systems. Matching Style with Sustainability. *Sustainability* **2021**, *13*, 2108. [CrossRef]
21. Lucchi, E. Integration between photovoltaic systems and cultural heritage: A socio-technical comparison of international policies, design criteria, applications, and innovation developments. *Energy Policy* **2022**, *171*, 112203. [CrossRef]
22. Cabeza, L.F.; de Gracia, A.; Pisello, A.L. Integration of renewable technologies in historical and heritage buildings: A review. *Energy Build.* **2018**, *177*, 96–111. [CrossRef]
23. United Nations (UN). Available online: https://www.un.org/en/climatechange/what-is-renewable-energy (accessed on 8 February 2023).
24. Falagas, M.E.; Pitsouni, E.I.; Malietzis, G.A.; Pappas, G. Comparison of PubMed, Scopus, Web of Science, and Google Scholar: Strengths and weaknesses. *FASEB J.* **2008**, *22*, 338–342. [CrossRef] [PubMed]
25. Cabeza, L.F.; Chàfer, M.; Mata, É. Comparative analysis of web of science and Scopus on the energy efficiency and climate impact of buildings. *Energies* **2020**, *13*, 409. [CrossRef]
26. Andersen, N.; Swami, V. Science mapping research on body image: A bibliometric review of publications in Body Image, 2004–2020. *Body Image* **2021**, *38*, 106–119. [CrossRef]
27. PV Accept. Available online: https://www.pvaccept.de (accessed on 8 February 2023).
28. IEA-SHC T41. Solar Energy and Architecture. Available online: https://task41.iea-shc.org (accessed on 8 February 2023).
29. IEA-PVPS T15. Enabling Framework for the Acceleration of BIPV. Available online: https://www.iea-pvps.org (accessed on 8 February 2023).
30. IEA-SHC T37. Advanced Housing Renewal with Solar & Conservation. Available online: https://task37.iea-shc.org (accessed on 8 February 2023).
31. New4Old. New Energy for Old Buildings. Available online: http://www.new4old.eu (accessed on 8 February 2023).
32. SECHURBA. Sustainable Energy Communities In Historic Urban Areas. Available online: www.sechurba.eu (accessed on 8 February 2023).
33. 3ENCULT. Efficient Energy for EU Cultural Heritage. Available online: https://www.3encult.eu (accessed on 8 February 2023).
34. ENBUAU. Energie und Baudenkmal Project. Available online: https://repository.supsi.ch/7836/1/20121001-EnBau-FinalReport.pdf (accessed on 8 February 2023).
35. Gracia, A.D.; Tarragona, J.; Pisello, A.L.; Cotana, F.; Rodríguez, X.F.; Burgués, J.M.; Cabeza, L.F.; Fernández, C. REHIB: Renewable Energies in Historical Buildings. In Proceedings of the Building Simulation 2019, Rome, Italy, 2–4 September 2019.
36. IEA-SHC T47. Solar Renewal of Non-Residential Buildings. Available online: https://task47.iea-shc.org (accessed on 8 February 2023).
37. EFFESUS: Energy Efficiency for EU Historic Districts' Sustainability. Available online: https://www.effesus.eu (accessed on 8 February 2023).
38. IEA-SHC T51. Solar Energy in Urban Planning. Available online: https://task51.iea-shc.org (accessed on 8 February 2023).
39. IEA-SHC T59. Deep Renovation of Historic Buildings towards Lowest Possible Energy Demand and CO2 Emission (nZEB). Available online: https://task59.iea-shc.org (accessed on 8 February 2023).
40. BIPV. Meets History: Value-Chain Creation for the Building Integrated Photovoltaics in the Energy Retrofit of Transnational Historic Buildings. Available online: http://www.bipvmeetshistory.eu (accessed on 8 February 2023).
41. Land Integrated Photovoltaics. Available online: https://sgp-a.com/single/land-integrated-photovoltaics-research-project/?msclkid=29336a29c20911ec9aa9d99ce7c203c2 (accessed on 8 February 2023).
42. Solarise. Available online: https://www.interregsolarise.eu (accessed on 8 February 2023).
43. Inayatullah, J.; Waheed, U.; Muhammad, A. Social acceptability of solar photovoltaic system in Pakistan: Key determinants and policy implications. *J. Clean. Prod.* **2020**, *274*, 123140.
44. Chen, T.; Heng, C.K. Analysis of the barriers to implementing building integrated photovoltaics in Singapore using an interpretative structural modelling approach. *J. Clean. Prod.* **2022**, *365*, 132652. [CrossRef]
45. Serraino, M.; Lucchi, E. Energy Efficiency, Heritage Conservation, and Landscape Integration: The Case Study of the San Martino Castle in Parella (Turin, Italy). *Energy Procedia* **2017**, *133*, 424–434. [CrossRef]

46. Rosa, F. Building-Integrated Photovoltaics (BIPV) in Historical buildings: Opportunities and constraints. *Energies* **2020**, *13*, 3628. [CrossRef]

47. Sánchez-Pantoja, N.; Vidal, R.; Pastor, M.C. Aesthetic impact of solar energy systems. *Renew. Sustain. Energy Rev.* **2018**, *98*, 227–238. [CrossRef]

48. Horváth, M.; Kassai-Szoó, D.; Csoknyai, T. Solar energy potential of roofs on urban level based on building typology. *Energy Build.* **2016**, *111*, 278–289. [CrossRef]

49. Peluchetti, A.; Guazzi, G.; Lucchi, E.; Dall'Orto, I.; López, C.S.P. Criteria for building types selection in preserved areas to pre-assess the Building Integrated Photovoltaics solar potential—The case study of Como land area. *IOP Conf. Series: Earth Environ. Sci.* **2021**, *863*, 012003. [CrossRef]

50. Tejedor, B.; Lucchi, E.; Nardi, I. Application of qualitative and quantitative infrared thermography at urban level: Potenctial and limitations. In *New Tecnhnologies in Building and Construction*; Bienvenido-Huertas, D., Moyano-Campos, J., Eds.; Lecture Notes in Civil Engineering; Springer: Singapore, 2022; Volume 258.

51. Chatzipoulka, C.; Compagnon, R.; Nikolopoulou, M. Urban geometry and solar availability on façades and ground of real. *Sol. Energy* **2016**, *138*, 53–66. [CrossRef]

52. Raybaud, B.; Desthieux, G. Adapted strategy for large-scale assessment of solar potential of facades in urban areas focusing on the reflection component. *Sol. Energy Adv.* **2022**, *2*, 10030.

53. Gassar, A.; Cha, S.H. Review of geographic information systems-based rooftop solar photovoltaic potential estimation approaches at urban scales. *Appl. Energy* **2021**, *291*, 116817. [CrossRef]

54. Rodríguez, L.R.; Nouvel, R.; Duminil, E.; Eicker, U. Setting intelligent city tilling strategies for urban shading simulations. *Sol. Energy* **2017**, *157*, 880–894. [CrossRef]

55. Peronato, G.; Rastogi, P.; Rey, E.; Andersen, M. A toolkit for multi-scale mapping of the solar energy-generation potential of buildings in urban environments under uncertainty urban forms: Using London as a case study. *Sol. Energy* **2018**, *173*, 861–874. [CrossRef]

56. Stendardo, N.; Desthieux, G.; Abdennadher, N.; Gallinelli, P. GPU-Enabled Shadow Casting for Solar Potential Estimation in Large Urban Areas. Application to the Solar Cadaster of Greater Geneva. *Appl. Sci.* **2020**, *10*, 5361. [CrossRef]

57. Camponovo, R.; Frei, A.; Desthieux, G.; Dubois, D.; Chognard, S.; Rinquet, L.; Frontini, F.; Polo, L.; Cristina, S.; Delucchi, A.; et al. *La Planification Solaire Globale, une Démarche au Service de la Transition Energétique et d'une Culture du Bâti de Qualité, Rapport d'étude*; FOC: Bern, Switzerland, 2018.

58. CitySim. Available online: https://www.epfl.ch/labs/leso/transfer/software/citysim (accessed on 8 February 2023).

59. Archelios Map. Available online: https://www.trace-software.com/archelios-pro/free-pv-design-software (accessed on 8 February 2023).

60. Grassopper. Available online: https://www.grasshopper3d.com (accessed on 8 February 2023).

61. Climate Studio. Available online: https://www.solemma.com/climatestudio (accessed on 8 February 2023).

62. Florio, P.; Probst MC, M.; Schüler, A.; Roecker, C.; Scartezzini, J.L. Assessing visibility in a multiscale urban planning: A contribution to a method enhancing social acceptability of solar energy in cities. *Sol. Energy* **2018**, *173*, 97–109. [CrossRef]

63. Munari Probst, M.C.; Roecker, C. Solar Energy promotion and Urban Context protection: LESO-QSV (Quality—Site—Visibility) method. In Proceedings of the 31th International PLEA Conference, Bologna, Italy, 9–11 September 2015.

64. Florio, P.; Peronato, G.; Perera AT, D.; Di Blasi, A.; Poon, K.H.; Kämpf, J.H. Designing and assessing solar energy neighborhoods from visual impact. *Sustain. Cities Soc.* **2021**, *71*, 102959. [CrossRef]

65. Lucchi, E.; Schito, E. Challenges and Opportunities for the Integration of Photovoltaic Modules in Heritage Buildings Through Dynamic Building Energy Simulations. In *The Future of Heritage Science and Technologies*; Furferi, R., Governi, L., Volpe, Y., Gherardini, F., Seymour, K., Eds.; Florence Heri-Tech 2022. Lecture Notes in Mechanical Engineering; Springer: Cham, Switzerland, 2023.

66. Schito, E. Integrating Different PV Roofs on a Heritage Building Considering Aesthetic, Technical, Energy and Environmental Aspects: A Multi-Perspective Approach. In Proceedings of the Mediterranean Green Forum, Florence, Italy, 20–22 July 2022.

67. *UNI 8290-1*; Edilizia Residenziale, Sistema Tecnologico. Parte 1: Classificazione e Terminologia. Unificazione Nazionale Italiana (UNI): Milano, Italy, 1981.

68. Lingfors, D.; Johansson, T.; Widén, J.; Broström, T. Target-based visibility assessment on building envelopes: Applications to PV and cultural-heritage values. *Energy Build.* **2019**, *204*, 109483. [CrossRef]

69. Dessì, V. *Visibility Assessment of the Integration of Technologies from R.E.S. in Sensitive Urban Environments. Proposal for a Simplified Graphical Tool*; University of Chieti-Pescara: Chieti, Italy, 2016.

70. Lucchi, E.; Baiani, S.; Altamura, P. Design criteria for the integration of active solar technologies in the historic built environment: Taxonomy of international recommendations. *Energy Build.* **2023**, *278*, 112651. [CrossRef]

71. Lucchi, E. Regenerative Design of Archaeological Sites: A Pedagogical Approach to Boost Environmental Sustainability and Social Engagement. *Sustainability* **2023**, *15*, 3783. [CrossRef]

72. Lucchi, E.; Polo, C.S.; Franco, G. A conceptual framework on the integration of solar systems in heritage sites and buildings. *IOP Conf. Ser. Mater. Sci. Eng.* **2020**, *949*, 012113. [CrossRef]

73. Hubinský, T.; Hajtmanek, R.; Šeligová, A.; Legény, J.; Špaček, R. Potential and limits of photovoltaic systems in historic urban structures: The case study of Monument Reserve in Bratislava, Slovakia. *Sustainability* **2023**, *15*, 2299. [CrossRef]

74. Lucchi, E.; Tiozzo Pezzoli, S.; Durante, A. Landscape Integrated Photovoltaic System for a Solar Island in the Venetian Lagoon. In *Sustainability in Energy and Buildings 2021, Smart Innovation Systems and Technologies 263*; Littlewood, J.R., Howlett, R.J., Jain, L.C., Eds.; Springer: Singapore, 2022.

75. Baiani, S.; Altamura, P.; Lucchi, E.; Romano, G. Solar Architecture in Rome: The Refurbishment of Historic Buildings with Active Solar Technologies. In Proceedings of the Mediterranean Green Forum, Florence, Italy, 20–22 July 2022.

76. Lin, Q.; Kensek, K.; Schiler, M.; Choi, J. Streamlining sustainable design in building information modeling BIM-based PV design and analysis tools. *Arch. Sci. Rev.* **2021**, *64*, 467–477. [CrossRef]

77. Ning, G.; Junnan, L.; Yansong, D.; Zhifeng, Q.; Qingshan, J.; Weihua, G.; Geert, D. BIM-based PV system optimization and deployment. *Energy Build.* **2017**, *150*, 13–22. [CrossRef]

78. Díaz, H.; Loughney, S.; Wang, J.; Soares, C.G. Comparison of multicriteria analysis techniques for decision making on floating offshore wind farms site selection. *Ocean. Eng.* **2022**, *248*, 1107751. [CrossRef]

79. Caceoğlu, E.; Yildiz, H.K.; Oğuz, E.; Huvaj, N.; Guerrero, J.M. Offshore wind power plant site selection using Analytical Hierarchy Process for Northwest Turkey. *Ocean Eng.* **2022**, *252*, 111178. [CrossRef]

80. Maxwell, S.M.; Kershaw, F.; Locke, C.C.; Conners, M.G.; Dawson, C.; Aylesworth, S.; Loomis, R.; Johnson, A.F. Potential impacts of floating wind turbine technology for marine species and habitats. *J. Environ. Manag.* **2022**, *307*, 114577. [CrossRef] [PubMed]

81. Doorga, J.R.; Hall, J.W.; Eyre, N. Geospatial multi-criteria analysis for identifying optimum wind and solar sites in Africa: Towards effective power sector decarbonization. *Renew. Sustain. Energy Rev.* **2021**, *144*, 110994. [CrossRef]

82. Islam, M.; Hassan, N.; Rasul, M.; Emami, K.; Chowdhury, A.A. Assessment of solar and wind energy potential in Far North Queensland, Australia. *Energy Rep.* **2022**, *8*, 557–564. [CrossRef]

83. Wieduwilt, P.; Wirth, P. Cultural heritage and wind turbines: A method to reduce conflicts in landscape planning and management. *Eur. Countrys.* **2018**, *10*, 652–672. [CrossRef]

84. Weydt, J. *World Heritage and Wind Energy Planning*; United Nations Educational, Scientific and Cultural Organization (UNESCO): Paris, France, 2021.

85. Watts, L. *Energy at the End of the World: An Orkney Island Saga*; MIT Press: Boston, MA, USA, 2019.

86. Kloos, M. Planning and Heritage Consultancy, Heritage Impact Assessment, 'Heumarkt Neu' construction Project and Development of the 'Historic Centre of Vienna' World Heritage Property. Available online: https://www.bundeskanzleramt.gv.at/service/publikationenaus-dem-bundeskanzleramt/publikationen-zu-kunst-und-kultur/berichte-studien-kultur.html (accessed on 8 February 2023).

87. Shadman, M.; Amiri, M.M.; Silva, C.; Estefen, S.F.; La Rovere, E. Environmental impacts of offshore wind installation, operation and maintenance, and decommissioning activities: A case study of Brazil. *Renew. Sustain. Energy Rev.* **2021**, *144*, 11094.

88. Massarotti, N.; Mauro, A.; Normino, G.; Vanoli, L.; Verde, C.; Allocca, V.; Calcaterra, D.; Coda, S.; De Vita, P.; Forzano, C.; et al. Innovative Solutions to Use Ground-Coupled Heat Pumps in Historical Buildings: A Test Case in the City of Napoli, Southern Italy. *Energies* **2021**, *14*, 296. [CrossRef]

89. Emmi, G.; Zarrella, A.; De Carli, M.; Moretto, S.; Galgaro, A.; Cultrera, M.; Di Tuccio, M.; Bernardi, A. Ground source heat pump systems in historical buildings: Two Italian case studies. *Energy Procedia* **2017**, *133*, 183–194. [CrossRef]

90. Pacchiega, C.; Fausti, P. A study on the energy performance of a ground source heat pump utilized in the refurbishment of an historical building: Comparison of different design options. *Energy Procedia* **2017**, *133*, 349–357. [CrossRef]

MDPI AG
Grosspeteranlage 5
4052 Basel
Switzerland
Tel.: +41 61 683 77 34

Buildings Editorial Office
E-mail: buildings@mdpi.com
www.mdpi.com/journal/buildings